Chemistry on Modified Oxide and Phosphate Surfaces: Fundamentals and Applications

INTERFACE SCIENCE AND TECHNOLOGY

Series Editor: ARTHUR HUBBARD

In this series:

Vol. 1: Clay Surfaces: Fundamentals and Applications
Edited by F. Wypych and K.G. Satyanarayana

Vol. 2: Electrokinetics in Microfluids
By Dongqing Li

Vol. 3: Radiotracer Studies of Interfaces
Edited by G. Horányi

Vol. 4: Emulsions: Structure Stability and Interactions
Edited by D.N. Petsev

Vol. 5: Inhaled Particles
By Chiu-sen Wang

Vol. 6: Heavy Metals in the Environment: Origin, Interaction and Remediation
Edited by H.B. Bradl

Vol. 7: Activated Carbon Surfaces in Environmental Remediation
Edited by Teresa J. Bandosz

Vol. 8: Tissue Engineering: Fundamentals and Applications
By Yoshito Ikada

Vol. 9: Particles at Interfaces: Interactions, Deposition, Structure
By Zbigniew Adamczyk

Vol. 10: Interface Science in Drinking Water Treatment: Theory and Applications
Edited by G. Newcombe and D. Dixon

Vol. 11: Surface Complexation Modelling
Edited by Johannes Lützenkirchen

Vol. 12: Theory of Colloid and Interfacial Electric Phenomena
By Hiroyuki Ohshima

Vol. 13: Sorbent Deformation
By A.V. Tvardovskiy

Vol. 14: Advanced Chemistry of Monolayers at Interfaces: Trends in Methodology and Technology
Edited by Toyoko Imae

Vol. 15: Macromolecules in Solution and Brownian Relativity
By Stefano A. Mezzasalma

Vol. 16: The Properties of Water and their Role in Colloidal and Biological Systems
By Carel Jan van Oss

Vol. 17: Chemistry on Modified Oxide and Phosphate Surfaces: Fundamentals and Applications
By Robson Fernandes de Farias

INTERFACE SCIENCE AND TECHNOLOGY – VOLUME 17

Chemistry on Modified Oxide and Phosphate Surfaces: Fundamentals and Applications

Robson Fernandes de Farias

Universidade Federal do Pará,
Belém, Pará, Brazil

ELSEVIER

Amsterdam • Boston • Heidelberg • London • New York • Oxford
Paris • San Diego • San Francisco • Singapore • Sydney • Tokyo
Academic Press is an imprint of Elsevier

Academic Press is an imprint of Elsevier
Radarweg 29, PO Box 211, 1000 AE Amsterdam, The Netherlands
Linacre House, Jordan Hill, Oxford OX2 8DP, UK
32, Jamestown Road, London, NWI 7BY, UK
30 Corporate Drive, Suite 400, Burlington, MA 01803, USA
525 B Street, Suite 1900, San Diego, CA 92101-4495, USA

First edition 2009

ISBN: 978-0-12-372554-7
ISSN: 1573-4285

For information on all Academic Press publications
visit our website at elsevierdirect.com

Printed and bound in The Netherlands
09 10 11 12 10 9 8 7 6 5 4 3 2 1

Contents

Preface

Oxides and phosphates are among the most important classes of inorganic compounds. As illustrative examples, silicon dioxide, which constitutes the most part of the earth globe and is used in the production of glasses, and calcium phosphate, which is present in bone, can be mentioned.

To modify a surface is to promote, through chemical reactions, or some kind of physical process, the modification of its chemical composition. Such modification is generally performed to enhance some "natural" property of the compound or, on the contrary, to give to a specific surface a property that it does not naturally exhibit. The best result can be achieved when the modifying species introduced on the surface interact, in a synergic way, with the chemical species on the host compound.

To promote the modification of the surface, any kind of chemical reaction/process can be used. However, the sol–gel process has played a very special role among the set of chemical methodologies for surface modification.

The term "surface" must be considered in a broad sense. If, for example, we are talking about an amorphous solid such as silica gel, one understands immediately what is being considered as the surface, but when considering a lamellar solid, such as molybdenum oxide, both "surfaces," that is, the external and internal (intralamellar) parts of the solid must be taken into account. So, to promote intercalation is also to modify a surface, of course.

As the title indicates, this little book is devoted specifically to modified oxides and phosphates. Each chapter consists of a general and short introduction to the specific subject, followed by some examples from the literature and by a specific example, or examples; this last part is organized, in some cases, with a "paper structure," that is, with an experimental part and a results and discussion section. I think that this is a good choice since not only "general" data are presented, but specific procedures to prepare the described compounds, adding to the "fundamental" knowledge presented. Hence, the present book is both a review about a specific theme, and also a preparative manual.

From the point of view of applications, as the reader can verify, modified oxide and phosphate surfaces can be used in various fields such as (a) sequestrating agents for metal cations, (b) active species in the production of electrochemical sensors, (c) new chromatographic phases, (d) new light conversion molecular devices, (e) molecular sieves, and (f) cation exchange matrices, among many others. In this context, it is not a surprise that it is practically impossible to read any issue of a magazine dedicated to materials or solid state or inorganic chemistry without finding, in every issue, some papers dealing with the preparation and applications of some modified oxide or phosphate. At the end of each chapter, in the "references" section are listed the articles from which specific information have been extracted. In the "bibliography" are shown a list of the works from which general information has been obtained to provide the "backbone" of each chapter, as well as an overview of the state of the art.

I hope that this little book will help to increase the number of professionals interested in so fascinating compounds such as oxides and phosphates.

Robson F. de Farias
Natal, Brazil. March, 2008.

CHAPTER ONE

Oxides and Phosphates

Contents

Oxides can be defined as binary compounds formed between various elements and oxygen, while phosphates can be defined as salts based formally on phosphorus (V) oxoacids and in particular salts of phosphoric (V) acid, H_3PO_4. Both oxides and phosphates are among the most important classes of inorganic compounds. For example, silicon dioxide (SiO_2) is the main compound in the Earth's crust, and apatite, a complex mineral form of calcium phosphate, $Ca_5(PO_4)_3$ (OH, F, Cl) is the main compound on the enamel of teeth [1–3].

Besides their uses as pigments, many oxides exhibit catalytic properties, making them a very important class of compounds, for both basic and technological research. Phosphates, on the other hand, have also a lot of practical uses, for example, fertilizers based on ammonium phosphate.

While "pure" oxides and phosphates exhibit many possible chemical interactions, modified oxide and phosphates could be considered as exhibiting "expanded" chemical interactions, with very interesting possible consequences for basic and applied chemistry.

Concerning the scope of this book, we are interested not in "pure" oxides and phosphates, but in modified ones, specifically those with chemical modifications that affect their superficial chemical composition (the bulk composition may or may not be affected).

Due to the presence of Lewis and/or Brønsted acidic and basic sites on their surfaces, both oxides and phosphates can interact (with or without the occurrence of chemical reactions) with a lot of substances, specially with organic ones, as will be shown in this book.

If the chemical species that is modifying the oxide or phosphate surface is organic, it forms a so-called inorganic–organic hybrid. A lot of modified oxides and phosphates are examples of that class of compounds.

Interface Science and Technology, Volume 17
ISSN: 1573-4285, doi: 10.1016/S1573-4285 (08) 00401-8

In a very interesting connection between modified oxides and phosphates, it is possible to modify lamellar phosphates with inorganic oxides, a process called pillaring [3]. This last example shows how rich is the chemistry of modified oxides and phosphates.

REFERENCES

1. Minidictionary of Chemistry, Oxford University Press, Oxford, 1988.
2. M.L. Occelli, H.E. Robson (Eds.), Expanded Clays and Other Microporous Solids, Van Nostrand Reinhold, New York, 1992.
3. E.F. Vansant, P. Van Der Voort, K.C. Vrancken, Characterization and Chemical Modification of the Silica Surface, Elsevier, Amsterdam, 1997.

CHAPTER TWO

SURFACES AND MODIFIED SURFACES

Contents

Many very important chemical and physical processes occur at solid surfaces, as is well illustrated by a large number of heterogeneous catalysts used in industry. Their heterogeneous nature is a remarkable characteristic of the chemical reactions that take place at surfaces/interfaces and so the principles usually followed to explain the chemical reactions that occur in the gaseous phase, liquid phase, or in solution cannot be straightforwardly applied to them.

In considering solid surfaces, properties such as superficial area, porosity (number and mean diameter of pores), and the existence of acidic and/or basic (from both types: Lewis and Brønsted) sites on the surfaces are of prominent importance to understand and predict the chemical behavior of such solids.

In considering surfaces and modified surfaces, the techniques and procedures usually used to characterize solid and surfaces are of prominent importance, such as Brunauer–Emmett–Teller (BET) isotherms, nuclear magnetic resonance (NMR) of solids (in our case, generally ^{29}Si and ^{31}P), scanning electron microscope (SEM), and transmission electron microscope (TEM).

Despite the fact that the surface is part of a given compound sample, its chemical properties can be very different from those observed in the bulk. Silica gel is a very good example: whereas the bulk is constituted by silicon atoms bonded only to oxygen, OH groups can be found in the surface [1], which are responsible for its most useful characteristic, namely, its reactivity toward organic species, producing organofunctionalized surfaces.

As we will discuss in the following chapters, among the most interesting classes of solids that can have their superficial chemistry changed by chemical modifications are oxides and phosphates.

REFERENCE

1. R.K. Iler, The Chemistry of Silica, John Wiley, New York, 1979.

Interface Science and Technology, Volume 17
ISSN: 1573-4285, doi: 10.1016/S1573-4285 (08) 00402-X

CHAPTER THREE

Chemistry on Organofunctionalized Amorphous Oxides Surfaces

Contents

1. Introduction

Amorphous oxides like silica can be conveniently organofunctionalized to produce surfaces having the ability to act, for example, as sequestrating agents toward transition metal cations.

In silica gel, for example, the functionalization process requires the presence of silanol (—Si—OH) groups on the oxide surface, to promote the reaction between the organofunctional agent and the oxide surface. In a typical synthesis, amorphous silica gel can react with 3-aminopropiltriethoxysilane, producing a modified surface, as shown in Fig. 3.1. Despite the fact that silica is the most commonly used

Interface Science and Technology, Volume 17
ISSN: 1573-4285, doi: 10.1016/S1573-4285 (08) 00403-1

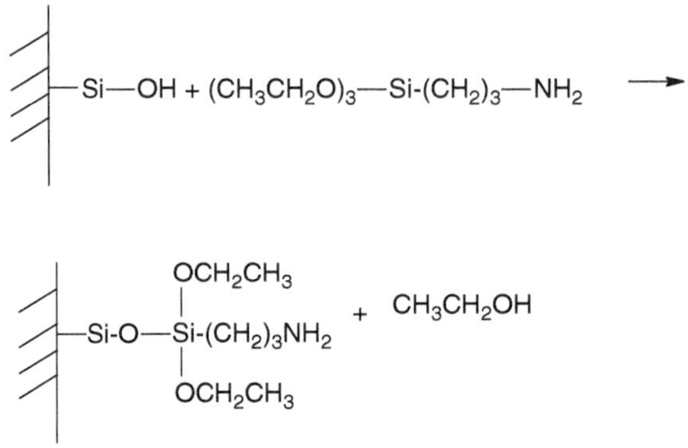

Figure 3.1 Schematic representation of the reaction of amorphous silica gel with 3-aminopropiltriethoxysilane.

compound to produce such modified surfaces, any other compound having surface OH groups does not show such type of chemical modification. In such a reaction, the number of available silanol groups on the surface is a critical factor, of course, and for silica, the number (medium value) is 5.0 OH groups per nm^2 [1,2]. As indicated in Fig. 3.1, of the three alkoxide groups of the functionalizing agent, only one or two are involved in the formation of covalent bonds with the surface, due to the fact that the tetrahedral nature of silicon, as well as the mean distance between each silanol group on the silica surface implies that the alkoxide groups are not far enough apart to allow reaction with three surface OH groups.

In the example shown in Fig. 3.1, since the nitrogen atom on the organofunctionalizing agent has a free electron pair, it can coordinate to transition metal cations, for example, producing supported coordination compounds.

One alternative, to promote the surface modification, is to react the silanol groups on the surface with an organofunctionalizing agent having no coordinative sites and then react such a modified surface with an organic compound having coordinative sites. A typical example is shown in Fig. 3.2. The organofunctionalizing method shown in Fig. 3.2 is called "heterogeneous" as opposed to a homogeneous method, for example, the previous reaction between the silicon alkoxide and urea, and the subsequent reaction between the product of such reaction and the amorphous silica.

Both, homogeneous and heterogeneous methods produce surfaces with a relatively low amount of organic groups, typically about 0.5 mmol of organic groups per gram of modified silica.

An alternative method to produce surface amorphous modified oxides (silica, titania, alumina, etc.) is to promote the reaction of the oxide precursor with the organic functionalizing agent. A typical example is shown in Fig. 3.3.

This kind of "sol–gel only" experimental approach exhibits two main advantages in comparison with the two others previously mentioned methods: (1) the preparation of modified surfaces with a higher degree of functionalization and (2) the

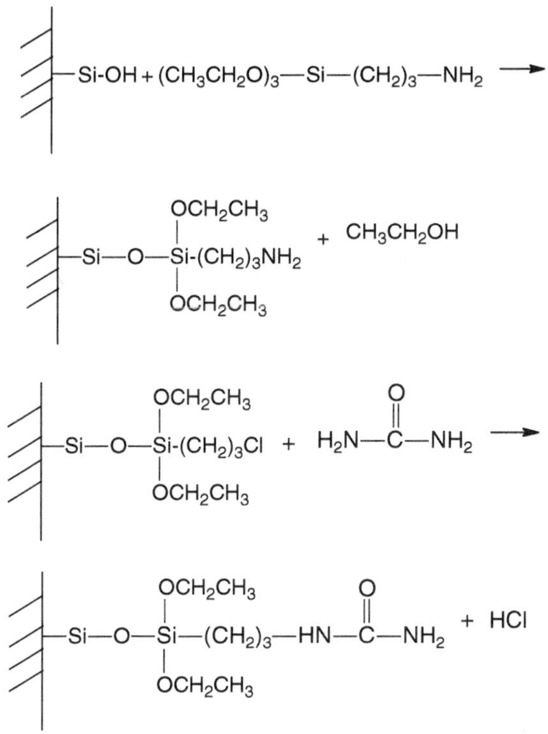

Figure 3.2 Schematic representation for the urea functionalization of silica gel surface by using 3-chloropropyltriethoxysilane as precursor.

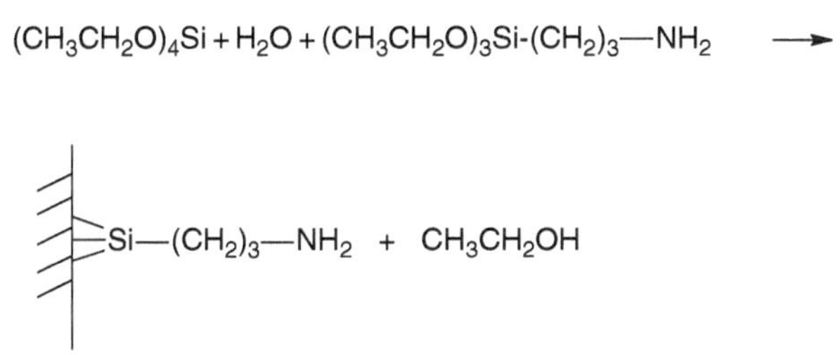

Figure 3.3 Schematic representation of the reaction between TEOS and 3-aminopropyltriethoxysilane, producing an organomodified surface.

formation of not only one or two, but three covalent bonds between the organic moiety and the oxide surface, and, consequently, the formation of a most stable (from chemical and thermal point of view) organomodified surface.

It is worth noting that in all reactions involving alkoxides, special care must be taken regarding the presence of water into the reaction system, since such compounds are very sensitive to hydrolysis processes.

2. Modified Amorphous Oxide Surfaces as Metal Cation Sequestrating Agents

Many modified surfaces do not have coordinating species, as in the case of the well-known modified surfaces, for example, C18, used in gaseous or liquid chromatography. However, if the organomodifing agent has one or more coordinative sites (and this is frequently the case) the modified surface can be successfully used to promote the withdrawl of metal cations from aqueous and nonaqueous solutions [3]. Solution calorimetric studies [3] have shown that the total amount of adsorbed (chemisorbed = coordinated) cations is a function of two main factors: the acid–base interactions between the metal cation (Lewis acid) and the coordinative site on the modified surface (Lewis base) and the metal cation–solvent interactions.

3. Organofunctionalized Metal-Modified Surfaces for Chromatographic Applications

Organically modified silica has been used as the stationary phase in chromatography for many years. Generally, long carbon chains like C18 bonded to silica are used. On the other hand, the search for supports more stable in basic solutions than silica (under aggressive conditions, using high pH mobile phases and elevated temperatures, for example) for use in high-performance liquid chromatography (HPLC) is still of importance. In this connection, the first reports on the preparation of more pH-stable reversed phases by alkylsilanization of zirconia, titania, and alumina date from the early 1990s. However, these approaches were somewhat limited due to the difficulty of obtaining these alternative supports in a suitable range of particle and pore sizes. In addition, the surface concentration of the C18 ligands on titania was low compared to silica.

The preparation of organically modified oxide surfaces for chromatographic purposes (some of them very specific) is a very rich and productive field of research. A few examples are presented in the following sections.

It has been shown that organofunctionalized silica surfaces further modified with *in situ* synthesized copper complexes can be used to produce new stationary phases for HPLC [4]. The immobilized copper complexes provide new sites in the stationary phase that can interact strongly with basic organic compounds. A test mixture of compounds such as benzene, toluene, naphthalene, anthracene, pyrene, and nitrobenzene shows that the presence of copper ion complexes on the modified surface strongly affects the retention factor (k) of the stationary phases.

A new stationary phase containing embedded urea groups ($NH_2C(O)NH_2$) was prepared by a procedure based on the synthesis of a trifunctional C18 urea–alkoxysilane, followed by modification of titanized silica and further endcapping to evaluate if the embedded group would minimize the higher retention and tailing for basic compounds seen with C18 titanized silica phases [5]. For this new stationary phase, chromatographic evaluations used hydrophobic, polar, and basic compounds to verify the effects of the polar urea groups embedded in the C18 urea phase. The chromatographic parameters, yielded favorable results.

4. Modified Amorphous Oxides for Luminescent Devices

Organically modified surfaces of oxides such as silica [6] or silica–titania [7] can be successfully used as a support for the adsorption of luminescent complexes, resulting in the production of new light conversion molecular devices (LCMDs). It is important to mention that, in these cases, the organically modified oxide is not merely an inert substrate, but interacts synergically with the adsorbed complexes affecting their optical properties, that is, lifetime emission, emission intensity, and so on.

A good example of the application of modified oxides for optical purposes is the use of a surface produced from 3-mercaptopropyltrimethoxysilane and ethylenediamine (en) as precursors. The compound 3-trimethoxypropylthioethylamine (mptt) was synthesized, and then used as a silylating agent to obtain, through a sol–gel process, the modification of a silica gel sample.

Such modified surface was then used as a substrate for the adsorption of the luminescent complex $Eu(fod)_3.2H_2O$ (fod is the anion of the 2,2-dimethyl-6,6,7,7,8,8,8-heptafluoro-3,5-octanedione). $Eu(fod)_3.2H_2O$ and the silicon hybrid were mixed in stoichiometric amounts to produce samples with 1%, 5%, and 10% (m/m) of adsorbed complex. The adsorbed complex was then studied from the spectroscopic point of view. The measured lifetimes for the emission process are 476, 542, and 566 μs for the 1%, 5%, and 10% samples, respectively. Furthermore, comparison with experimental data for *n*-[3(trimethoxysilyl) propyl]-ethylenediamine-modified silica gel surfaces [8] emphasizes the prominent role of the chemical composition of the silica-modified surface in the spectroscopic properties of the adsorbed complex.

4.1 A 3-trimethoxypropylthioethylamine-modified surface

Silica gel (Aldrich) (particle size 70-230 mesh) with mean diameter of 6 nm and $0.75\,m^3\,g^{-1}$ pore volume was used. The silica sample (10.0 g) was stirred with H_2SO_4–HNO_3 ($2.0\,mol\,dm^{-3}$) in a 9:1 proportion. The suspension was left standing for 24 h. The solid was filtered and extensively washed with double-distilled water until the filtrate had the same pH as water. After this treatment, the silica was activated by heating at 423 K for 10 h in vacuum under a nitrogen atmosphere.

The silylating agent 3-mercaptopropyltrimethoxysilane from Aldrich was used as supplied.

Ethylenimine (etn) was synthesized following a described procedure, by reacting 2-amino-hydrogen sulfate (Aldrich) in a hot alkali medium.

The silylating agent mptt was synthesized as follows: a $5.5\,cm^3$ volume (29.0 mmol) of 3-mercaptopropyltrimethoxysilane was reacted with $6.0\,cm^3$ (116.0 mmol) of etn under reflux at 323 K, with magnetic stirring, under an inert dry nitrogen atmosphere for 24 h. Excess of etn was eliminated from the cooled solution in vacuum.

The mptt-modified silica gel surface was prepared as follows: the synthesized mptt was allowed to react with 5.0 g of the previously activated silica, in dry xylene at 343 K and the mixture was mechanically stirred for 72 h. Then, the solid was filtered, washed with xylene and acetone until the filtered solution gave no positive test for thiol groups. The sequence of reactions involved in the synthesis of the modified silica is summarized in Fig. 3.4.

The activated and modified silica samples were characterized by the BET surface area in a Flowsord II 2300 Micromeritics apparatus. Carbon, nitrogen, sulfur, and hydrogen contents were determined by using a Perkin-Elmer microelemental analyzer model PE 2400.

The Eu(fod)$_3$.2H_2O complex was of analytical grade (Aldrich) and was used after recrystallization from ethanol.

The europium-doped inorganic–organic hybrid samples were prepared as follows: the Eu(fod)$_3$.2H_2O was dissolved in dry ethanol and to the solution obtained was added to the insoluble hybrid.

The resulting suspension was stirred for 30 min and the solvent evaporated. The resulting powder was dried at 60°C under vacuum in a stove for 12 h. Eu(fod)$_3$ and the silica gel hybrid were mixed in stoichiometric amounts to produce samples with 1%, 5%, and 10% (m/m) of the adsorbed complex. The X-ray diffraction (XRD) patterns were recorded on Shimadzu equipment, model XD3A, with nickel-filtered Cu K_α radiation (35 kV, 25 mA).

The emission spectra were analyzed using a Jobin Yvon model double monochromator, model U-1000, and the fluorescence signal detected by a water-cooled RCA C31034-02 photomultiplier was processed by a Jobin Spectralink system. The excitation wavelength was 340 nm. To ensure the reliability of the comparison of the spectra main line intensities, the geometry of the counting system was maintained constant, that is, with the same sample container, the same excitation wavelength, and the same slit width (0.05 mm).

The lifetime measurements were performed at 298 K exciting the sample with the third harmonic of a Nd–YAG laser ($\lambda_{exc} = 355$ nm) and detecting the emitted light, after dispersion through a Jobin Yvon model H-10 monochromator, with a P28 photomultiplier tube.

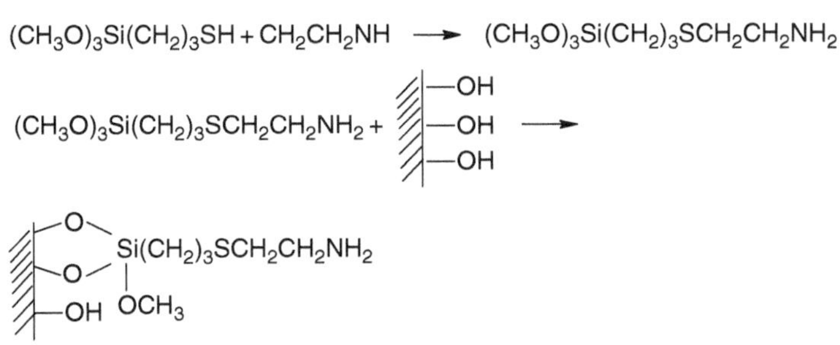

Figure 3.4 Schematic representation of the sequence of the reactions involved to produce the mptt-modified silica surface.

The samples for analysis were pressed on a uniaxial press under 380 ± 19 MPa for 30 s. The prepared samples will be named here as SilSNH$_2$Eu, SilSNH$_2$5Eu, and SilSNH$_2$10Eu, respectively.

The elemental analysis indicated a total amount of 0.8 mmol of S—(CH$_2$)$_2$NH$_2$ groups per gram of hybrid matrix. The XRD pattern confirms that the synthesized matrix is amorphous.

All emission spectra exhibited very similar profiles. For example, the spectrum of the SilSNH$_2$10Eu sample is shown in Fig. 3.5.

The emission bands from 1 to 5 are, respectively, $^5D_0 \rightarrow {}^7F_0$, $^5D_0 \rightarrow {}^7F_1$, $^5D_0 \rightarrow {}^7F_2$, $^5D_0 \rightarrow {}^7F_3$, and $^5D_0 \rightarrow {}^7F_4$. It was observed that the fluorescence intensity decreases with an increase of the total amount adsorbed complex. Such fact was previously observed for another europium complex-doped hybrid surfaces and could be explained based on both, the organic moiety/complex ratio, as well as concentration quenching effect.

The measured lifetimes (for lifetime measurements, the main emission band, $^5D_0 \rightarrow {}^7F_2$, was considered) for the emission process are 476, 542, and 566 μs for the samples SilSNH$_2$Eu, SilSNH$_2$5Eu, and SilSNH$_2$10Eu, respectively. So, it is verified that the lifetime increases with an increased amount of adsorbed complex. Hence, it can be inferred that proximity between the Eu centers at the surface (higher complex concentrations) favors longer lifetimes.

This behavior is opposite to that observed for Eu(fod)$_3$.2H$_2$O adsorbed on Si—Ti [6] or silicon inorganic–organic hybrids. For those samples, the organic moiety was the group —(CH$_2$)$_3$—NH—(CH$_2$)$_2$NH$_2$, and lifetimes of 414, 384, and 380 μs were measured for the 1%, 5%, and 10% complex adsorbed samples, respectively. So, the experimental results obtained in this work, compared with those previously reported [7,8] provide evidence for the prominent role of the chemical composition of the silica-modified surface in the spectroscopic properties of the adsorbed complex.

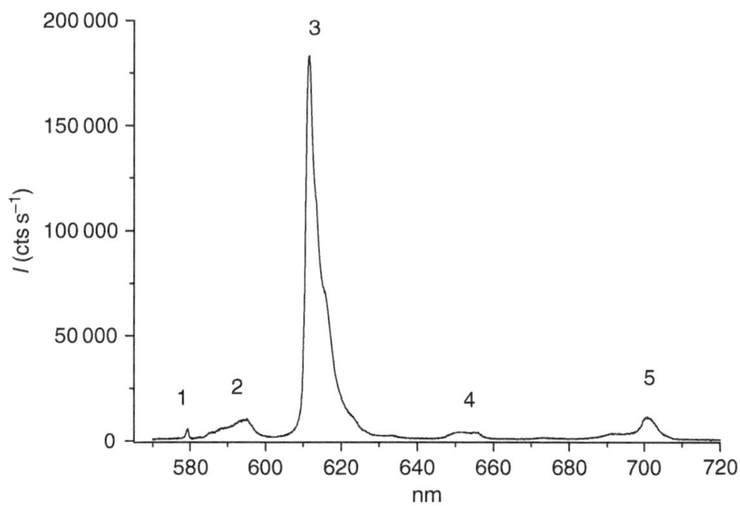

Figure 3.5 Emission spectra for the sample SilSNH$_2$10Eu.

It is worth remembering that the Eu(fod)$_3$.2H_2O complex has two water molecules in its coordination sphere, and it is well known that such H_2O molecules can be replaced by other ligands, as exemplified by 1,10-phenanthroline (phen) and 2,2′-dipyridyl (both with nitrogen as donor atoms). Hence, in previously *n*-[3(trimethoxysilyl) propyl]-ethylenediamine-modified surfaces [6,7] it was supposed that some water molecules in the coordination sphere of the complex were replaced by the nitrogen atom of the organic moiety. As the complex/hybrid matrix ratio increases, from 1% to 5% and finally 10%, the number of nitrogen atoms available per complex molecule is reduced, and it can be supposed that, because of this, the symmetry of the adsorbed compound is undergoing modifications that make the absorption and transfer of energy (from ligands to the lanthanide) less effective. The fact that, for other matrices [7,8] the fluorescence intensity exhibited irregular variations depending on the duration of the applied pressure [7,8], reinforces this hypothesis.

Comparing the samples for which the organic moiety is the group —$(CH_2)_3$—S—$(CH_2)_2NH_2$ with those with the group —$(CH_2)_3$—NH—$(CH_2)_2NH_2$, one can verify that the only difference from the point of view of chemical composition, is the replacement of a nitrogen atom by a sulfur atom. However, such a difference was enough to modify (increase or decrease) the lifetimes for the emission process as well as to increase the lifetimes of the sulfur-containing samples. Such phenomena could be, as a first hypothesis, attributed to the different hardnesses of N and S, and so, it could be concluded that softer bases increase the lifetime, whereas hard bases reduce the lifetime. So, it could be inferred that when adsorbed on a inorganic–organic hybrid surface, the lifetime of the fluorescence process will be favored by polarizable (soft) atoms on the surface and, on the contrary, will be shortened by the presence of hard (less polarizable) atoms on the surface.

However, the role of the N—H vibration modes of en could also be considered to explain the less efficient quenching of the S—$(CH_2)_2NH_2$ group. Furthermore, it can also be inferred that the progressive substitution of water molecules for S atoms on the coordination sphere of Eu(fod)$_3$.2H_2O exerts a significant effect resulting in the observed phenomena.

So, it can be concluded that the experimental results show that a mptt-modified silica gel surface can be successfully used as a substrate for luminescent lanthanide compounds. Furthermore, one could suppose that, controlling the total amount of available coordination sites in the surface, as well as the external pressure suffered by the doped samples, a complete control of the intensity and lifetime of the emitted light could be achieved. Furthermore, comparison with experimental data for *n*-[3(trimethoxysilyl)propyl]-ethylenediamine-modified silica gel surfaces indicates the prominent role of the chemical composition of the silica-modified surface on the spectroscopic properties of the adsorbed complex.

4.2 *In situ* synthesis of luminescent complexes on amorphous modified silica surfaces

Besides the adsorption of previously prepared luminescent complexes on modified amorphous surfaces, it is also possible to synthesize the luminescent species *in situ*, that is, an organic ligand that will coordinate to the lanthanide cation can be covalently bonded to the oxide surface, which will react with the metal cation, forming an *in situ* coordinated complex. A specific example [9] is the *in situ* synthesis on organofunctionalized silica and spectroscopic properties of europium complexes with eta (eta = ethyl 4,4,4-trifluoroacetoacetate) and phen. The optical properties of the *in situ* synthesized and "free" complexes are compared. The choice of a β-ketoester ligand is based on the fact that this class of compounds resembles the well-known β-diketones which are very good ligands to prepare new LCMDs.

The complex $[Eu(eta)_2(H_2O)_4]Cl$ was synthesized as follows (all reactions were performed under an argon atmosphere): 2.0 mmol of $EuCl_3$ was dissolved in triethylorthoformate under magnetic stirring, after which dried THF (tetrahydrofuran) was added. The ligand was deprotonated by using NaI. The resulting solution was stirred for 24 h. The obtained complex was dried, neutralized with NH_4Cl, filtered off, and washed with deionized water and ethanol to remove the excess ligand. The obtained solid was then dried under vacuum at room temperature for 8 h.

The complex $[Eu(eta)_2(phen)(H_2O)_2]Cl$ was synthesized by dissolving the previously prepared complex in a solution of phen (1:1 mol:mol) in chloroform. The resulting solution was maintained in reflux for 24 h, and the resulting complex washed with deionized water and dried under vacuum at room temperature for 8 h.

Organofunctionalized silica was prepared as follows: 50 cm^3 of tetraethylorthosilicate (TEOS) was dissolved in 38 cm^3 of 3-chloro(propyl)trimethoxysilane. To the resulting solution, 400 cm^3 of a 0.3 mol dm^3 HCl solution was added. The obtained gel was mechanically stirred for 30 min. The resulting mixture was aged for 24 h, the supernatant was discarded, and 400 cm^3 was deionized water was added. This sequence of operations was repeated three times. The obtained powder was then dried at room temperature in a fume hood and then dried under vacuum at 40°C for 4 h.

The eta-modified silica was prepared by reacting the 3-chloropropyl silica with eta by using the same procedure used to the synthesis of eta complex.

The *in situ* organofunctionalized silica complexes were prepared by reacting the eta-modified silica with an ethanolic solution of $EuCl_3$. Schematic representations of the *in situ* synthesized complexes are shown in Fig. 3.6.

The CHN elemental analysis was performed on a Perkin-Elmer 240 analyzer. The Eu (III) contents were determined by EDTA complexometric titration.

The ultraviolet–visible (UV–Vis) spectra were obtained on a Perkin-Elmer lambda 6 spectrophotometer with a tungsten lamp for excitation in the visible wavelength range and a deuterium lamp for excitation in the ultraviolet wavelength range.

The excitation and emission spectra were obtained on a Jobin Ivon Ramanor U1000 model H-10 equipment. The emission detection was performed by using a RCA C31034-02 photomultiplier and the register and processing of the emission signals were performed by using a Spectralink interface.

The luminescence decay curves were registered on a SPEX Fluorolog F212 spectrofluorimeter linked to a 1934 D phosphorimeter with a 150 W pulsed xenon lamp.

The excitation and emission data were collected at a 22.5° "front face" angle (with respect to the incident light) with a SPEX-Fluorolog 2 spectrofluorimeter

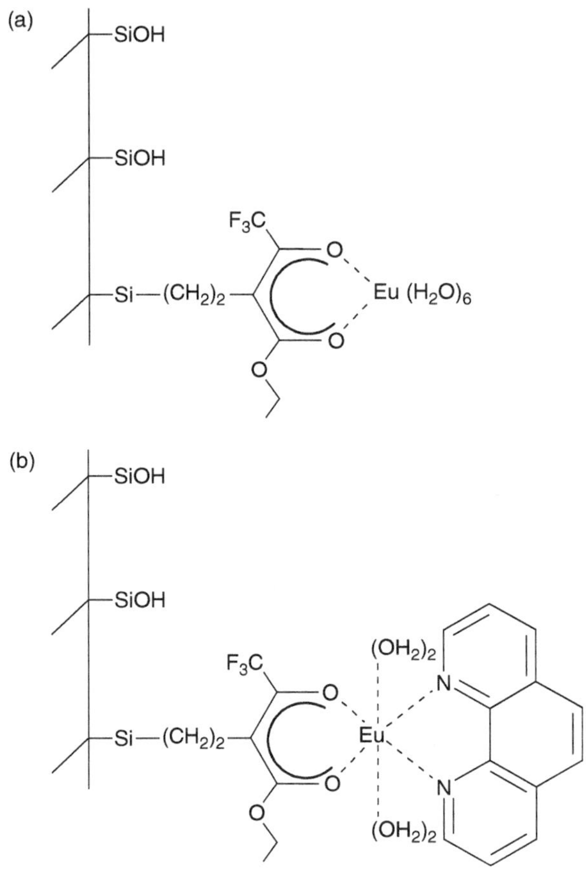

Figure 3.6 Schematic representation of eta (a) and eta–phen (b) of *in situ* organofunctionalized silica europium complexes.

with double monochromators of 0.22 m of SPEX 1680 type by using a "continuous" 450 W xenon lamp as excitation source.

The experimental intensity parameters Ω_2 and Ω_4 for the trivalent europium ion are expressed as:

$$I_{J-J'} = \hbar\omega_{J-J'} A_{J-J'} N_J$$

where N_J is the population of the emission level J, $\hbar$ is the energy of the J–J' transition, and $A_{J-J'}$ is the Einstein's coefficient of spontaneous emission. The transition $^5D_0 \rightarrow {}^7F_1$, is used as reference for the intensity parameters determination, since it is not dependent of the Ω_λ values [9]. From the equation

$$A_{J-J'} = \frac{S_{J-J'}}{S_{0-1}} \frac{\omega_{0-1}}{\omega_{J-J'}} A_{0-1}$$

We can correlate frequencies (ω) and areas (S), obtained from the emission spectra of the complex and the emission rate of the reference band (0–1), which is defined as

$$A_{0-1} = 0.31\times10^{-11}\eta^3\tilde{\nu}^3$$

where η is the refraction index (used value = 1.5) and $\tilde{\nu}$ denotes the wave number of the 0–1 transition. From this last equation, it is possible to obtain the experimental intensity parameters in combination with the following equation [9]:

$$\Omega_\lambda = \frac{3\hbar C^3 A_{0-1}}{4e^2\omega^3\chi\langle^7\mathrm{F}_\lambda||U^{(\lambda)}||^5\mathrm{D}_0\rangle^2}$$

where

$$\chi = \frac{n(n^2+2)^2}{9}$$

The CHN elemental analysis results for the 3-chloro(propyl)-modified silica show that the functionalization degree is of 20%. The CHN and Eu (III) elemental analysis results are in good agreement with the proposed formula shown in Table 3.1 for "free" complexes.

The thermogravimetric curves for the "free" complexes are shown in Fig. 3.7. For [Eu(eta)$_2$ (H_2O)$_4$]Cl (Fig. 3.7a) the first two mass loss steps are as follows: [Eu(eta)$_2$ (H_2O)$_4$]Cl (s) $\rightarrow$ [Eu(eta)$_2$ (H_2O)$_2$]Cl (s) + 2H_2O (g); [Eu(eta)$_2$ (H_2O)$_2$]Cl (s) $\rightarrow$ [Eu(eta)$_2$]Cl (s) + 2H_2O (g). So, it is inferred that in the complex [Eu(eta)$_2$ (H_2O)$_4$]Cl there are, from the energetic point of view, two distinct coordination sites: two water molecules are closer to the Eu(III) cation and, consequently, exhibit a higher bond enthalpy, being released at a higher temperature (125°C) than the other two (75°C).

The release of gaseous eta molecules and the formation of stable Eu_2O_3 (possibly with some $EuCl_3$) are the next observed steps.

For [Eu(eta)$_2$ (phen) (H_2O)$_2$]Cl (Fig. 3.7b) the sequence is as follows: [Eu(eta)$_2$ (phen) (H_2O)$_2$]Cl (s) $\rightarrow$ [Eu(eta)$_2$ (phen)]Cl (s) + 2H_2O; [Eu(eta)$_2$ (phen)]Cl (s) $\rightarrow$ [Eu(eta)$_2$]Cl (s) + phen (g); [Eu(eta)$_2$]Cl (s) $\rightarrow$ 2eta (g) + $EuCl_3$/Eu_2O_3 (mixture) is proposed.

As an illustrative example, the infrared spectra for [Eu(eta)$_2$ (phen) (H_2O)$_2$]Cl (s) and its respective silica-supported complex, Silica/Eu(eta) phen (H_2O)$_4$ are shown in Fig. 3.8.

Table 3.1 Ω_2 and Ω_4 experimental intensity parameters for "free" and *in situ* organofunctionalized silica complexes

Complexes	Ω_2 (10^{-20} cm^2)	Ω_4 (10^{-20} cm^2)
[Eu(eta)$_2$ (H_2O)$_4$]Cl	10.2	3.7
Silica/Eu(eta) (H_2O)$_6$	6.4	3.8
[Eu(eta)$_2$ (phen) (H_2O)$_2$]Cl	17.2	2.8
Silica/Eu(eta) phen (H_2O)$_4$	7.1	0.2

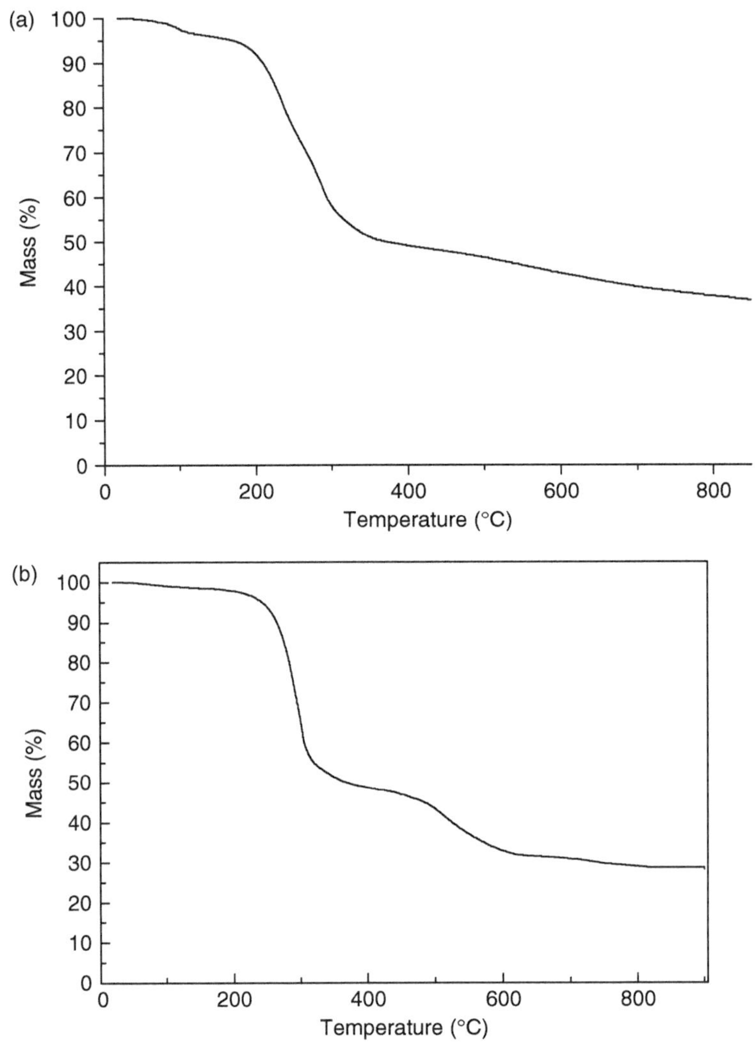

Figure 3.7 Thermogravimetric curves for (a) $[Eu(eta)_2\ (H_2O)_4]Cl$ and (b) $[Eu(eta)_2\ (phen)\ (H_2O)_2]Cl$.

As can be verified for the silica-supported sample, the characteristic bands observed for the free complex, such as those at (free complex) 1675/1645 cm^{-1} due to the ν C=O vibration mode and 1430 cm^{-1} associated with the ν C=N mode. In the silica-supported complex, the C=O bands are shifted to 1699/1656 cm^{-1}, indicating that the metal–ligand bond order is higher in this case. So, it is possible to suppose that the silica-supported complex is more stable than the free one.

The excitation spectra were obtained in the 200–400 nm range, checking the emission at 612 nm. The solid samples were excited at 298 K.

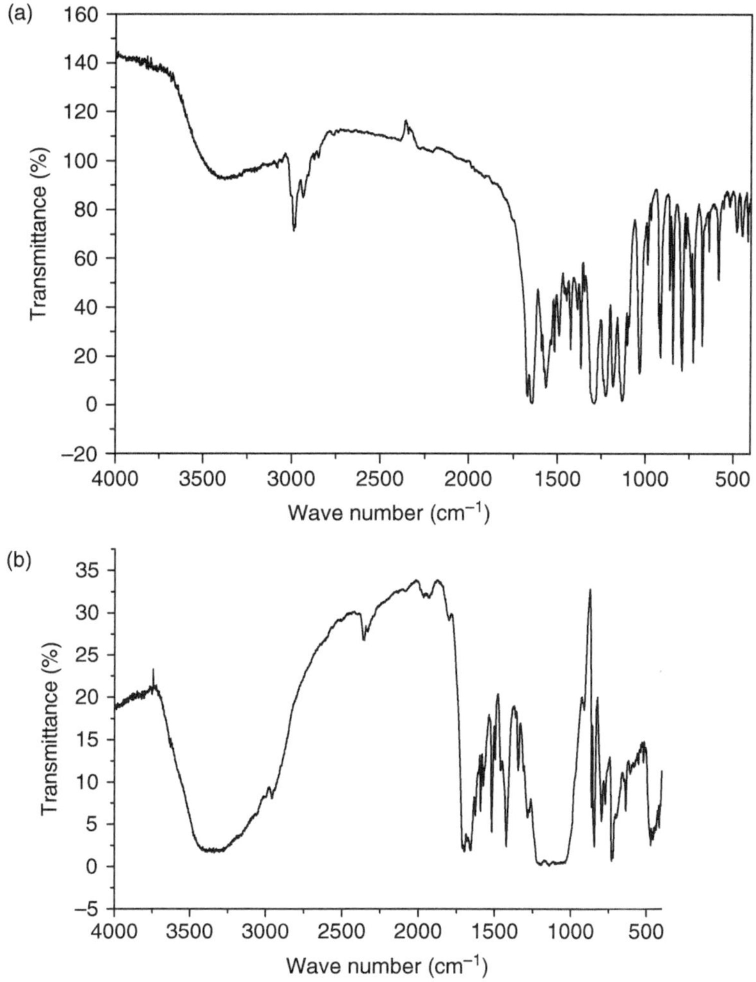

Figure 3.8 Infrared spectra for: (a) [Eu(eta)$_2$ (phen) (H$_2$O)$_2$]Cl and (b) Silica/Eu(eta)$_2$phen (H$_2$O)$_4$.

The emission spectra were obtained at 298 K and 77 K in the 570–720 nm range. Transitions from the $^5D_0$ to the ^7F_J ($J=0$, 1, 2, 3, and 4) were observed, which are consistent with a low symmetry around the europium cation which could be occupying a C_{nv}, C_n, or C_s symmetry site. The substitution of two water molecules by a phen molecule (by comparing the complexes [Eu(eta)$_2$ (H$_2$O)$_4$]Cl and [Eu(eta)$_2$ (phen) (H$_2$O)$_2$]Cl) intensify the $^5D_0 \rightarrow {}^7F_2$ transition. This indicates that the chemical environmental has been changed, since such a transition is a hypersensitive one. The spectra exhibit a number of states higher than $2J+1$, suggesting that the compounds consist of a mixture of isomers.

The emission spectra for "free" and *in situ* organofunctionalized silica prepared complexes are shown in Figs. 3.9 and 3.10. As can be seen from Figs. 3.9a and 3.9b,

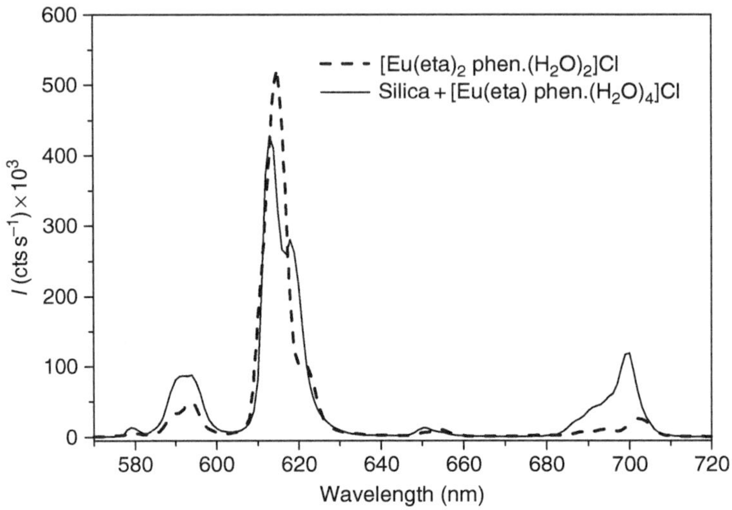

Figure 3.9 Emission spectra for "free" and *in situ* organofunctionalized silica prepared Eu–eta–phen complexes.

the intensity of the emission is higher (about 26 times higher) for the *in situ* prepared eta complex. This can be explained by taking into account that the silica complex exhibits a lower degree of freedom in comparison with the free compound. This fact (decreased degree of freedom) overwhelms the possible decrease of the intensity due to an increase in the number of water molecules into the coordination sphere. However, for the phen complex (Fig. 3.9), a lowering of intensity is observed. This suggests that, in this case, the effect of the increased number of water molecules is the predominant one.

Furthermore, it is possible to suppose that the efficiency of the actuation of phen as a chromophor has been reduced by its immobilization on the silica surface.

This hypothesis is reinforced by the fact that even minor modifications in the chromophor composition and structure can exert remarkable effects on energy transfer, as recently verified by a combinatory chemistry study [9].

The Ω_λ and Ω_4 values are shown in Table 3.1. Higher values of Ω_λ are related to transitions occurring mainly by dynamic coupling. The differing Ω_λ values of "free" and silica-based complexes indicate that in the "free" complexes the chemical environment is most polarizable, in comparison with the silica-based complexes. Furthermore, the differing Ω_λ values between "free" and silica-based complexes could be associated with a decrease of the metal–ligand bond energies for the silica-based species.

The above hypothesis regarding the effect of deposition of silica on the ability of phen to act as a chromophor is also reinforced by the fact that, for the eta–phen complex, a lowering of the Ω_4 value is observed when comparing the "free" and silica-based complexes, whereas for eta complex such lowering is not observed, since Ω_4 is related with orientation of the ligand molecule around the europium cation.

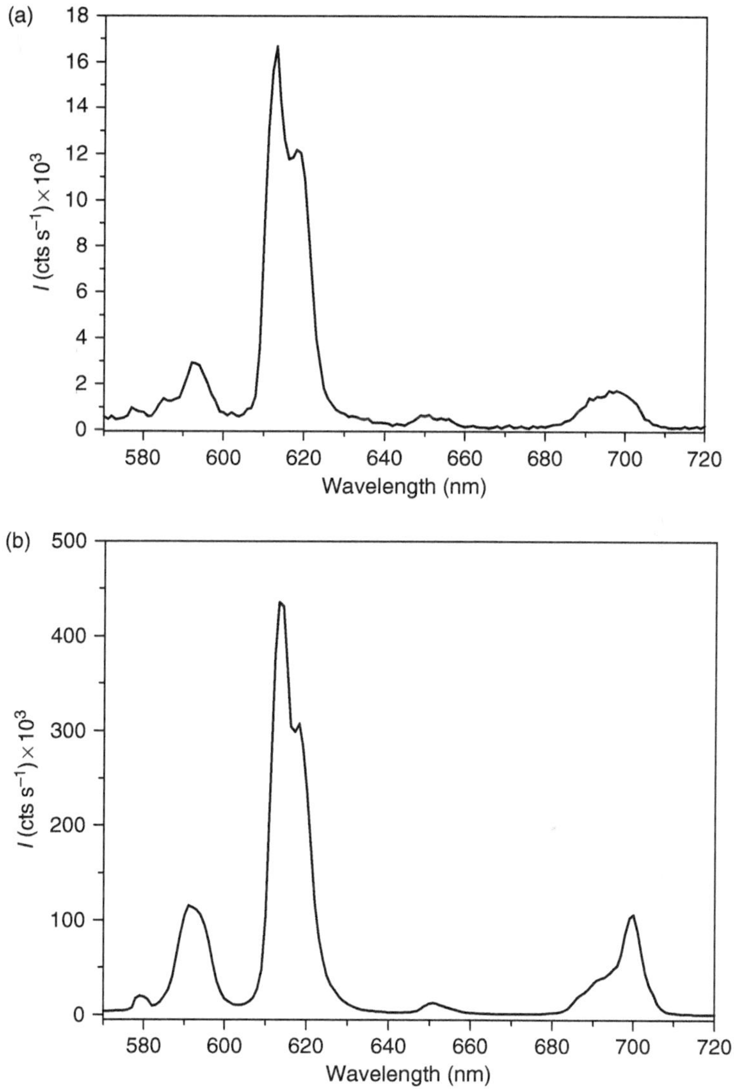

Figure 3.10 Emission spectra for (a) $[Eu(eta)_2 (H_2O)_4]Cl$ and (b) silica/$Eu(eta) (H_2O)_6$.

The experimental results indicate that the substitution of water molecules by a phen molecule into the coordination sphere of a Eu–eta complex enhances the luminescence intensity due to the lowering of the number of OH oscillators (which suppress the luminescence), as well as the ability of phen to act as a good chromophor. Furthermore, the *in situ* organofunctionalized silica–eta complex exhibits an optimization of optical properties, in comparison with the "free" complex, mainly an enhancement of the emission intensity. The experimental parameters show that the dynamic coupling mechanism is predominant.

5. Modified Amorphous Oxide Surfaces as Electrochemical Sensors

Due to their ability to coordinate metal cations, as well as interact with organic species, the amorphous oxide-modified surfaces can be successfully used to produce electrochemical sensors [10], as will be illustrated by two specific examples.

5.1 Metal cation electrochemical sensors

In this example, the mptt-modified surface previously described [11] will be used. Such a silica-modified surface can successfully used to promote the electrochemical determination (by redissolution voltammetry) of Co^{2+} and Ni^{2+} in aqueous solutions. It is found that both species can be simultaneously investigated in the concentration range 10^{-5}–10^{-7} mol dm^{-3}.

The detailed preparations and experimental conditions will be not reported here, but the electrochemical determinations were performed by using a modified carbon paste (CP) electrode produced by using the modified silica surface as "sensitive" species. The main experimental conditions and results are summarized in Figs. 3.11 and 3.12 and Table 3.2.

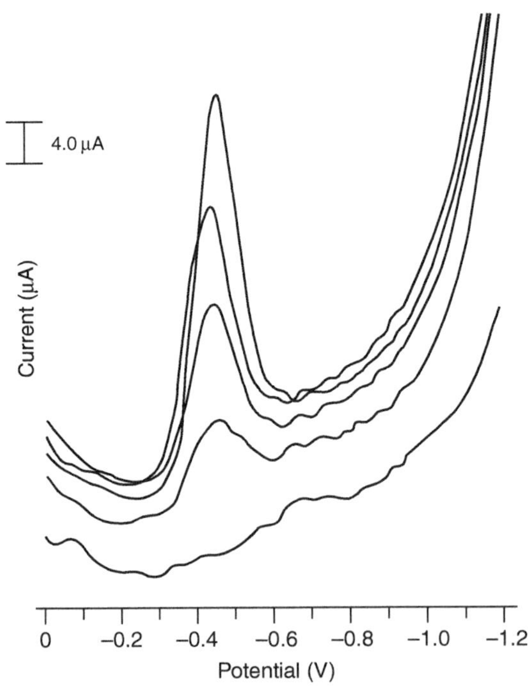

Figure 3.11 Characteristic voltammogram for the voltammetric determination of Co^{2+}. $E_i = -1200$, $E_f = 0$, and $V = 10$ mV s^{-1}. Co^{2+} concentration: 6875×10^{-4} mol dm^{-3}. Standard additions: 100, 200, 300, and 400 μl.

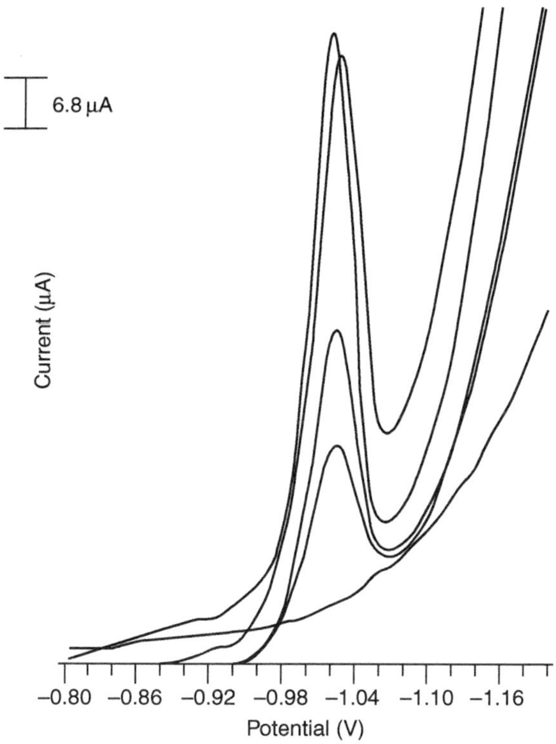

Figure 3.12 Characteristic voltammogram for the voltammetric determination of Ni^{2+}. $E_i = -800$, $E_f = 1150$, and $V = 10\,mV\,s^{-1}$. Co^{2+} concentration: $8.5 \times 10^{-5}\,mol\,dm^{-3}$. Standard additions: 30, 60, 90, and 120 μl.

Table 3.2 Experimental parameters to obtain the experimental voltammetric curves

Parameters	Results
Velocity	$10\,mV\,s^{-1}$
Direction of scanning	Anodic (positive)
Sensibility	$10\,\mu A\,V^{-1}$
Nitrogen time purge	900 s
Percentage of modified silica surface on the electrode	3%
Pulse amplitude	25 mV
Preconcentration potential (Co)	0–1.2 V
Preconcentration potential (Ni)	–800 to –1200 V
Preconcentration time	150 s

Based on the obtained experimental results, it is possible to verify that the mptt-modified surface can be successfully used to produce modified CP electrodes for qualitative and quantitative determinations of Co^{2+} and Ni^{2+} (even simultaneously) in aqueous solutions in the 10^{-5}–$10^{-7}\,mol\,dm^{-3}$ concentration range.

5.2 Paracetamol sensors

Paracetamol (*N*-acetyl-*p*-aminophenol (APAP), acetaminophen), is a well-known analgesic present in many pharmaceutical formulations. This work reports the use of an organofunctionalized silica–titania as a modifier to produce a CP electrode sensitive to paracetamol. An amino-modified surface is used as modifying agent, and anodic redissolution voltammetry is the technique used.

The organofucntionalized silica–titania was synthesized as follows: 0.15 mol of TEOS and 0.15 mol of titanium tetrabutoxide were dissolved into 0.1 mol of *N*-[3-(trimethoxysilyl)propyl]-ethylenediamine. The resulting solution was magnetically stirred by 30 min under nitrogen atmosphere and then aged for 24 h. Hence, to the previously prepared solution 8.0 mol of deionized water was added. The obtained gel was mechanically stirred for 5 h and the resulting powder aged for 48 h. The final product was then washed with deionized water and dried under vacuum at 35°C for 12 h.

Carbon, hydrogen, and nitrogen elemental analyses were performed in a Perkin-Elmer microanalyzer. XRD powder diffraction analysis was performed with a Shimadzu XD-3A instrument (Cu K_{α}, 35 kV, 25 mA). The voltammetric measurements were performed on a BAS equipment, model CV-50 W, in a 20 cm^3 Pyrex glass cell with a modified electrode (work electrode), a platinum wire (auxiliary electrode), and a Ag/AgCl electrode (reference electrode). As electrolyte, 0.1 mol dm^{-3} KCl solution was used.

The modified electrode was a CP electrode with the organofunctionalized silica–titania as modifier and was prepared as follows: 1.0 g of graphite (powder), three different modifier masses (0.002, 0.004, and 0.005 g), as well as 0.32 g of mineral oil were ground in a mortar. Better results were observed for the 0.005 g electrode.

The CHN elemental analysis results are 24% C, 11.1% H, and 5.2% N, giving a total of 4.0 mmol of groups —$(CH_2)_3$—NH—$(CH_2)_2$—NH_2 per gram of hybrid matrix. The XRD powder diffraction patterns (not shown) are characteristic of an amorphous compound.

Before the anodic redissolution experiments, a preliminary cyclic voltammetric study was performed and it was verified that paracetamol exhibits irreversible behavior, with potential peaks into the range 0.0–1.0 V. Similar results were reported [11] with the 750 mV potential used for paracetamol quantitative determinations.

Based on the previous results, the following work conditions were established: $E_i = +900$ mV, $E_f = +400$ mV, speed = 20 mV s^{-1}, pulse amplitude = 50 mV, equilibrium time = 15 s, sensibility = 100 μA, scanning direction: anodic (positive), and preconcentration time = 120 s. It was also verified that the maximum activity of the electrode occurs at pH 2.0. Under these experimental conditions, no electrochemical signal due to the organic or inorganic fractions of the modifier are observed.

By addition of paracetamol standard samples, it was verified that the electrochemical response is proportional to the total amount of analyte in solution. Hence, with all these conditions previously established, the modified electrode was evaluated.

The results obtained by analyzing four paracetamol commercial samples (each one from different suppliers) are summarized in Table 3.1. As can be verified, the results show the reliability of the modified electrode. Furthermore, it was verified that the paracetamol detection limit is 1.4×10^{-6} mol dm^{-3}. For example, the voltammograms for sample A (see Table 3.3) are shown in Fig. 3.13 Similar results were obtained for other samples [12].

Based on the experimental results, it is possible to conclude that an organofunctionalized silica–titania surface CP electrode can be used for both qualitative and quantitative determinations of paracetamol in pharmaceutical formulations, providing a low cost, quick, and reliable analytical methodology for such analyses.

Table 3.3 Concentration (ppm) of paracetamol in four commercial samples

Sample	Theoretical[a]	Experimental
A	25	25.1
B	500	413
C	750	780
D	500	434

[a] Values provided by the respective suppliers.

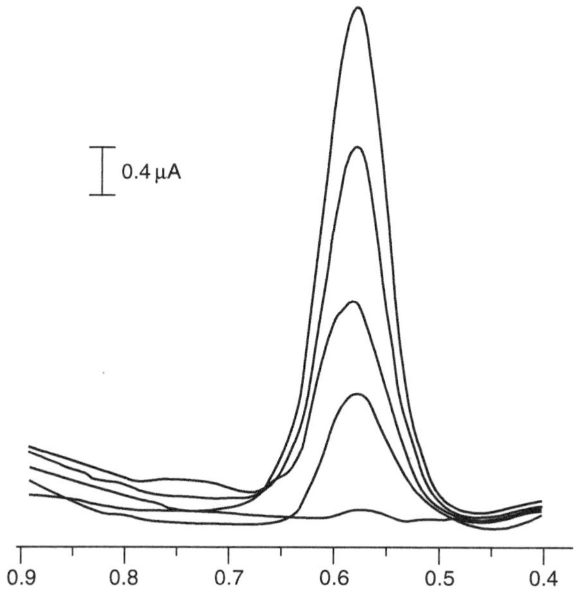

Figure 3.13 Characteristic voltammograms for paracetamol (sample A; Table 3.1). Solution concentration = 1.3×10^{-3} mol dm^{-3}.

5.3 Chemically modified oxide surfaces capable of molecular recognition

The title of present topic is the same of the review by Lisichkin, Novotortsev, and Bernadyuk [13], and is presented only to remember one of the possible applications of chemically modified oxides. The mentioned review is focused mainly on organo-functionalized silicas, which were, to a large extent, previously discussed in the present chapter.

6. The Influence of Magnesium Trisilicate and Silicon Dioxide on the Thermal Degradation of Ascorbic Acid

In the pharmaceutical industry, the use of excipients is of fundamental importance, since they serve several important functions such as agglutination, and they facilitate tablet decomposition. However, the excipients must not interact chemically with the chemical therapeutic substance, since such interaction could substantially affect the desired therapeutic properties. Magnesium trisilicate ($2MgO.3SiO_2.xH_2O$) and SiO_2 are two commonly used excipients. Hence, the purpose of the present study is to investigate, from the point of view of surface chemistry, the possible effects of magnesium trisilicate and SiO_2 on the thermal degradation of ascorbic acid. In this case, ascorbic acid is viewed as a surface modifier for $2MgO.3SiO_2.xH_2O$ and SiO_2. For comparison, magnesium stearate, $C_{36}H_{70}MgO_4$ is also used. The thermal degradation processes are studied by thermogravimetry (TG) and differential scanning calorimetry (DSC) analysis. The thermal study was performed on standard ascorbic acid samples and binary mixtures of ascorbic acid and excipients, under nitrogen and air atmospheres.

Ascorbic acid (Sigma), magnesium stearate (CIA-Estearina Parananheci), magnesium trisilicate (Farmar), and SiO_2 (Cab-O-Sil) were used without further purification.

Ascorbic acid samples as well as binary mixtures of ascorbic acid and the considered excipients were investigated. The binary mixtures were prepared taking into account the shoichiometry generally used for the preparation of ascorbic acid tablets: ascorbic acid (500 mg, of a 100 mesh sample), magnesium stearate (6.7 mg), magnesium trisilicate (13.3 mg), and SiO_2 (1.0 mg). The compounds were mixed in a mortar.

The thermogravimetric curves were obtained in the range 30–1200°C on a Shimadzu TGA-50) apparatus under dry air, air ("humid" air), and nitrogen. Samples of 4 mg, a gas flow of $50\,cm^3\,min^{-1}$, and the following three heating rates were used: $5°C\,min^{-1}$, $10°C\,min^{-1}$, and $20°C\,min^{-1}$. The DSC curves were obtained in the range 30–500°C in a Shimadzu DSC-50H apparatus under nitrogen atmosphere with a heating rate of $10°C\,min^{-1}$ and a gas flow of $50\,cm^3\,min^{-1}$.

The infrared spectra were obtained in KBr disks in the range 4000–400 cm^{-1} by using a FTIR BOMEM apparatus model MB 102. The XRD patterns (powder method) were obtained in a Phillips PW 1710 apparatus by using Cu K_α radiation. The diffraction patterns were compared with standard ones [13].

The infrared and X-ray powder diffraction data are in agreement with the standard nature of the studied ascorbic acid and excipients samples.

The thermogravimetric data obtained for the thermal degradation of ascorbic acid samples are summarized in Tables 3.4 and 3.5. Based on these results, the heating rate of 10°C min^{-1} was chosen for further analysis. The thermogravimetric data for thermal degradation of ascorbic acid samples at a heating rate of 10°C min^{-1} in the three studied atmospheres are summarized in Table 3.2. The three thermogravimetric curves are shown in Fig. 3.14.

As can be seen, there is no difference in the thermal degradation profile from dry to humid air. On the other hand, the elimination of the carbonaceous residue occurs at lower temperatures in oxidizing atmospheres than in an inert one (N_2). Furthermore, it can be seen that ascorbic acid is stable until almost 200°C. The visual observation performed in an electric furnace of the thermal degradation residues at 190°C, 220°C, and 240°C shows that they exhibit characteristics typical of fired sugar [16]. Furthermore, the infrared spectra obtained for each residue shows that they

Table 3.4 Thermogravimetric data for the degradation of standard ascorbic acid samples at different heating rates

Heating rate (°C min^{-1})	T_1 (°C)	T_2 (°C)	T_p (°C)	Δm (%)
5	212	806	228	97.6
			304	
			753	
10	221	262	240	
	274	481	323	
	466	845	772	100.0
20	219	800	250	
			324	
			673	92.3

T_1, T_2, and T_p are the initial, final, and "peak" temperatures for the thermal degradation processes, respectively. Δm is the observed mass loss.

Table 3.5 Summary of the thermogravimetric data for the thermal degradation of acid ascorbic samples at a heating rate of 10°C min^{-1} in the three studied atmospheres

Atmosphere	T_1 (°C)	T_2 (°C)	T_p (°C)	Δm (%)
N_2	220	262	240	33.8
	274	461	323	33.2
	466	845	772	33.0
Dry air	274	547	232	33.9
			322	36.0
			524	30.0
Humid air	227	558	251	33.9
			325	34.0
			529	32.0

T_1, T_2, and T_p are the initial, final, and "peak" temperatures for the thermal degradation processes, respectively. Δm is the observed mass loss.

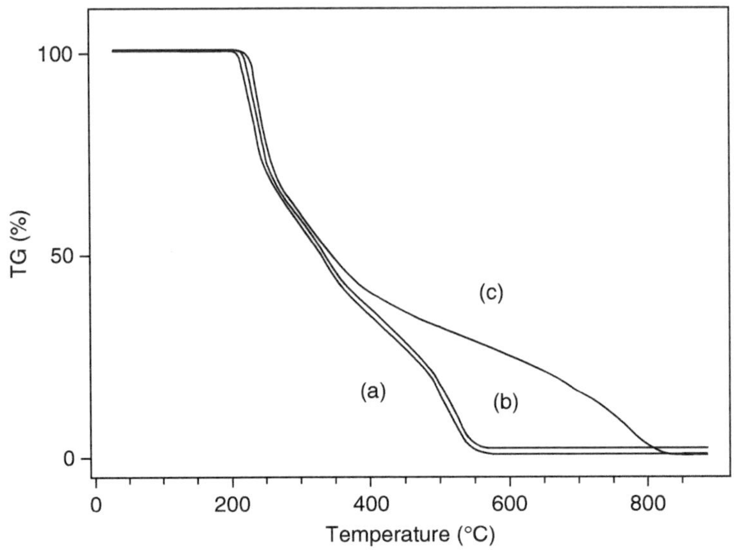

Figure 3.14 Thermogravimetric curves for the degradation of standard ascorbic acid samples under three different atmospheres: (a) N_2, (b) dry air, and (c) humid air.

are a mixture of nondecomposed ascorbic acid and carbonaceous residues; that is, all products of the thermal degradation of ascorbic acid are volatile and there are no solid intermediate products. Gas chromatography–mass spectrometry (GC–MS) data previously obtained for the volatile products [14] shown that they are a mixture of 32 different compounds.

The DSC curves for ascorbic acid are shown in Fig. 3.15.

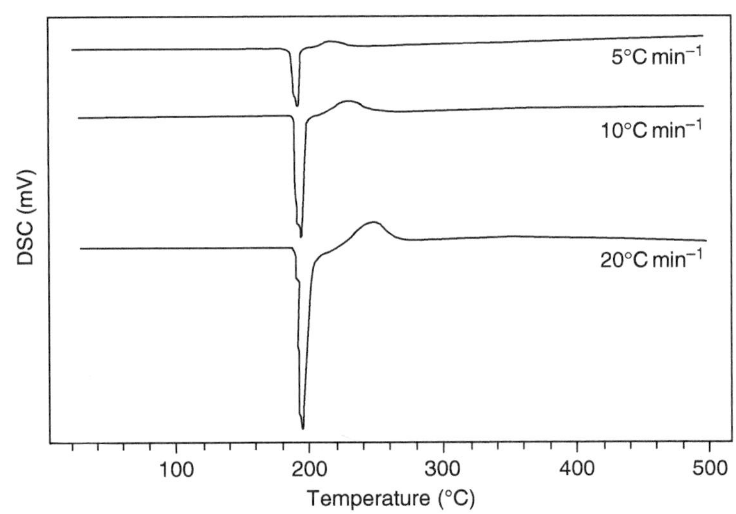

Figure 3.15 DSC curves for standard ascorbic acid samples at three different heating rates.

The heating rate of 10°C min^{-1} leads to a melting temperature of 193°C, which is a value in agreement with those previously reported [14,15]. The first peak (endothermic) is associated with the melting of ascorbic acid and the second one (exothermic) is associated with the thermal degradation process, with the release of volatile compounds and the formation of a carbonaceous residue. For the heating rate of 10°C min^{-1} these two processes exhibit values of 252 and −112 J g^{-1}, respectively.

The TG and DSC data obtained for the magnesium stearate are summarized in Table 3.6. The TG–DTG/DSC curves for magnesium stearate are shown in Fig. 3.16. Magnesium trisilicate and SiO_2 have not exhibited any mass loss or

Table 3.6 TG and DSC data summary for magnesium stearate (N_2 atmosphere, 10°C min^{-1})

TG data	T_1	T_2	T_p (°C)	Δm (%)
	56	123	107	4.2
	332	461	384	88.6
DSC data	T_1	T_2	T_p (°C)	ΔH (J g^{-1})
	66	126	92	302
			116	
	305	380	350	233
	382	398	395	1
	399	458	441	10

T_1, T_2, and T_p are the initial, final, and "peak" temperatures for the thermal degradation processes, respectively. Δm is the observed mass loss. ΔH is the enthalpy variation.

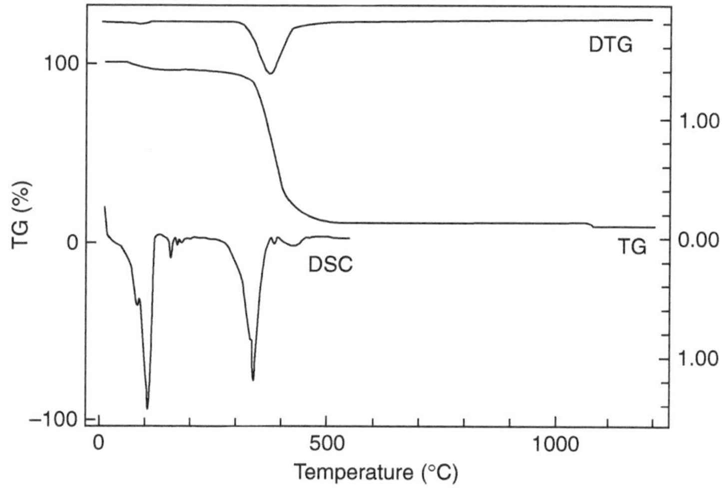

Figure 3.16 TG–DTG/DSC curves for magnesium stearate.

enthalpic variations and so their TG–DTG (differential thermal gravimetry)/DSC curves and data are not shown.

For magnesium stearate, the first mass loss step (4.2%) is associated with dehydration and the second one (88.6%) with thermal degradation of the organic moiety. In the DSC curve, the first and second observed endothermic peaks are associated with the dehydration and thermal degradation processes, respectively.

The composition of binary mixtures are described in the experimental section. The comparison of the infrared and X-ray powder diffraction data obtained for pure ascorbic acid and excipients with those obtained for the respective binary mixtures strongly suggests that there are no interactions between the ascorbic acid and excipient molecules, since the main infrared bands for all compounds remains unchanged and the X-ray powder diffraction is merely the superposition of individual patterns, with no new diffraction peaks associated with the formation of a new crystalline phase. The main TG and DSC data obtained for the binary mixtures are summarized in Tables 3.7 and 3.8, respectively. The TG and DSC curves for the binary mixtures are shown in Figs. 3.17–3.22. From the TG and DSC curves it is evident that the presence of the excipients has no influence on the thermal degradation process observed for ascorbic acid samples, except for minor dislocation of the temperature range observed for the thermal degradation processes.

Table 3.7 TG data for binary mixtures involving ascorbic acid and various excipients (N_2 atmosphere, 10°C min^{-1})

Binary mixture	T_1	T_2	T_p (°C)
Ascorbic acid/magnesium stearate	206	248	225 313 628
Ascorbic acid/magnesium trisilicate	215	–	236 315 633
Ascorbic acid/SiO_2	214	–	230 312 665

T_1, T_2, and T_p are the initial, final, and "peak" temperatures for the thermal degradation processes, respectively.

Table 3.8 DSC data summary for the studied binary mixtures involving ascorbic acid and excipients (N_2 atmosphere, 10°C min^{-1})

Binary mixture	T_1	T_2	T_p (°C)	ΔH (J g^{-1})
Ascorbic acid/magnesium stearate	183	199	190	229.4
	208	246	223	−106.8
Ascorbic acid/magnesium trisilicate	191	207	193	215.8
	210	250	232	−74.4
Ascorbic acid/SiO_2	188	203	193	230.7
	204	251	232	−142.2

T_1, T_2, and T_p are the initial, final, and "peak" temperatures for the thermal degradation processes, respectively.

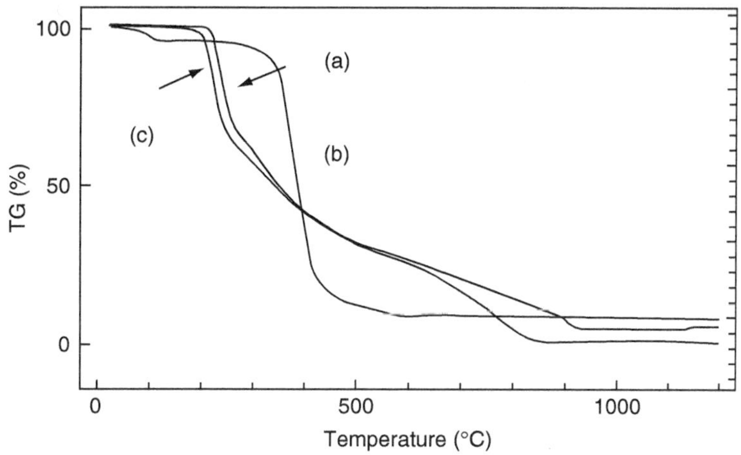

Figure 3.17 TG curves for (a) ascorbic acid, (b) magnesium stearate, and (c) ascorbic acid/magnesium stearate binary mixture.

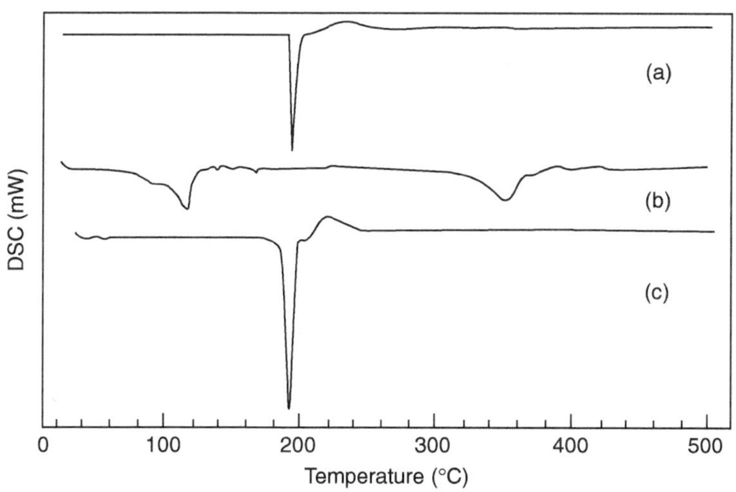

Figure 3.18 DSC curves for (a) ascorbic acid, (b) magnesium stearate, and (c) ascorbic acid/magnesium stearate binary mixture.

Based on the TG and DSC experimental results it can be concluded that the excipients, that is, magnesium stearate, magnesium trisilicate, and SiO_2, have no substantial influences on the thermal stability of ascorbic acid. This result shows that the three compounds can be used (at least from a thermal point of view) as excipients, with no restrictions, for ascorbic acid tablet formulations, since they will not provoke or accelerate the thermal degradation of ascorbic acid. However, by comparison of Tables 3.5 and 3.7, it is evident that the temperature of the

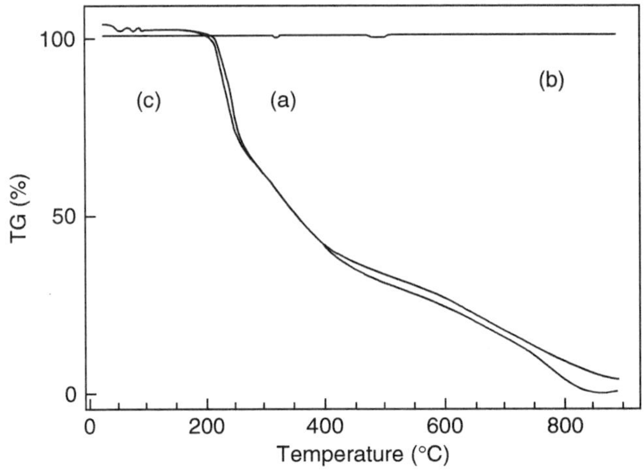

Figure 3.19 TG curves for (a) ascorbic acid, (b) magnesium trisilicate, and (c) ascorbic acid/ trisilicate binary mixture.

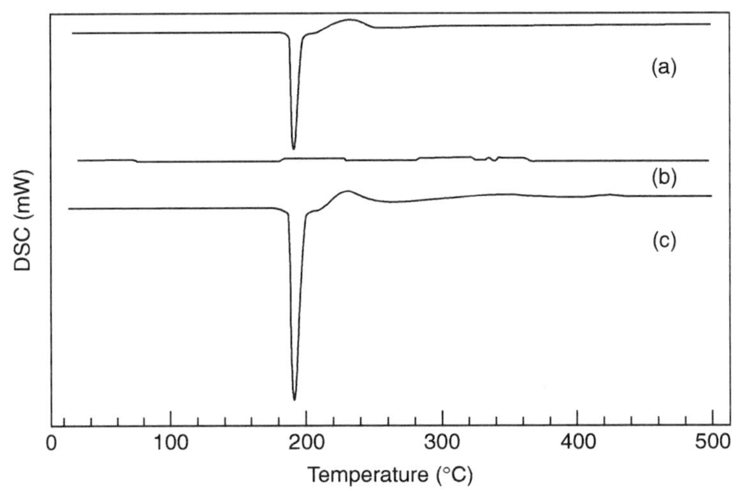

Figure 3.20 DSC curves for (a) ascorbic acid, (b) magnesium trisilicate, and (c) ascorbic acid/ trisilicate binary mixture.

beginning of thermal degradation of ascorbic acid is lowered by about 5°C for magnesium trisilicate and SiO_2, and about 15°C for magnesium stearate mixtures. Furthermore, the melting point of ascorbic acid is lowered from 193°C to 183°C, 191°C, and 188°C for magnesium stearate, magnesium trisilicate, and SiO_2 binary mixtures, respectively. Hence, it can be inferred that ascorbic acid exhibits a surface interaction with magnesium trisilicate and SiO_2, showing that such excipients are not completely "inert" from a physico-chemical point of view. Such facts must be considered in ascorbic acid tablet storage planning.

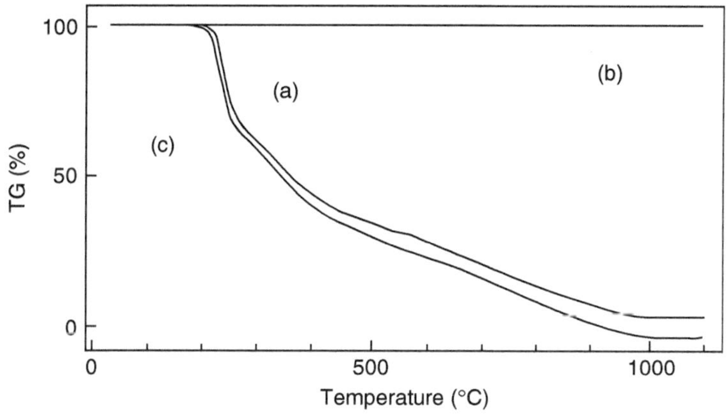

Figure 3.21 TG curves for (a) ascorbic acid, (b) silicon oxide, and (c) ascorbic acid/silicon oxide binary mixtures.

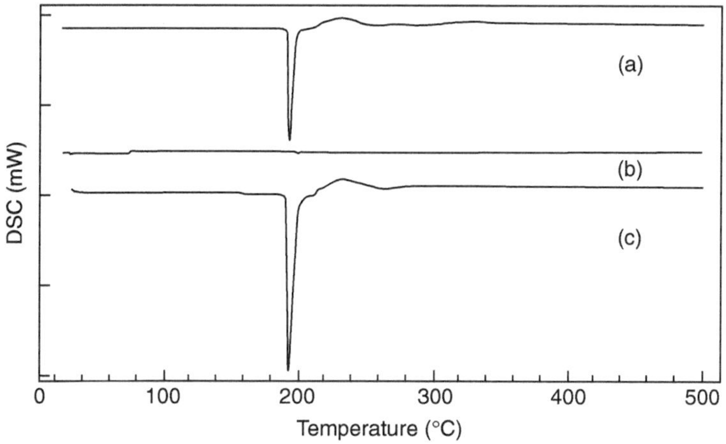

Figure 3.22 DSC curves for (a) ascorbic acid, (b) silicon oxide, and (c) ascorbic acid/silicon oxide binary mixtures.

REFERENCES

1. R.F. de Farias, C. Airoldi, J. Thermal Anal. 53 (1998) 751–756.
2. R.K. Iler, The Chemistry of Silica, John Wiley, New York, 1979.
3. C. Airoldi, R.F. de Farias, Quím. Nova 23 (2000) 496–503.
4. C.R. Silva, I.C.S.F. Jardim, C. Airoldi, J. Sep. Sci. 24 (2001) 197.
5. C.R. Silva, C. Airoldi, K.E. Collins, C.H. Collins, J. Chromatogr. A 1087 (2005) 29.
6. R.F. de Farias, S. Alves Jr., M.F. Belian, M.R.S. Vieira, J.M. de Souza, G.G. Pedrosa, G.F. de Sá, J. Serb. Chem. Soc. 69 (2004) 387.
7. R.F. de Farias, S. Alves Jr., M.F. Belian, G.F. de Sá, J. Colloid Interface Sci. 243 (2001) 523.
8. R.F. de Farias, S. Alves Jr., M.F. Belian, G.F. de Sá, Opt. Mater. 18 (2002) 431.

9. A.P.S. de Freitas, M.F. Belian, O.L. Malta, G.F. de Sá, R.F. de Farias, S. Alves Jr., J. Alloy Compounds, 459 (2008), 543.
10. R.O. Freire, R.Q. Albuquerque, S. Alves Jr., G.B. Rocha, M.E. Mesquita, Chem. Phys. Lett. 405 (2005) 123.
11. R.F. de Farias, J.A. Rocha, E.P. Marques, Quim. Bras. 1(1) (2007) 25.
12. D.J. Miner, J.R. Rice, R.M. Riggin, P.T. Kissinger, Anal. Chem. 53 (1981) 2259.
13. G.V. Lisichkin, R.Yu. Novotortsev, S.Z. Bernadyuk, Colloid J. 66 (2004) 387.
14. Powder Diffraction File of the Joint Committee on Powder Diffraction Data, International Centre for Diffraction data, Swarthmore, PA, 1995.
15. G. Vernin, S. Chakib, C. Parkanyi, Carbohydr. Res. 305 (1998) 1.
16. S. Lerdkanchanapon, D. Dollimore, J. Thermal Anal. 49 (1997) 887.

CHAPTER FOUR

Chemistry on Organofunctionalized Nanostructured Oxides' Surfaces

Contents

1. Introduction

The use of alkoxides as precursors to produce new materials [1] has increased in the last decade due to the advantages of the so-called sol–gel process [2] such as the preparation of compounds with high homogeneity, at room temperatures, in contrast with the typical high-temperature syntheses in solid-state chemistry. Such an experimental approach can lead not only to modified amorphous oxide surfaces, as discussed in Chapter 1, but also to nanostructured modified surfaces, as will be shown here.

Interface Science and Technology, Volume 17
ISSN: 1573-4285, doi: 10.1016/S1573-4285 (08) 00404-3

Figure 4.1 Scanning electron micrography of nanostructured, organomodified zirconia–titania.

2. Inorganic–Organic Hybrids

As the first example, we mention the organomodified zirconia–titania [3] obtained using 1,12-diaminedodecane, zirconium, and titanium butoxides as precursors. That organofunctionalized compound exhibits a hexagonal nanostructure and particles with spherical morphology, as shown in Fig. 4.1.

This illustrates the very important effect of experimental conditions on the nature of the modified surface. In this case, specifically, when alkoxides are used, the total amount and speed of water addition are critical factors affecting the final properties of the modified surface.

3. Sol–Gel Lamellar Silica

The sol–gel process can be used to obtain lamellar silica (LS) samples by using neutral amines as template molecules. Its is found that such LS samples are able to act as sequestrating agents toward transition metal cations and that the coordination of such metal cations of the three dimensional network structure of the silica exerts remarkable effects on its nanostructure and thermal stability [4]. Furthermore, such metal sequestrating ability and the consequent nanostructure modifications are observed even if the metal–silica reaction is performed in the solid state and room temperature [5]. By performing a solution calorimetric study, it is possible to verify that the total amount of metal cations that the lamellar matrix is able to sequester as well as its affinity by the metal cations, for example, $Ni^{2+} > Cu^{2+} > Co^{2+}$ [6] is a consequence of the influence of main two factors: the metal–nitrogen coordination enthalpies and the structural disorder provoked into the lamellar network by the metal–nitrogen coordination.

For LS obtained by the neutral amine route, it is also found that an increase of the external pressure can provoke a 16% reduction of the interlayer distance [7] as well as a decrease of the emission intensity of europium complexes adsorbed on such surface [8], making the matrix suitable for the production of "pressure sensors."

Through the sol–gel process, single oxides such as silica can be synthesized with an organically modified surface, as well as double or triple oxides [9] such as Si—Al, Si—Ti, Si—Zr, Al—Zr, Al—Ti, and Al—Ti—Zr. For such syntheses, 1,10-diaminedecane and 1,12-diaminedodecane, as well as a surfactant such as dodecyltrimethylammonium bromide, can be used, with the final modified oxides exhibiting lamellar or hexagonal nanostructure. Such compounds with modified surface are also able to coordinate transition metal cations.

Due to its ability to coordinate metal cations, the sol–gel amine-modified LS can be used as precursor for the synthesis of ceramic powders [10]. It has been found that the surface area of the powders are a function of the nature (charge and radius) or the coordinated cations.

3.1 Eu(II)-doped lamellar silica

LS was synthesized as follows: 1.0×10^{-2} mol of 1,12-diaminododecane was dissolved in a mixture of water (3.3 mol) and propanol (0.50 mol). To the resulting solution, 1.0×10^{-2} mol of TEOS was added. The formed gel was aged for 24 h and then dried under vacuum at 80°C for 8 h. The resulting powder was characterized as described in detail elsewhere [2–6], confirming the lamellar nature of the matrix, which exhibits an interlayer distance of 2.3 nm. The synthesized lamellar matrix exhibits the following chemical composition, as found by TG: 53% SiO_2 and 47% 1,12-diaminedodecane.

The $Eu(fod)_3.2H_2O$ complex (fod is the anion of the 2,2-dimethyl-6,6,7,7,8,8,8-heptafluoro-3,5-octanedione) was of analytical grade (Aldrich) and used after recrystallization from ethanol.

Europium-doped LS samples were prepared as follows: the $Eu(fod)_3$ was dissolved in dried ethanol and to the resulting solution the insoluble LS was added. The suspension was stirred for 30 min and the solvent evaporated. The resulting powder was dried at 60°C under vacuum in a stove. $Eu(fod)_3$ and LS were mixed in stoichiometric amounts to produce samples with 1%, 5%, and 10% (m/m) of adsorbed complex.

The emission spectra were analyzed using a Jobin Yvon model double monochromator, model U-1000, and the fluorescence signal detected by a water-cooled RCA C31034-02 photomultiplier was processed by a Jobin Spectralink system. The excitation wavelength was 370 nm. To ensure the reliability of the comparison of the spectra main line intensities, the geometry of the counting system was maintained constant, that is, with the same sample container, the same excitation wavelength, and the same slit width 0.05 mm. The lifetime measurements were performed at 298 K exciting the sample with the third harmonic of a Nd–YAG laser ($\lambda_{exc} = 355$ nm) and detecting the emitted light, after dispersion through a Jobin Yvon model H-10 monochromator, with a P28 photomultiplier tube.

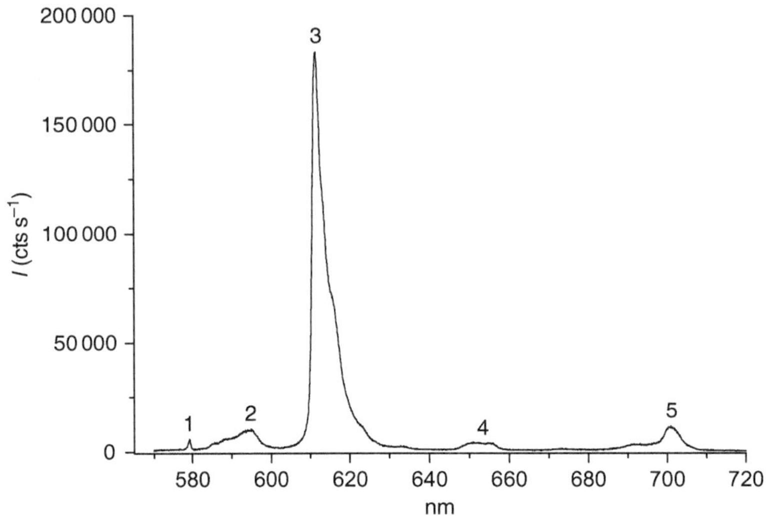

Figure 4.2 Emission spectrum of the LS sample with 10% of $Eu(fod)_3$, pressed at 380 ± 19 MPa for 60 min. The emission bands from 1 to 5 are named in the text.

The samples for analysis were pressed on an uniaxial press under 380 ± 19 MPa for 0.5, 30, and 60 min. The product will be denoted *x*EuSLy, with *x* being the percentage of Eu(fod) in the matrix [1%, 5%, and 10%] and *y* being the time of applied pressure [0.5, 30, and 60 min].

All of the emission spectra are very similar; as a typical example, the spectrum of the sample 10EuSL60 is shown in Fig. 4.2. The emission bands from 1 to 5 are, respectively, $^5D_0 \rightarrow {}^7F_0$, $^5D_0 \rightarrow {}^7F_1$, $^5D_0 \rightarrow {}^7F_2$, $^5D_0 \rightarrow {}^7F_3$, and $^5D_0 \rightarrow {}^7F_4$ [7]. For discussion purposes only the most intense $^5D_0 \rightarrow {}^7F_2$ band will be considered. The intensity values for the $^5D_0 \rightarrow {}^7F_2$ band for all samples are shown in Table 4.1.

As can be seen, the intensity increases, as the percentage of $Eu(fod)_3$ in the solid sample increases. However, the minor increase in intensity from 5%

Table 4.1 Intensities for the $^5D_0 \rightarrow {}^7F_2$ emission band for solid samples of LS with adsorbed $Eu(fod)_3$ complex

Sample	Intensity (cts s^{-1}) × 10^4
1EuSL0.5	5.54
1EuSL30	3.94
1EuSL60	3.10
5EuSL0.5	15.44
5EuSL30	14.31
5EuSL60	14.45
10EuSL0.5	16.14
10EuSL30	16.22
10EuSL60	18.23

to 10% of the complex-adsorbed sample suggests a concentration quenching. Furthermore, as a general behavior, it can be observed that the intensity of the considered emission band decreases with time, at the constant applied pressure.

The increase of the intensity of the $^5D_0 \rightarrow {}^7F_2$ emission band with the increase of the percentage of the adsorbed complex is a predictable result, taking into account that the measured intensity is proportional to the complex amount. For samples with 1% and 5% of the europium complex, it is found that the intensity suffers a decrease with increasing time of applied pressure. Since a decrease of the interlayer distance is observed with time, under a constant pressure [4], it is reasonable to suppose that the intensity of the $^5D_0 \rightarrow {}^7F_2$ emission band decreases as the interlayer distance decreases, with both phenomena very closely related. The observed intensity increase for the 10EuSL60 sample, in comparison with the 10EuSL0.5 and 10EuSL30 ones, can be explained by supposing that for a concentrated europium sample, a large reduction of the interlayer distance provokes the reorientation of the diamine molecules in a way that favors energy transfer to the metal.

Since the complex $Eu(fod)_3$ has two water molecules in its coordination sphere, which can be replaced by other ligands, including those with nitrogen as coordination site, like phen, and taking into account that the synthesized LS have "free" nitrogen basic sites, able to coordinate metals, it is possible to suppose that $Eu(fod)_3$ molecules are coordinated to the LS sample (chemisorption) and not only weakly adsorbed (physisorption). However, this is presented only as a hypothesis, since the infrared spectra are not conclusive, due to the presence of large and intense bands attributed to the inorganic substrate, which masks the infrared bands due to the organic moiety. The presence of the $^5D_0 \rightarrow {}^7F_0$ band is indicative of the presence of a center of inversion in the complex molecule, in agreement with D_{3h} symmetry.

To verify the possible effects of the total amount of adsorbed complex on the luminescent lifetimes of the samples, lifetime measurements were performed for the 1%, 5%, and 10% matrices pressed for 0.5 min. The measured luminescent lifetimes are 507, 551, and 581 μs, respectively. So, it is evident, for the pressed hybrid samples, the luminescence lifetime is increased with an increased amount of adsorbed luminescent complex. This suggests that not only adsorption but also coordination occurs between the europium complex and the nitrogen atoms of the organic moiety, since a large complex/matrix ratio implies a minor amount of nitrogen atoms per complex molecule, that is, a minor amount of nitrogen atoms enters the europium coordination sphere.

Taking into account that the emission spectrum of "free" $Eu(fod)_3$ is almost identical to that obtained for the $Eu(fod)_3$-doped LS, it is possible to claim that LS could be used as an inert substrate to support $Eu(fod)_3$ and, possibly, other LCMDs. Furthermore, a relationship between interlayer decrease and intensity emission band decrease was observed for some doped samples, suggesting the possibility of uses of such doped compounds as some king of "pressure sensors," which, of course, will require most extensive and deep investigations, which is now on progress.

4. Titania Foam and Thiol-Functionalized Nanoporous Titania–Silica

Titanium dioxide is a well-known photocatalyst for air purification, exhibiting other possible applications [11]. In this connection, the sol–gel process has been successfully used for the preparation of a series of titania containing single, double, or triple oxides with regular nanostructure [12–17].

In an attempt to develop new, easy, and cheaper experimental procedures to prepare crystalline titania, it has been shown that anatase phase can be synthesized at room temperature without any previous or further thermal treatment [17,18]. Furthermore, it has been shown that, for sol–gel-prepared titania samples, a doping process with cerium or copper oxides, as well as silica, can enhance the thermal stability of the anatase phase, producing very stable samples (up to 1000°C).

However, to improve and/or extend the possible applications of TiO_2, it is necessary to develop porous high surface area samples, with the control of nanostructure being a main purpose. In this connection, the dissolution of TiO_2 into hydrogen peroxide has been used to produce TiO_2 foams [19]. However, in such preparations the foam obtained is sponge-like and not a really porous matrix, for example, as in the sense used for active carbon.

A specific example is the synthesis of a titania–silica inorganic–organic hybrid matrix, functionalized with thiol groups [20]. Such a matrix exhibits a porous nanostructure, with pores from 200 to 500 nm. The matrix is obtained by a sol–gel process, using titanium and thiol-functionalized silicon alkoxides. The porous nanostructure is achieved by using polymeric spheres as templates. The main purpose of such a preparation is to obtain a hybrid matrix with several functions, able to act as a photocatalyst, as well as a soft metal-sequestering agent (e.g., mercury). The preparation of a titania foam exhibiting regular pores has also been reported.

4.1 Thiol-functionalized nanoporous titania–silica

In a typical preparation, 50 mg of methyl methacrylate copolymer was dissolved in a mixture of 10 cm^3 of titanium butoxide and 2 cm^3 of mercaptopropyltriethoxysilane. Another synthesis with higher amounts of the thiol-functionalized alkoxide was also performed. To evaluate the role of the thiol silicon alkoxide on nanoporous hybrid formation, some syntheses were performed using only the methyl metacrylate copolymer and the titanium alkoxide.

SEM micrographs were obtained with a JEOL JSM 6360 LV SEM, with an accelerating voltage of 15 kV. XRD patterns were obtained with a Shimadzu XD-3A instrument (Cu K_α radiation, 35 kV, 25 mA). The SEM micrographs are shown in Fig. 4.2.

Inspecting the SEM micrographs shown in Fig. 4.3, some conclusions can be pointed out:

a. The methyl methacrylate spheres were dissolved in a mixture of titanium–thiol-functionalized silicon alkoxides, and the dissolved spheres led to the formation of nanopores with diameters from 200 to 500 nm.

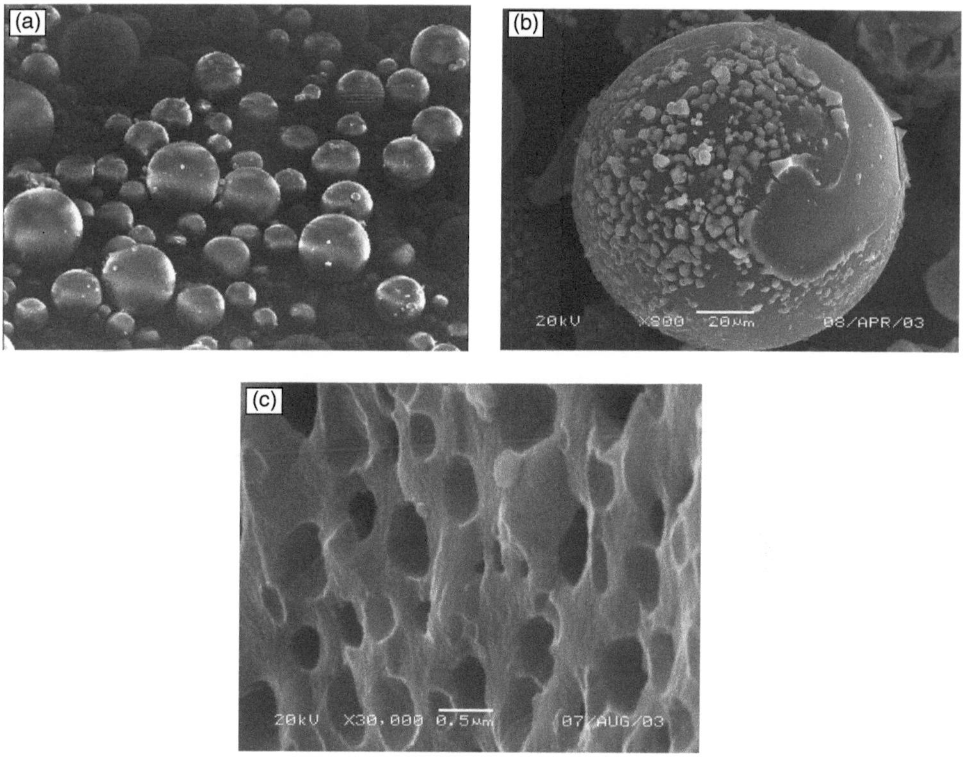

Figure 4.3 Scanning electron micrographs for methyl methacrylate spheres (a), methyl methacrylate sphere covered with TiO_2 (b), and nanoporous thiol-functionalized titania–silica (c).

b. The presence of the thiol-functionalized silicon alkoxide is of prominent importance to the dissolution of the methyl methacrylate spheres, producing nanoporous hybrids, since when only the titanium alkoxide was used, the obtained product was not a nanoporous hybrid but merely the methyl methacrylate spheres covered wth TiO_2.

The evidence for the formation of a new crystalline phase is also reinforced by the XRD patterns shown in Fig. 4.4. As can be verified, the methyl methacrylate spheres exhibit some degree of crystallinity. Furthermore, the titania-covered spheres exhibit the same XRD patterns as the uncovered spheres, indicating that the TiO_2 "coat" is amorphous. On the other hand, the nanoporous thiol-functionalized titania–silica hybrid exhibits a distinct XRD pattern, showing that a new crystalline phase was formed. By comparison with previously prepared double oxides [13,16] the diffraction peak at 5.5° observed in Fig. 4.2c could be attributed to the {100} diffraction plane of a hexagonal phase.

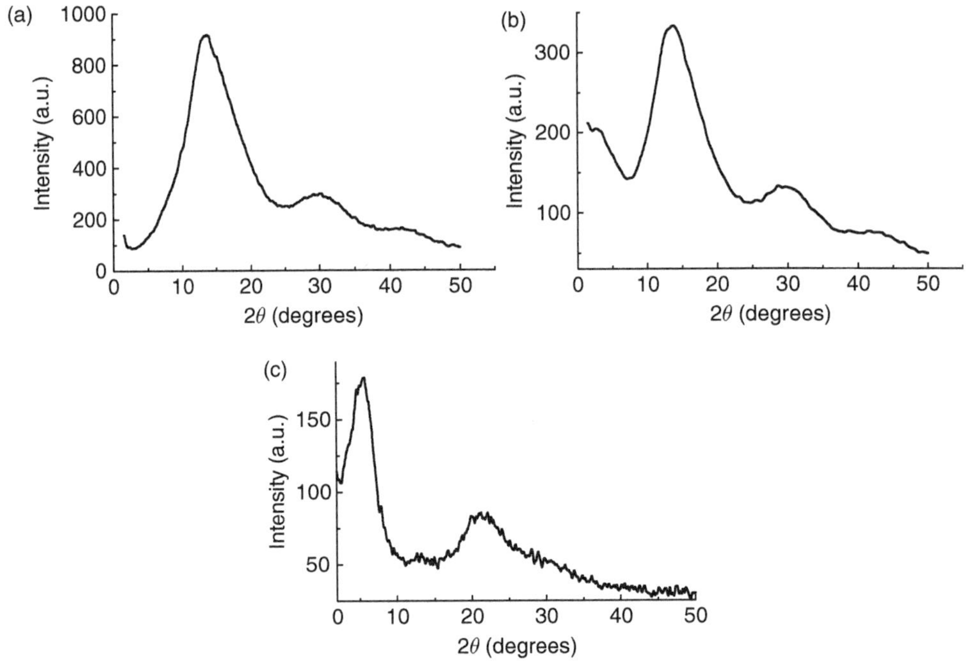

Figure 4.4 XRD patterns for methyl methacrylate spheres (a), methyl methacrylate spheres covered with TiO_2 (b), and nanoporous thiol-functionalized titania–silica (c).

Three of the SEM micrographs for the titania foams are shown in Fig. 4.5. The magnifications are shown in each figure. As can be observed, the TiO_2 foam exhibits a very porous microstructure with pore diameters ranging from 1.5 to 3.0 μm (mean values). Furthermore, the observed morphological features are quite different from those verified for a titania sample prepared by using a similar experimental approach previously reported [9].

4.2 Titania foam

In a typical preparation, 4.0 g of polyethyleneglycol 4000 was dissolved in 50 cm^3 of deionized water. To this solution, 2.0 g of amorphous TiO_2 (previously prepared by hydrolysis and polycondensation of titanium butoxide in deiozined water) was added, as well as 40 cm^3 of a 40% H_2O_2 solution. To increase the H_2O_2 degradation rate, with the consequent liberation of O_2 bubbles, 10 cm^3 of a 0.5 mol dm^{-3} KI solution was added, under magnetic stirring. Then a brown-yellow foam was formed, with an increase of 900% (in comparison with the initial volume) was observed. The foam was then dried at room temperature (27°C) for 2 days in a fume hood before SEM analysis. The SEM micrographs were obtained with a JEOL JSM 6360 LV SEM with an accelerating voltage of 15 kV.

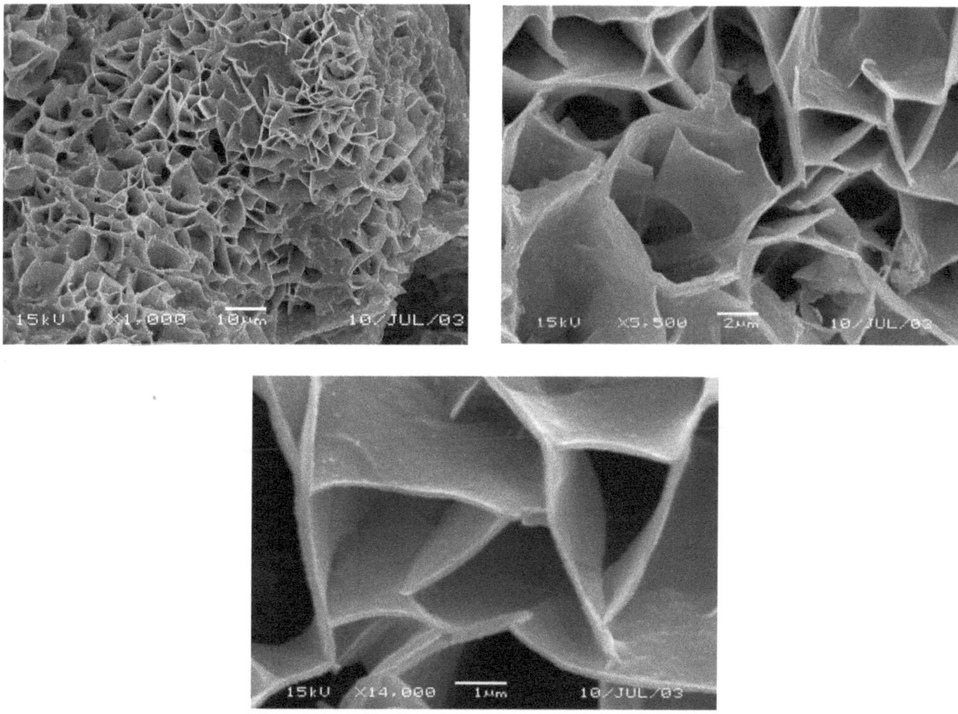

Figure 4.5 Scanning electron micrographs of TiO_2 porous foam at 1 000×, 5 500×, and 14 000×.

5. The Effects of Nanostructure on Adsorption Features: Metal Chemisorption on Hexagonal Templated Zirconia Obtained Thorough Sol–Gel Process

It was verified by investigation of the adsorption of Co(II), Ni(II), Cu(II), and Zn(II) from aqueous solutions on a templated hexagonal zirconia matrix obtained by a sol–gel process [21] that the nanostructure of the modified surface can exert remarkable effects on the adsorption features. Such a modified surface was prepared as follows:

About 20.0 mmol of 1,12-diaminodecane was dissolved in 100 cm^3 of 2-butanol. To this solution, 0.55 mol of zirconium butoxide was added dropwise under magnetic stirring, until a new solution was formed, after which, 76.0 mmol of 1,8-diaminooctane, previously dissolved in 300 cm^3 of deionized water, was immediately added, under magnetic stirring to the diaminodecane–zirconium butoxide solution. The powder was stirred for 3.0 min and then aged for 48 h (at room temperature, i.e., 27°C), after which it was washed with 300 cm^3 of deionized water, followed by 300 cm^3 of dry ethanol.

The adsorption experiments were performed as follows: 0.100 g of the hexagonal templated zirconia matrix was suspended, under magnetic stirring, for 60 min, into 20 cm^3 of an aqueous solutions of Co(II), Ni(II), Cu(II), and Zn(II) previously prepared by dissolution of the respective sulfates (with exception of cobalt, for which chloride was used). The flasks containing the suspensions were kept on a water bath at controlled temperature (25 ± 0.5°C). The total amount of metal cations before and after the adsorption process was measured for each individual cation, by volumetric titration with a 3.2×10^{-3} mol dm^{-3} EDTA solution.

The X-ray powder diffraction pattern for the modified matrix is shown in Fig. 4.6. The adsorption results are summarized in Fig. 4.7.

The XRD pattern for the hybrid zirconia sample exhibits a peak at 3.1°, corresponding to a d value of 2.9 nm, associated with the 100 diffraction plane of a hexagonal phase. The obtained scanning micrographs (not shown) showed that the synthesized powder exhibits shapeless grains.

The adsorption experimental results, summarized in Fig. 4.6, indicate that the adsorption affinity of the synthesized hexagonal templated zirconia toward Co(II), Ni(II), Cu(II), and Zn(II) follows the sequence Cu(II) > Zn(II) >> Ni(II) > Co(II). The adsorption affinities sequence is in total agreement with the calculated values for the adsorption constants, K_{ads} = 611.8 (Cu), 228.0 (Zn), 208.5 (Ni), and 66.1 (Co).

It is found that the adsorption of the considered cations follows the Langmuir and not the Freundlich isotherm. All isotherms are of type I according to the IUPAC classification [22]. This is as expected, since the adsorption of the metal cations to the inorganic substrate occurs through coordination with the nitrogen atoms of the template molecules, that is, through the formation of a chemical bond (coordination). So, there is a chemisorption process, with the formation of a "monolayer" of adsorbed (in this case coordinated) cations.

These adsorption affinities of the zirconia matrix could, as a first hypothesis, be attributed to the different coordination features of Co(II), Ni(II), Cu(II), and

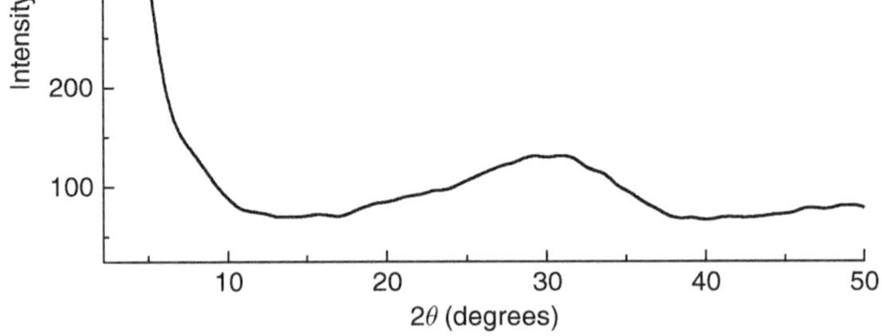

Figure 4.6 XRD pattern for the hexagonal templated zirconia sample.

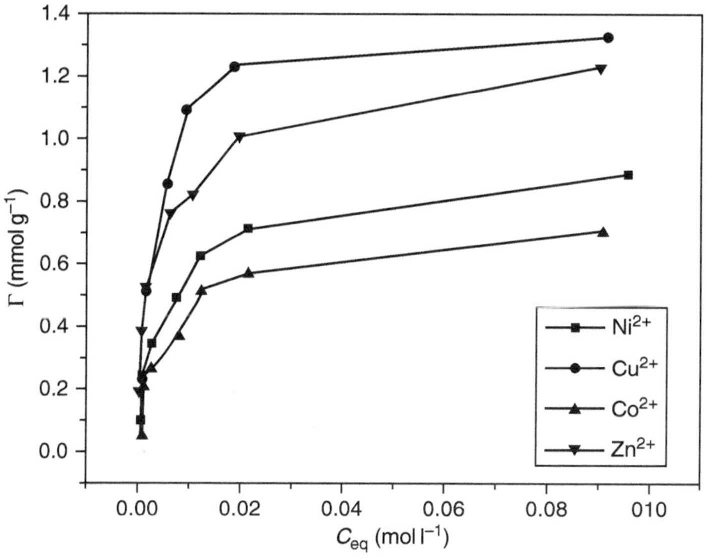

Figure 4.7 Adsorption isotherms for the adsorption of Co(II), Ni(II), Cu(II), and Zn(II) on the synthesized hexagonal templated zirconia.

Zn(II): Zn(II) generally produces four coordinated tetrahedral compounds [23]. On the other hand, Ni(II) and Cu(II) generally forms four- or five-coordinate complexes, with tetrahedral (copper), square planar (nickel), or bipyramidal trigonal (nickel and copper) geometries, whereas coordination number 6, with octahedral geometry is the most common for cobalt compounds [23]. So, Co(II) is able to coordinate to a higher number of nitrogen atoms, in comparison with Zn(II), Ni(II), and Cu(II). Hence, each Co(II) cation is "taking" a higher number of possible coordination sites in the hybrid matrix, giving a minor amount of coordinated cations in comparison with Zn(II), Cu(II), and Ni(II).

The adsorption affinities toward Ni and Co could also be understood taking into account their ligand field stabilization enthalpies (kJ mol^{-1}) to the formation of a $M(H_2O)_6$ complex: 111 for Co and 102 for Ni [24]. So, it is most difficult to remove a water molecule from the coordination sphere of cobalt, allowing the formation of a Co—N bond.

The adsorption affinities of the hexagonal zirconia matrix toward Ni(II) and Cu(II), Cu(II) >> Ni(II), are opposite to those observed for LS samples [6]: Ni(II) > Cu(II) (in this case the respective chlorides were used for both Ni and Co). This indicates that the nanostructure of the templated substrate can exert remarkable effects not only on the total amount of adsorbed cations, but also on the selectivity of the adsorption process. It could be inferred that hexagonal nanostructures, as well as lamellar ones, do not favor cobalt adsorption (or the adsorption of other cations with coordination number 6, forming octahedral compounds). However, lamellar nanostructures favor Ni(II) adsorption, whereas Cu(II) adsorption is favored by templated hybrid matrices with hexagonal nanostructures.

This hypothesis is schematically presented in Fig. 4.8. In hexagonal matrices, the distorted octahedral coordination sphere of the cation is completed by coordination

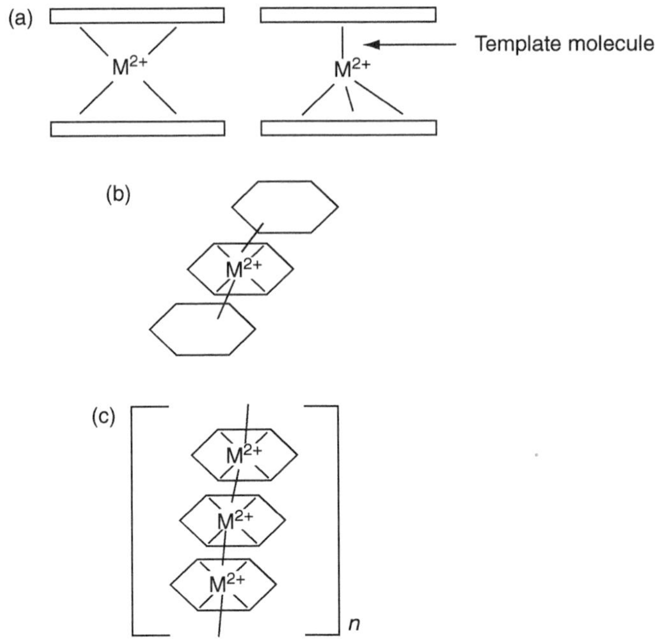

Figure 4.8 Schematic representation of the possible coordination features of transition metal cations of a hybrid matrix with lamellar (a) or hexagonal (b) and (c) nanostructure.

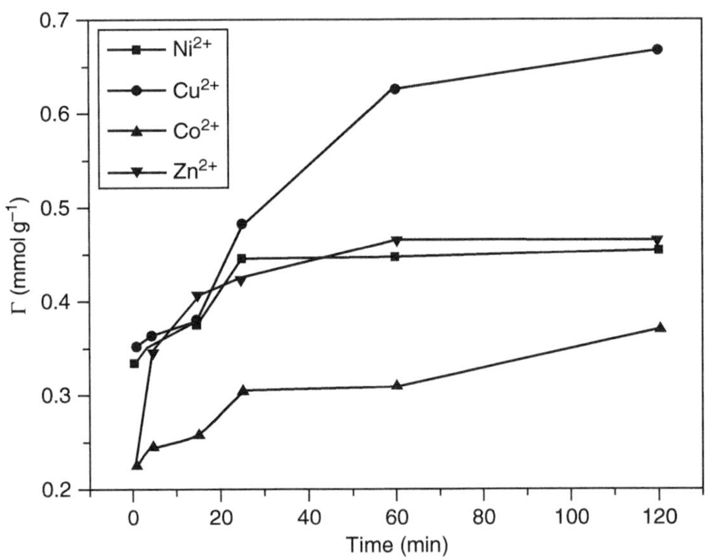

Figure 4.9 Γ values as a function of time for the adsorption of Co(II), Ni(II), Cu(II), and Zn(II) on the synthesized hexagonal templated zirconia.

with template molecules of adjacent hexagonal "rings." Since the template molecules have two nitrogen atoms, it can be supposed that the same diamine molecule could be linked to two different cations, as shown in Fig. 4.8c.

The "equilibration" curve, that is, a curve in which the Γ values are plotted as a function of time is shown in Fig. 4.9 (for all cations, the initial concentration was $1.0\times10^{-2}\,\text{mol}\,\text{dm}^{-3}$). The observed sequence is $\text{Ni} < \text{Zn} < \text{Co} \approx \text{Cu}$. As can be observed, the equilibrium concentration, that is, the concentration value at which the adsorption and desorption rates are equal, is reached firstly by Ni and Zn. So, it can be concluded that the kinetic and thermodynamic features of the adsorption of cations are very different, since the cation with maximum adsorbed amount (Cu) exhibits practically the same equilibration time as the cation with minimum adsorbed amount (Co). Such facts must be carefully taken into account if the hybrid matrix is to be used for metal adsorption purposes in a dynamic process (as a filter, for example).

6. Surface Modifications Promoted by Not Bonded Species

In this section, by considering a specific example, it will be shown that even in minor amounts, metals salts and organic species such as dimethylformamide (DMF) and dimethylsulfoxide (DMSO) can act, during the synthetic step, as surface modifiers and exert remarkable effects on the oxide.

6.1 Effects of metal salts, DMF, and DMSO on the anatase–rutile transition in sol–gel-synthesized TiO_2

The sol–gel process is one of the so-called "*chimie douce*," routes. It has been widely used a as synthetic route to the preparation of new materials, providing very homogeneous samples, even at low temperatures.

Taking into account its possible applications such as gas sensors, dielectric ceramics, and photocatalysts, titanium dioxide (TiO_2) has been extensively studied, and it has been shown that titanium oxide as anatase phase can be synthesized at room temperature without use of any previous or further thermal or hydrothermal treatment, performing hydrolysis and polycondensation processes in saturated metal chloride solutions.

Titania has three naturally occurring polymorphs: anatase, brookite, and rutile. Anatase and brookite are considered to be kinetic products despite the fact that depending on the particle size, anatase becomes more stable than rutile. The anatase–rutile transition is exothermic and irreversible, occurring in the range 400–1200°C [25,26].

The aim of this chapter is to report thermal studies of anatase–rutile structural transitions of anatase powders obtained by sol–gel process. The titania syntheses were performed in metal chloride saturated aqueous solutions, as well as DMF and DMSO solutions.

To obtain the titania powders, hydrolysis and polycondensation of titanium (IV) butoxide (Aldrich) was performed in water and aqueous saturated solutions of

KCl, $CaCl_2$, $NiCl_2$, $CoCl_2$, and $MnCl_2$. In a typical synthesis, 5 cm^3 of titanium butoxide was added to 50 cm^3 of metal chloride saturated solution at 25°C, and the resulting mixture stirred for 30 s. The resulting powders were aged for 24 h and then washed several times with deionized water and dried under vacuum at room temperature for 20 h.

Hydrolysis and polycondensation of titanium tetrabutoxide was also performed in DMF or DMSO. In a typical synthesis, 5 cm^3 of titanium tetrabutoxide were added to 50 cm^3 of DMF or DMSO, under stirring. The resulting solutions were aged for 48 h, and then 1.5 cm^3 of deionized water was added. The resulting powders were aged for 24 h and then washed several times with deionized water and dried under vacuum at room temperature (25°C) for 24 h.

The XRD patterns were recorded with a Shimadzu apparatus using Cu K_α radiation. The TG–DTA (differential thermal analysis) curves were obtained under argon atmosphere with a heating rate of 10°C min^{-1} using a DuPont instrument.

All titania powders synthesized in saturated aqueous solutions of metal chlorides are crystalline. From the broadening of the $2\theta = 25°$ peak, due to the {101} diffraction plane ($d = 0.36$ nm), the mean crystallite size was calculated as 11 nm for all samples. Titania powders synthesized in DMF or DMSO solutions are also crystalline. From the peak broadening of the anatase {101} diffraction, the mean crystallite sizes were calculated as 6 nm for both samples. This indicates that the mean crystallite size is strongly affected by the composition of the reaction media and that organic species induce the formation of crystallites of low dimensions.

It is worth noting that all titania samples are crystalline as synthesized, without needing any previous or further treatment such as heating or hydrothermal aging, as observed for other sol–gel-derived titania powders [25,26]. This phenomenon is probably related to the formation of OTi_3 groups, which could favor the formation of an anatase phase, with a localized structure having four edges shared by TiO_6 octahedral units, instead of the amorphous phase, with only two shared edges.

In the XRD patterns, all diffraction peaks, including the one near $2\theta = 25°$ confirm that anatase is the crystalline phase formed [25,26]. On the other hand, the diffraction peak near 30° can be attributed to the presence of a minor amount of brookite phase [25,26]. Obviously, the presence of the organic molecules (DMF or DMSO), or the ions of metal chlorides, provides new paths for the hydrolysis and polycondensation processes, leading to ordered matrices, instead of amorphous ones, as observed, if the hydrolysis and polycondensation processes are performed in water. For example, the XRD pattern for titania powder synthesized in DMF is shown in Fig. 4.10.

For the titania powder synthesized in $CaCl_2$ solution, the anatase–rutile structural transition, associated with a well-defined exothermic peak, occurs at 512°C, as verified by DTA analysis. For the titania powders synthesized in solutions of K, Ni, Co, and Mn chlorides, the transition occurs at 460°C, 455°C, 510°C, and 570°C, respectively. So, variations of 55°C can be observed for such structural transitions, depending on the metal chloride used.

For DMF- or DMSO-synthesized titania samples, an endothermic peak from 30°C to 250°C in the DTA curves is observed, which is associated with a significant mass loss step (TG curves), attributed to the removal of physisorbed water and DMF or DMSO molecules, as well as unreacted titanium alkoxide.

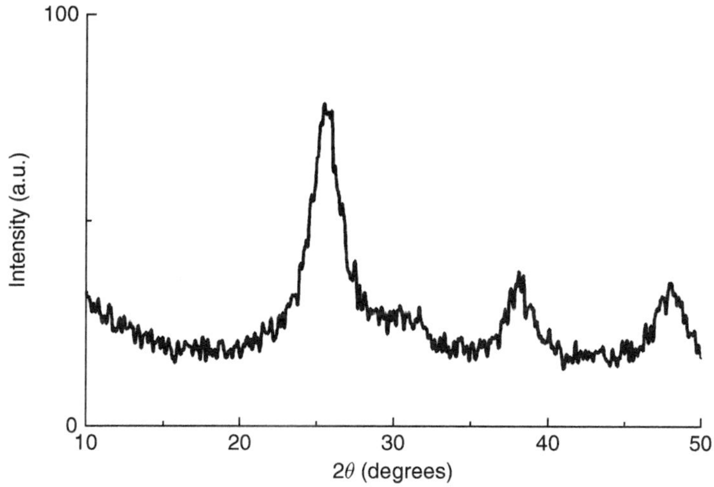

Figure 4.10 XRD patterns for titania powder obtained by sol–gel process in DMF.

For both samples, the anatase–rutile structural transition, associated with a well-defined exothermic peak, occurs at 485°C. Such fact suggests that the temperature of the structural transition is associated with the mean crystallite sizes, which have, for both samples, the same value. From 600°C to 800°C, an endothermic peak is observed for both samples, associated with a minor mass loss step, probably due to the sublimation of a minor amount of powder or the thermal degradation of the last alkoxide molecules, trapped into the three dimensional oxide structure. As an example, the TG–DTA curve for the DMF-synthesized powder is shown in Fig. 4.11.

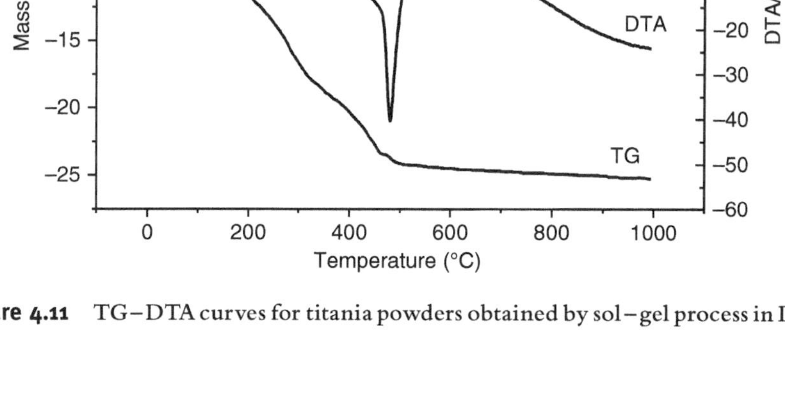

Figure 4.11 TG–DTA curves for titania powders obtained by sol–gel process in DMF.

It was observed that amorphous and hydrothermally treated titania powders begin rutile formation in the range 700–820°C [25]. So, the titania powders synthesized in this work allow the formation of a rutile phase at temperatures about 200°C below those observed otherwise. This is probably related to the mean crystallite sizes, which were smaller for these samples.

The experimental results show that organic species (DMF and DMSO, in this case) induce the formation of titania crystallites of low dimensions (6 nm). Furthermore, the temperatures needed to promote the anatase–rutile structural transition can be decreased by 200°C, when compared with other sol–gel-prepared titania powders [27], using the experimental procedures reported in this work.

7. Hexagonal Mesoporous Silica

An instructive example of the attainment of organically modified mesoporous silica is the synthesis [27] of hexagonal mesoporous silicas chemically modified with en moieties via co-condensation of TEOS with two different silylating agents: (i) *N*-[3-(trimethoxysilyl)propyl]-ethylenediamine (SiNN) and (ii) the new agent prepared from the incorporation of the en molecule into the epoxide group of the precursor 3-glycidoxypropyltrimethoxysilane (Siepox).

In such preparations, the silica source of the inorganic framework was TEOS. The silylating agents Siepox, SiNN, the surfactant *n*-octylamine, en, methanol, and ethanol were all reagent grade. A mixture of 5.0 g of Siepox and 2.66 g of en was slurried in 20.0 cm^3 of methanol in the presence of 1.0 cm^3 of tributylamine as deprotonating agent. The solution was mechanically stirred under dry nitrogen at 323 K for 24 h, to yield the new silylating agent denoted as SiGNN.

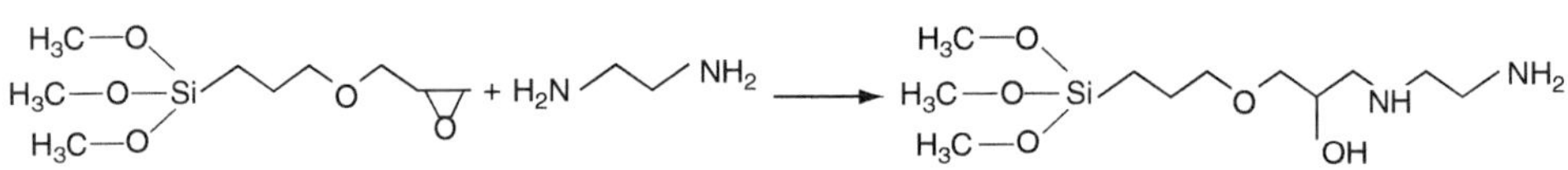

The parent mesostructure was prepared by stirring *n*-octylamine (5.6 mmol) in 200 cm^3 of water for 30 min until an opalescent solution was obtained, as a consequence of micelle formation (Fig. 4.12). To this micellar suspension, TEOS (23.2 mmol) and the organosilane SiGNN (4.8 mmol) were then added. Each suspension was stirred for 24 h at room temperature, resulting in a precipitate. The *n*-octylamine inside the pores of the synthesized compounds was extracted with ethanol at reflux temperature for 72 h.

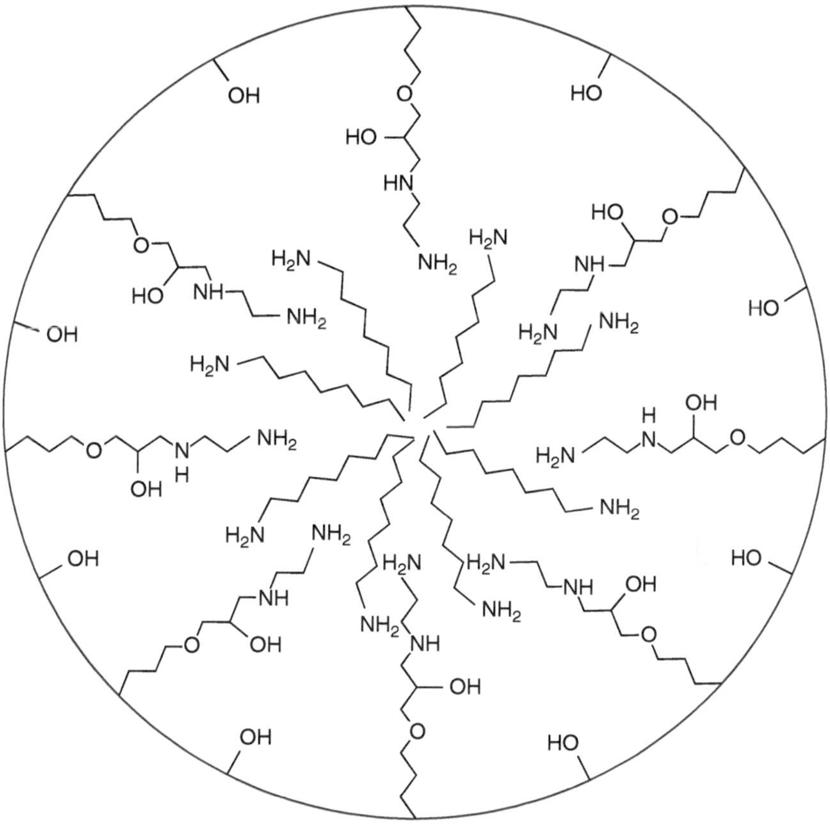

Figure 4.12 Co-condensation betweenTEOS and SiGNN around the micelle.

REFERENCES

1. C. Airoldi, R.F. de Farias, Quím. Nova 27 (2004) 84.
2. C.J. Brinker, G.W. Scherer, Sol-Gel Science, Academic Press, San Diego, 1990.
3. R.F. de Farias, C. Airoldi, J. Colloid Interface Sci. 220 (1999) 255.
4. R.F. de Farias, C. Airoldi, J. Solid State Chem. 149 (2000) 113.
5. R.F. de Farias, C. Airoldi, J. Non-Cryst. Solids 261 (2000) 181.
6. R.F. de Farias, C. Airoldi, Colloids Surf. A 172 (2000) 145.
7. R.F. de Farias, L. Martinez, C. Airoldi, J. Non-Cryst. Solids 276 (2000) 56.
8. R.F. de Farias, S. Alves Jr., M.F. Belian, G.F. de Sá, Bull. Chem. Technol. Macedonia 23 (2004) 7.
9. R.F. de Farias, C. Airoldi, J. Solid State Chem. 158 (2001) 134.
10. R.F. de Farias, J. Non-Cryst. Solids 288 (2001) 218.
11. A. Fujishima, K. Hashimoto, T. Watanabe, TiO_2 Photocatalysis: Fundamentals and Applications, Bkc Inc., Tokyo, 1999.
12. R.F. de Farias, C. Airoldi, J. Non-Cryst. Solids 351 (2005) 84.
13. R.F. de Farias, et al., J. Phys. Chem. Solids 64 (2003) 2385.
14. R.F. de Farias, Int. J. Inorg. Mater. 3 (2001) 825.
15. R.F. de Farias, J. Non-Cryst. Solids 288 (2001) 218.
16. R.F. de Farias, C. Airoldi, J. Solid State Chem. 158 (2001) 134.

17. R.F. de Farias, Quím. Nova 25 (2002) 1027.
18. R.F. de Farias, J. Colloid Interface Sci. 239 (2001) 584.
19. I.M. Arabatzis, P. Falaras, Nano Lett. 3 (2003) 249.
20. R.F. de Farias, C. Airoldi, Synth. React. Inorg. Met. Org. Nano-Met. Chem. 35 (2005) 1.
21. R.F. de Farias, A.A.S. do Nescimento, C.W.B. Bezerra, J. Colloid Interface Sci. 277 (2004) 19.
22. K.S.W. Sing, D.H. Everett, R.A.W. Haul, L. Moscou, R.A. Pierotti, J. Rouquerol, T. Siemieniewska, Pure Appl. Chem. 57 (1985) 603.
23. J.E. Heheey, E.A. Keiter, R.L. Keiter, Inorganic Chemistry, Harper Collins, New York, 1993.
24. D.M.P. Mingos, Essential Trends in Inorganic Chemistry, Oxford University Press, Oxford, 1998.
25. J. Ovenstone, K. Yanagizawa, Chem. Mater. 11 (1999) 2770.
26. K. Yanagizawa, Y. Yamamoto, Q. Feng, N. Yamasaki, J. Mater. Res. 13 (1998) 825.
27. J.A.A. Sales, A.G.S. Prado, C. Airoldi, Surf. Sci. 590 (2005) 51–62.

CHAPTER FIVE

Chemistry on Conducting Polymer-Modified Oxide Surfaces

Contents

1. Introduction

Conducting polymers, for example, polyaniline, polypyrrol, have been extensively investigated taking into account its possible technological applications. From the specific point of view of modified surfaces, it has been shown that, when adsorbed on oxide surfaces such as molybdenum [1], arsenic [2,3], and iron [3] oxides, polyaniline exerts remarkable effects on their redox properties [4] and/or thermal stability of the respective oxides. This chapter presents a brief outline about the achievements on this field of investigation.

2. Synthetic Approach

Once the conducting polymer has been synthesized by the usual chemical or electrochemical approaches [5], the adsorption of the polymer on an oxide surface it is very simple. As illustrated by polyaniline [1–4], it is only a question of finding a solvent for the polymer, for example, DMF, and then dispersing the investigated oxide in solution, promoting the doping of the conducting polymer with some inorganic oxide (if desired), promoting the aging of the new product, for example, for 24 h, filtering the polymer adsorbed oxide, drying it, and performing desired measurements.

Interface Science and Technology, Volume 17
ISSN: 1573-4285, doi: 10.1016/S1573-4285 (08) 00405-5

3. Effects of Adsorbed Conducting Polymer on the Oxide Properties

For molybdenum and arsenic oxides [1–3], it has been shown that the modification of the oxide surface by adsorption of a conducting polymer (polyaniline, in these cases) can exert remarkable effects on the electrochemical properties of the oxides.

For molybdenum oxide, for example, it is shown that the adsorption of 4% (in mass) of polyaniline is enough to enhance the electrode response, affecting the potential in which the formation of the so-called "molybdenum blue" is formed. For arsenic oxide, it is shown that the modification of the oxide surface by 13% polyaniline adsorption was responsible for enhancement of the oxide response to applied potential as well as for a change in the potentials observed for oxidation–reduction processes occurring in the unmodified oxide.

From the point of view of thermal stabilization, it is seen that the sublimation temperature of As_2O_3 is increased by 41°C (from 165°C to 206°C) when the oxide surface is modified with polyaniline. Furthermore, for Fe_3O_4 the surface modification with Pani (polyaniline) avoids the oxide sublimation making it thermally stable up to 1000°C (the unmodified oxide sublimate at 780°C). For comparison, the obtained thermogravimetric curves are shown in Fig. 5.1.

4. Effects of the Oxide on the Adsorbed Polymer Properties

The modification of an oxide surface with conducting polymer can exert remarkable effects not only on the properties of the oxide but also on the properties of the conducting polymer, as illustrated for Ba, Mg, Sn, Ge, and Al oxides [6].

For this investigation, polyaniline was obtained by using a 0.50 mol dm^{-3} solution of double-distilled aniline (Aldrich) in 1.0 mol dm^{-3} HCl (Aldrich) as starting

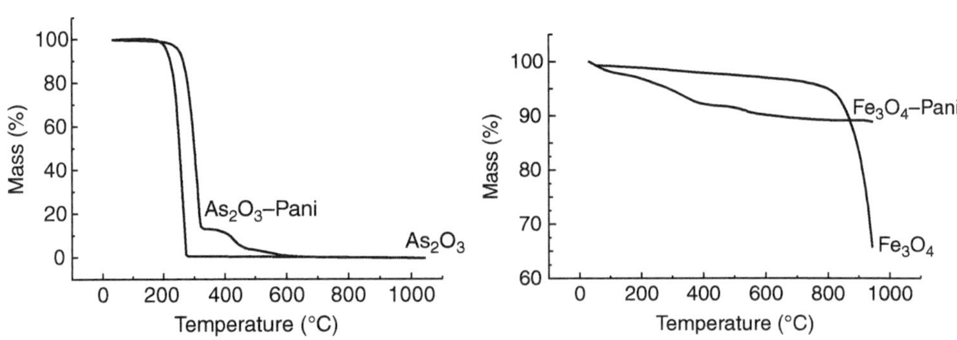

Figure 5.1 Thermogravimetric curves for Pani-modified and unmodified iron and arsenic oxides.

reagents. To this mixture, maintained at 5°C, was added dropwise, during a period of 2 h, 100 cm^3 of a 0.80 mol dm^{-3} of $(NH_4)_2S_2O_8$ solution. After the complete addition of the oxidant solution, the mixture was stirred for 10 h, until the polymerization was completed. The final polymer was washed several times with distilled water and acetonitrile to remove oligomers as well another possible impurities. The polyaniline thus obtained was then treated with a 1.0 mol dm^{-3} NH_4OH solution to promote deprotonation.

In the next stage, 1.6 g of the polyaniline was dissolved in 180 cm^3 of *N*,*N*-dimethylformamide (Merck). This solution was used to promote the deposition of polyaniline on BaO, GeO_2, MgO, SnO_2, and Al_2O_3. In a typical synthesis, 2 g of the oxide were suspended in 40 cm^3 of the previously prepared solution, under stirring. Then, 0.5 cm^3 of concentrated (85%) H_3PO_4 was added to the system, to promote doping. The mixture was aged for 48 h at room temperature under static conditions before withdrawing the excess of solution and drying at 30°C overnight. The final products were isolated as green powders.

The oxide–polyaniline powders were characterized by Fourier transform infrared (FTIR), using a Bomem spectrometer (KBr disks), by TG in a Shimadzu TGA-50 instrument and by DSC in a DuPont 2000 instrument. TG and DSC curves were obtained under Ar atmosphere with a heating rate of 5°C min^{-1}. The SEM images were obtained in a JEOL, JSM T-300 microscope, using an accelerating voltage of 15 kV. No success was achieved in attempting to obtain$^{\alpha\text{-}13}$C cross-polarization-magic angle spinning (CP-MAS) spectrum for oxide–Pani samples. UV–Vis spectra for the solid samples were recorded with a Cary 5G spectrometer.

The infrared spectra for the oxide–polyaniline samples confirms the presence of the conducting polymer, exhibiting some of its characteristic bands.

Based on thermogravimetric data, percentages of 3–10% of conducting polymer on the oxide surfaces can be calculated. The DSC curves exhibit endothermic peaks associated with the respective mass loss steps. As examples, the TG–DTG and DSC curves for MgO–Pani are shown in Fig. 5.2.

The adsorption of conducting polymer on the oxide surfaces did not change the grain morphology of the oxides, except for MgO. For example, the SEM

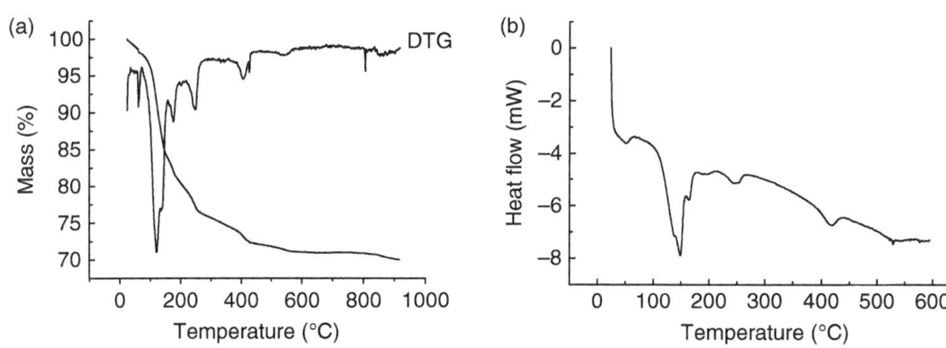

Figure 5.2 TG–DTG (a) and DSC (b) curves for MgO–Pani.

Figure 5.3 SEM micrographs for SnO_2, 750× (a) and SnO_2-Pani, 750× (b).

micrographs for SnO_2/SnO_2–Pani and MgO/MgO–Pani are shown in Figs. 5.3 and 5.4, respectively.

For all samples, a very interesting phenomenon was observed: the Brønsted acid-doped samples remain doped after treatment with a NH_4OH 1.0 mol dm^3 solution. This fact can be verified by inspection of the oxide sample with the naked eye or with an optical microscope (the doped polyaniline samples exhibit a characteristic green color, whereas the undoped samples are blue) or by UV–Vis spectra of the samples before and after treatment with 1.0 mol dm^{-3} NH_4OH solution. As an example, the UV–Vis spectra for BaO–polyaniline samples are shown in Fig. 5.5.

The spectrum shown in Fig. 5.5a is for the doped BaO–polyaniline sample, which exhibits absorption bands at 420 nm and 780 nm, characteristic of Brønsted acid-doped polyaniline. The spectrum shown in Fig. 5.5b is for the same sample after treatment with a NH_4OH 1.0 mol dm^3 solution. As can be seen, the 420 nm and 780 nm absorption bands remain at the same positions, even after treatment in basic solution. This is not observed for the "free" polymer, for which a reaction with base promotes deprotonation, which causes the appearance of an absorption band at 625 nm.

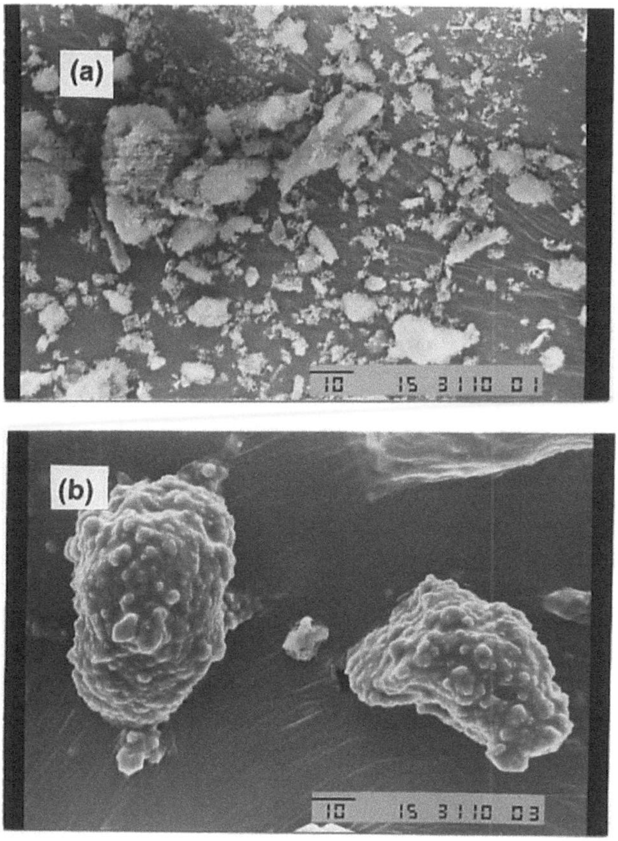

Figure 5.4 SEM micrographs for MgO, 1000× (a) and MgO–Pani, 1000× (b).

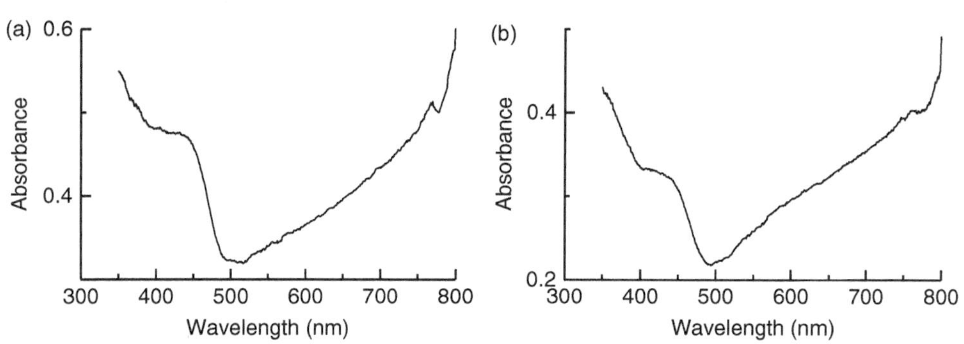

Figure 5.5 UV–Vis spectra for (a) BaO–polyaniline-doped sample (a) and BaO–polyaniline-doped sample after reaction with a NH_4OH 1.0 mol dm^{-3} solution (b).

The UV–Vis spectrum of Pani has two absorption bands, at 325 nm and 625 nm, which are due to π–π^* transitions of the benzenoid rings and excitation absorption of the quinoid rings, respectively. However, when nitrogen atoms in imine groups are protonated, N and its neighboring quinoid ring become a semiquinoid radical cation, causing a decrease in the exciton absorption peak intensity and a generation of the absorption peaks about 420 nm and 865 nm due to the presence of polaron/bipolaron. When Pani is doped with lithium salts, absorption bands at 420 and 865 nm are observed, suggesting a doping process similar to those caused by Brønsted acids.

However, when Pani samples are doped with $SnCl_4$ or $FeCl_3$ somewhat different results are obtained. $SnCl_4$-doped samples exhibit an absorption band at 370 nm, and a broad absorption band with maximum at 1530 nm. On the other hand, $FeCl_3$-doped samples exhibits an extremely broad absorption band extending toward NIR with a maximum around 1700 nm. This is evidence of the influence of the Lewis acid on the UV-Vis/NIR (near-infrared) spectra of Lewis acid-doped polyaniline. In the present case, the UV–Vis spectra for the oxide–Pani samples resemble those obtained for Pani doped with lithium salts.

Based on the experimental results as well as literature data, it can be concluded that the adsorption of Brønsted acid-doped polyaniline on Ba, Mg, Ge, Sn, and Al oxide surfaces enlarges the pH range in which the conducting polymer remains doped, that is, enlarges its pH range of maximum conductivity. Furthermore, the oxide–polymer samples exhibit higher processability, when compared with the polymer. Obviously, such facts could have useful influences on the possible technological applications of polyaniline.

REFERENCES

1. R.F. de Farias, J.M. de Souza, J.V. de Melo, C. Airoldi, J. Colloid Interface Sci. 212 (1999) 123–129.
2. R.F. de Farias, J.M. de Souza, J.V. de Melo, C. Airoldi, J. Colloid Interface Sci. 227 (2000) 147–151.
3. R.F. de Farias, Quím. Nova 23 (2000) 313–315.
4. R.F. de Farias, Effects of Adsorption on the Redox Process of Oxide Surfaces in Encyclopedia of Surface and Colloid Science, Marcel Dekker, New York, 2002.
5. H.S. Nalva, Handbook of Organic Conductive Molecules and Polymers, Wiley, New York, 1997.
6. R.F. de Farias, J.M. de Souza, Quim. Bras. 1(1) (2007) 37–42.

CHAPTER SIX

Chemistry on Modified Layered Oxides

Contents

1. Introduction

Inorganic layered oxides have attracted considerable attention due to their unique structural properties and applications, such as in intercalation reactions, ionic exchange processes, photochemical and semiconductor properties, and catalytic applications. In this chapter, titanates, niobates, tungstates, molybdates, and molybdenum oxide will be considered as specific examples. However, it is necessary to say that many other specific lamellar oxides are investigated, such as V_2O_5.

2. Titanates and Niobates

Layered titanates and niobates such as $K_2Ti_4O_9$, $Na_2Ti_3O_7$, KNb_3O_8, and $K_4Nb_6O_{17}$, as well as their acid forms, that is, $H_2Ti_4O_9$, $H_2Ti_3O_7$, HNb_3O_8, and $H_4Nb_6O_{17}$ can be used as cation exchangers toward Cu^{2+} [1]. Furthermore, the acidic compound $H_2Ti_4O_9.nH_2O$ was used for the preparation of intercalation compounds with n-alkyldiamines of formula $H_2N(CH_2)_nNH_2$ with $n=2$, 3, 4,

Interface Science and Technology, Volume 17
ISSN: 1573-4285, doi: 10.1016/S1573-4285 (08) 00406-7

6, 7, and 8 [2]. These specific systems will be discussed here, to provide a most profound insight about modified titanates and niobates.

From the point of view of cation exchange, alkaline titanates have been extensively investigated due to their high selectivity toward metal cations. In this respect, the layered compound $K_2Ti_4O_9$ had been investigated and it has been found that its corresponding acidic form exhibits a high ion exchange capacity.

The syntheses of the above-mentioned alkaline titanates and niobates are performed through a traditional solid-state chemistry approach at high temperatures: $K_2Ti_4O_9$ and $Na_2Ti_3O_7$ are prepared by reaction of K_2CO_3 or Na_2CO_3 with stoichiometric amounts of TiO_2 in a platinum crucible. The mixture is heated at 800°C for 20 h at room atmosphere. After being cooled and ground, the mixture is heated again at the same temperature for another 20 h.

3. Tungstates and Molybdates

It has been demonstrated that tungstates can act as effective corrosion inhibitors in low-level carbon stainless steel [3]. Furthermore, molybdates have attracted considerable interest due to their possible applications as catalysts [4].

The bulk and surface compositions of tungstates and molybdates can be modified through ionic exchange process, such as those used to obtain transition metal molybdate from sodium molybdate [5]. In this topic, as a specific example of preparation and uses of modified tungstates and molybdates, will be presented the preparation of Eu(III) compounds for optical purposes.

To produce new LCMDs [6] with high thermal stability and processibility in comparison with traditional lanthanide complexes, new strategies have been adopted, such as adsorbing the complex on amorphous inorganic–organic hybrid surfaces [7,8], or LS [9], or insert them into polysiloxane three dimensional networks [10]. The specific example reported here belongs to the above-mentioned context and is dedicated to the investigation of entirely inorganic (with no organic ligand molecules) LCMDs with high thermal stability and prepared in aqueous medium at room temperature (in contrast with the well-studied rare-earth-doped glasses). To achieve such purposes, the well-known ammonium molybdate $(NH_4)_6Mo_7O_{24}.4H_2O$ and a new sodium tungstate $Na_8WO_7.2H_2O$ are used as precursors [11].

Sodium tungstate, Na_8WO_3 was prepared by grinding stoichiometric amounts of $Na_2CO_3.10H_2O$ and WO_3 for 30 min. Then the resulting mixture was heated at 700°C and the liquid product was cooled to a solid at room temperature. This final solid was then ground to a powder in a mortar. So far as the authors are concerned this is a new tungstate.

The europium tungstate and molybdate were prepared by mixing stoichiometric amounts of aqueous solutions of Na_8WO_7 or $(NH_4)_6Mo_7O_{24}.4H_2O$ with aqueous solutions of $EuCl_3$. In both cases, with precipitates were obtained which were filtered off and dried under in a fume hood at room temperature for 8 h.

The XRD patterns were obtained with a Shimadzu XD-3A instrument, using Cu K_α radiation (35 kV, 25 mA). The infrared spectra were obtained in a Bomem spectrometers in the range 4000–400 cm^{-1} with a resolution of 4 cm^{-1}, by dispersing the samples in KBr disks. Thermogravimetric curves were obtained with a Shimadzu TGA 50 instrument under nitrogen atmosphere at a heating rate of 10°C min^{-1}. The sodium/tungsten ratio from which the sodium tungstate formula was established was measured by energy-dispersive X-ray (EDX) analysis in a JEOL instrument, model JSM T-300, with an accelerating voltage of 15 kV. The reliability of such analysis was previously verified by using a Pd (80%)/Au (20%) standard disk, and the ratio was as expected. The spectral analyses (emission spectra) were performed at room temperature and at 77 K (N_2) using a Jobin Yvon Ramanor U-1000 spectrometer, using a 450W Xe lamp. The detector was an RCA C31034 photomultiplier tube or an intensified diode array. Excitation spectra of the compounds were registered in an ISS K2 Multifrequency Phase Fluorometer.

The obtained thermogravimetric curves are shown in Fig. 6.1. Based on the experimental mass loss percentages, the formula $Na_8WO_7.2H_2O$ can be proposed to the new synthesized tungstate. For this compound, the first mass loss step, started at 65°C can be associated with the process: $Na_8WO_7.2H_2O$ (s) $\rightarrow$ Na_8WO_7 (s) + $2H_2O$ (g) (experimental value: 7.2%; calculated value: 7.0%). After this mass loss, a large plateau, in the range 105–550°C is observed. That is, the anhydrous compound is stable (form a thermal point of view) is this temperature interval. At 600°C a large mass loss is observed. However, as we know from the synthetic procedure, the

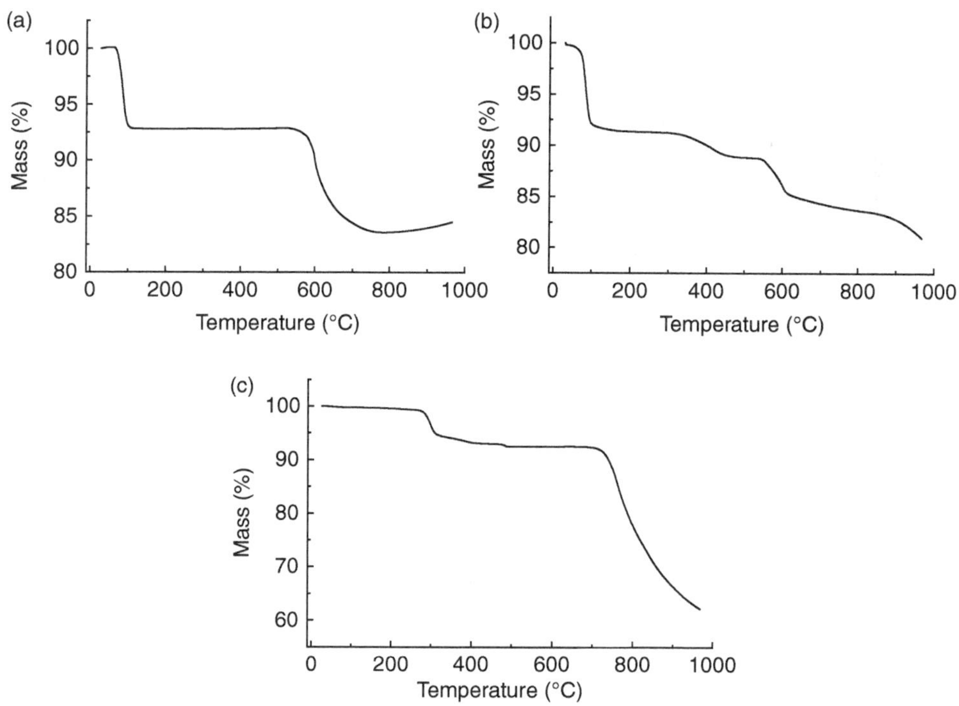

Figure 6.1 Thermogravimetric curves for $Na_8WO_7.2H_2O$ (a), $Eu_8(WO_7)_3.5H_2O$ (b), and $Eu_2Mo_7O_{24}.4H_2O$ (c).

anhydrous Na_8WO_7 melts at this temperature, and thus this observed loss is associated with the evaporation of the liquid tungstate. The mass "gain" observed near 800°C is probably due to a measuring error resulting from the "swinging" of the sample container due to the quick evaporation of the compound.

Based on the experimental mass loss of 8.0% observed for the first mass loss step for europium tungstate (Fig. 6.1b) the formula $Eu_8(WO_7)_3.5H_2O$ can be proposed, with the release of water molecules occurring as follows: $Eu_8(WO_7)_3.5H_2O$ (s) $\rightarrow$ $Eu_8(WO_7)_3$ (s) + $5H_2O$ (g) (calculated value: 8.0%). The anhydrous $Eu_8(WO_7)_3$ is stable in the 110–350° range.

Based on the experimental value obtained to the first mass loss step for the europium molybdate (5.2%) the formula $Eu_2Mo_7O_{24}.4H_2O$ can be proposed (calculated water content is 5.0%), for which the water molecules release process is: $Eu_2Mo_7O_{24}.4H_2O$ (s) $\rightarrow$ $Eu_2Mo_7O_{24}$ (s) + $4H_2O$ (g). So, an exchange of six NH_4^+ cations by two Eu^{3+} cations into the molybdate network does not require any "extra" water molecules in the network, to produce a stable structure from the interrelated structural and thermodynamic points of view. The anhydrous compound is stable up to 720°C.

The X-ray powder diffraction patterns for the new tungstate, $Na_8WO_7.2H_2O$ and the europium tungstate and molybdate are shown in Fig. 6.2. By comparison of Figs. 6.2a and 6.2b, it can be seen that $Na_8WO_7.2H_2O$ and $Eu_8(WO_7)_3.5H_2O$

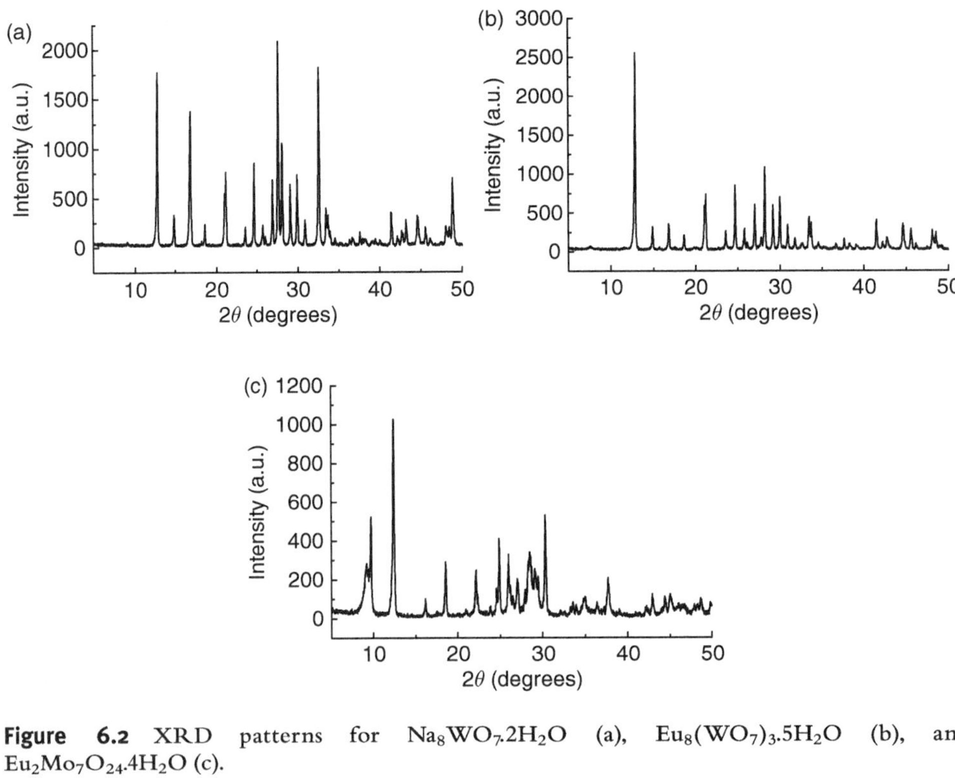

Figure 6.2 XRD patterns for $Na_8WO_7.2H_2O$ (a), $Eu_8(WO_7)_3.5H_2O$ (b), and $Eu_2Mo_7O_{24}.4H_2O$ (c).

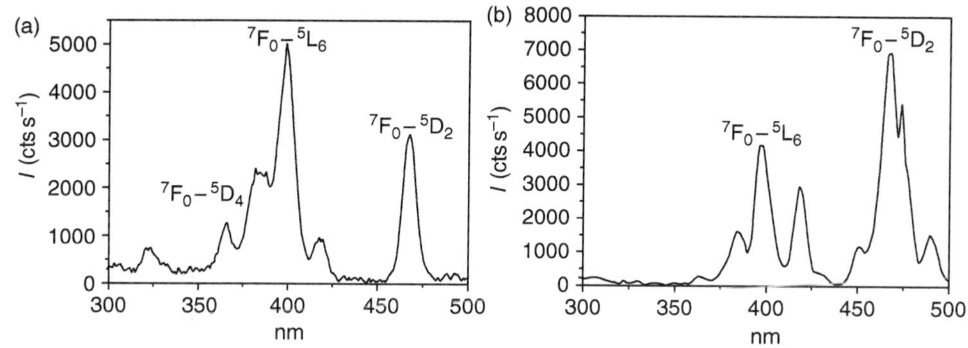

Figure 6.3 Excitation spectra for $Eu_8(WO_7)_3.5H_2O$ (a) and $Eu_2Mo_7O_{24}.4H_2O$ (b).

are isomorphs. Furthermore, it can be inferred, from the $Eu_2Mo_7O_{24}.4H_2O$ diffraction pattern (Fig. 6.1c) that this compound exhibits a structure similar to that exhibited by $Eu_8(WO_7)_3.5H_2O$.

The excitation spectra are shown in Fig. 6.3. These spectra were recorded in the 300–500 nm range and the emission monitored at ~612 nm (the hypersensitive $^5D_0 \rightarrow {}^7F_2$ transition of Eu^{3+}). Large bands associated with the allowed transitions ($S_0 \rightarrow S_1$) from ligands are not observed. Only the bands due to the low-intensity transitions ($^7F_0 \rightarrow {}^5D_4$; $^5L_6$; $^5D_2$) schematically shown in Fig. 6.4 are observed.

The emission spectra are shown in Fig. 6.5.

In both spectra, the characteristic $^5D_0 \rightarrow {}^7F_J$ bands are observed, with exception of the $^5D_0 \rightarrow {}^7F_1$ band. Comparison of the spectra, indicates that the most intense band ($^5D_0 \rightarrow {}^7F_2$), due to the characteristic red emission of europium complexes, is most intense (about three times) to $Eu_2Mo_7O_{24}.4H_2O$ in comparison with

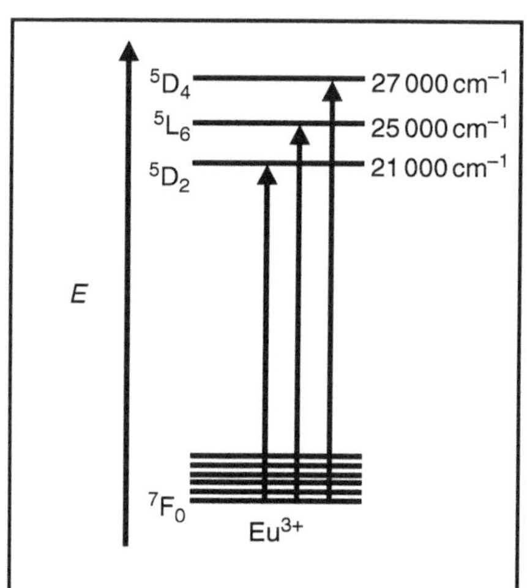

Figure 6.4 The Eu^{3+} transitions observed for the tungstate and molybdate.

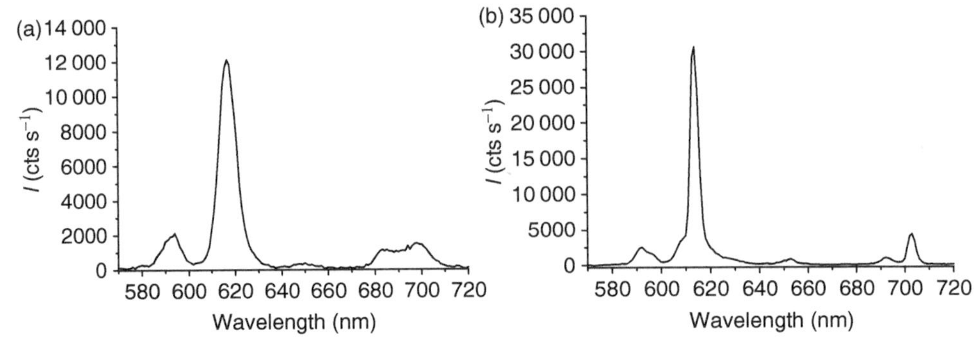

Figure 6.5 Emission spectra for $Eu_8(WO_7)_3.5H_2O$ (a) and $Eu_2Mo_7O_{24}.4H_2O$ (b).

$Eu_8(WO_7)_3.5H_2O$. The presence of five water molecules in the tungstate structure, that is, an "extra" molecule in relation to the molybdate, can be pointed out as the main factor leading to such quenching phenomena. Furthermore, the presence of a larger number of oxygen atoms in the molybdate structure can promote efficient energy absorption and transfer to the europium cation.

4. Molybdenum Oxide as a Molecular Sieve

Molybdenum trioxide (α-MoO_3) has a lamellar structure [12], being insoluble in water [13]. Its layered structure is constituted of parallel rows of distorted MoO_6 octahedra [12] as shown in Fig. 6.6.

Due to its many applications for catalytic purposes [14], redox effects on MoO_3 have been studied [15,16], by deposition of conducting polymer on its surface. Furthermore, taking into account that its sheets are separated by a van der Walls gap, lamellar MoO_3 can be used as host species to produce intercalation compounds.

The most recent studies of such compounds involves MoO_3–pyrazine [17], MoO_3–pyrrolidinedithiocarbamate [18] and MoO_3–phen [19] hybrids. The obtained results suggest that lamellar MoO_3 can be successfully used as a molecular sieve toward a series of nitrogen-containing organic species.

In this chapter, specific examples of the use of MoO_3 as a molecular sieve, will be discussed the synthesis and characterization of MoO_3 intercalation compounds with nicotinamide (nic) and hexamethylenetetramine (hmta) [20], as well the study of the adsorption of caffeine (caff), dimethylglyoxime (dmg), and rodamin-B (rod-B) on lamellar molybdenum oxide will be presented.

4.1 MoO_3 intercalation compounds with nic and hmta

The MoO_3 intercalation compounds were synthesized at room temperature. In a typical synthesis, MoO_3 (Aldrich) was suspended in 60 cm^3 of a 1:1 (v/v) water–ethanol solution of nic or hmta (both from Aldrich). The obtained suspension was

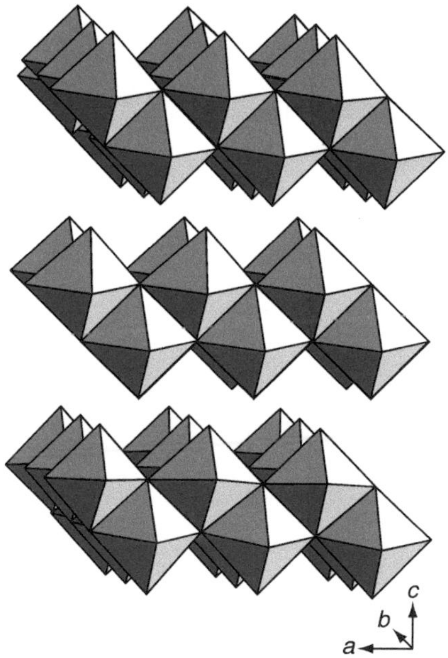

Figure 6.6 Schematic representation of the network structure of lamellar MoO_3.

aged for 5 days, and then, the solvent was slowly evaporated at 30°C for another 7 days. We have used a total of six MoO_3/nic and MoO_3/hmta stoichiometric ratios:

a. MoO_3 (0.60 g) + nic (0.10 g) = Mo1
b. MoO_3 (0.60 g) + nic (0.20 g) = Mo2
c. MoO_3 (0.60 g) + nic (0.30 g) = Mo3
d. MoO_3 (0.50 g) + hmta (0.20 g) = Mo4
e. MoO_3 (0.50 g) + hmta (0.35 g) = Mo5
f. MoO_3 (0.50 g) + hmta (0.55 g) = Mo6

To calculate the lengths of nic and hmta molecules, as well as the structural features of their intercalation compounds with MoO_3, quantum chemical calculations (Hartree–Fock, STO-3G) were performed by using the Spartan-Pro 1.0.3 package.

The elemental analysis results are summarized in Table 6.1. Each presented value is the average of three distinct determinations. For nic, the %C/%N ratio is 2.57. For Mo1, Mo2, and Mo3 samples this ratio is equal 2.50, 2.56, and 2.63, respectively. For hmta the %C/%N ratio is 1.29. For Mo4, Mo5 and Mo6 samples the values are 1.29, 1.28, and 1.28 respectively. So, for both, nic and hmta intercalation compounds the presence of ethanol molecules is excluded, since the presence of alcohol molecules increase the carbon percentage, and so, the %C/%N ratio. The proposed formula shown in Table 6.1 accounts for both, intercalated and surface adsorbed molecules, as will be discussed later.

Table 6.1 Elemental analysis results for MoO_3 intercalation compounds with nic and hmta

Matrix	C (%)	H (%)	N (%)	Organic content (mmol g^{-1})	Proposed formula
Mo1	8.63	0.75	3.45	1.20	$MoO_3.(nic)_{0.20}$
Mo2	14.77	1.21	5.77	2.05	$MoO_3.(nic)_{0.40}$
Mo3	18.93	1.91	7.19	2.63	$MoO_3.(nic)_{0.56}$
Mo4	14.70	2.72	11.43	2.04	$MoO_3.(hmta)_{0.4}$
Mo5	18.90	3.42	14.79	2.63	$MoO_3.(hmta)_{0.60}$
Mo6	23.56	4.03	18.39	3.72	$MoO_3.(hmta)_{1.12}$

The XRD patterns for the oxide and intercalation compounds are shown in Figs. 6.7a–6.7g. In the oxide diffraction pattern, peaks at 12.6° and 25.2° are associated with the {001} and {002} diffraction planes [17–19]. Taking into account the {001} diffraction peak position, the interlayer distance can be calculated as 0.70 nm.

For Mo1, the 001 diffraction peak of the pure oxide is observed. However, two peaks, at 7.9° (d = 1.12 nm) and 15.8° (d = 0.56 nm) can be attributed to the {001} and {002} diffraction of a new phase, with intercalated nic molecules. However, not all nic molecules are intercalated, as can be verified by the presence of a series of diffraction peaks due to "free" (merely adsorbed on the surface, not intercalated) nic molecules. For Mo2, {001} diffraction peak of MoO_3 is not observed. The {001} and {002} diffraction peaks due to the MoO_3–nic intercalation compound are observed. However, the diffraction peaks due to free nic molecules are of very low intensity, showing that almost all nic molecules are intercalated is this case. So, comparing Mo1 and Mo2 diffraction patterns, it is possible to verify that an increase of the nic/MoO_3 ratio induces the intercalation of the organic species. Hence, it can be inferred that the efficiency of MoO_3 as a molecular sieve toward nic, will be dependent on the relative nic concentration. For Mo3, the diffraction peaks at 7.9° and 15.8° observed in Mo1 and Mo2 samples are also observed. However, a new diffraction peak at 6.7° (d = 1.32 nm) is observed, and can be attributed to the 001 diffraction plane of a new MoO_3–nic intercalation compound. So, a further increase of the nic/MoO_3 ratio leads to the formation of a new intercalation compound.

The structural formulas for nic and hmta are shown in Fig. 6.8, with the calculated values for molecular length. Based on the molecular lengths, as well as the experimental X-ray data (interlayer distances), it could be proposed that Mo1 and Mo2 samples, nic is intercalated as presented in Fig. 6.9a. For the Mo3 sample this intercalation feature is also observed, as well as those shown in Fig. 6.5b. Is possible to infer, by simple geometric considerations, that the intercalation feature shown in Fig. 6.9a favors the interaction of the ring nitrogen with the Lewis acidic sites of the inorganic substrate, whereas the coordination feature shown in Fig. 6.9b favors the interaction of the nitrogen from the NH_2 group. So, the nic/MoO_3 ratio can induce different coordination features of the nic molecules with the acidic sites in the inorganic surface.

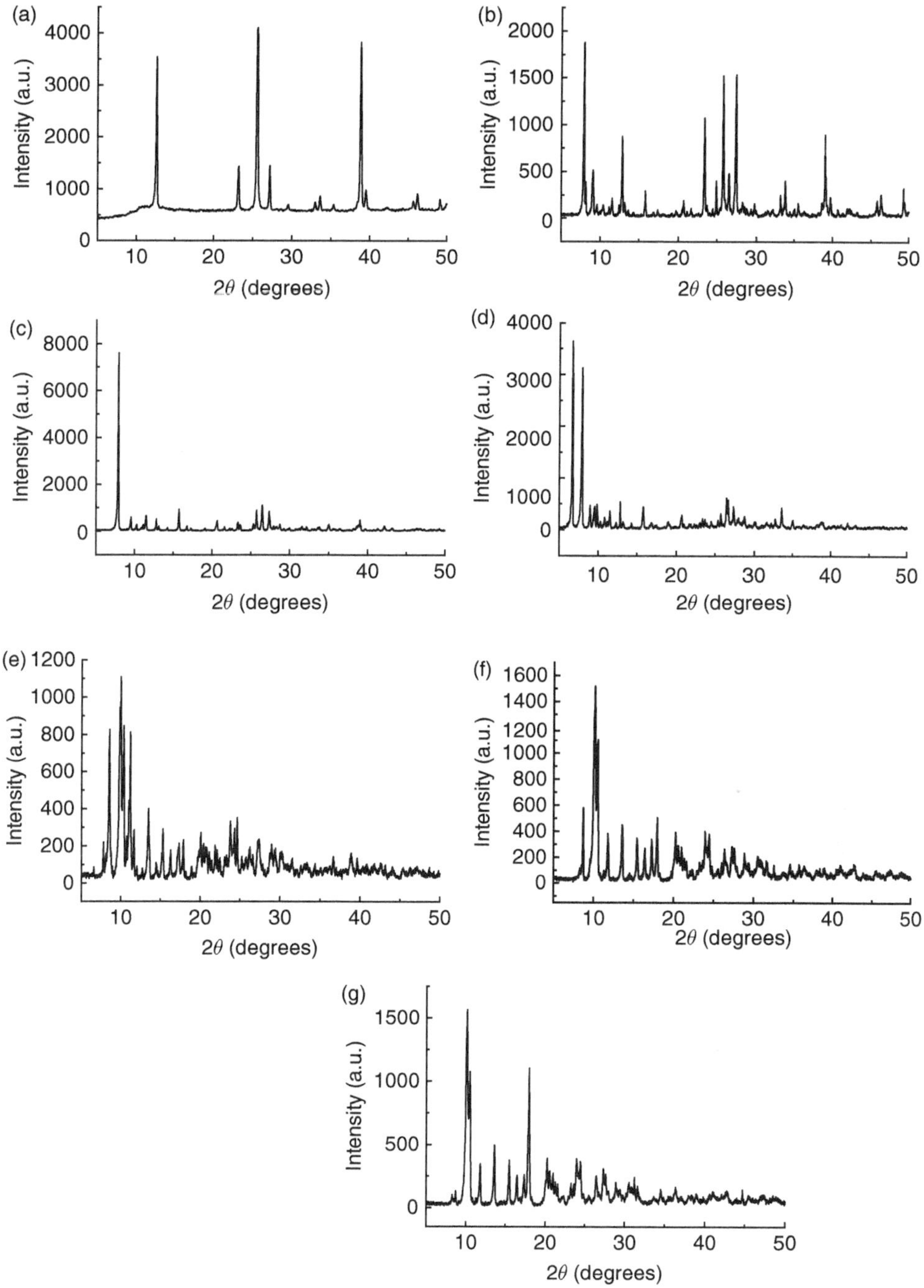

Figure 6.7 XRD patterns for (a) MoO_3, (b) Mo1 (c) Mo2, (d) Mo3, (e) Mo4, (f) Mo5, and (g) Mo6.

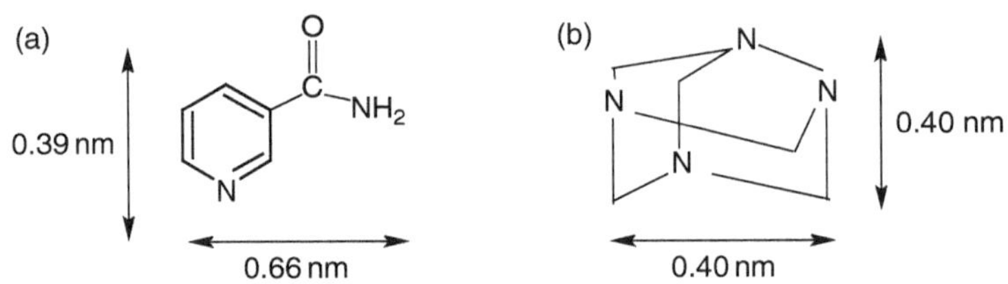

Figure 6.8 Structural formulas for nic (a) and hmta (b). The molecular lengths as calculated by using Spartan-Pro 1.0.3 package are shown.

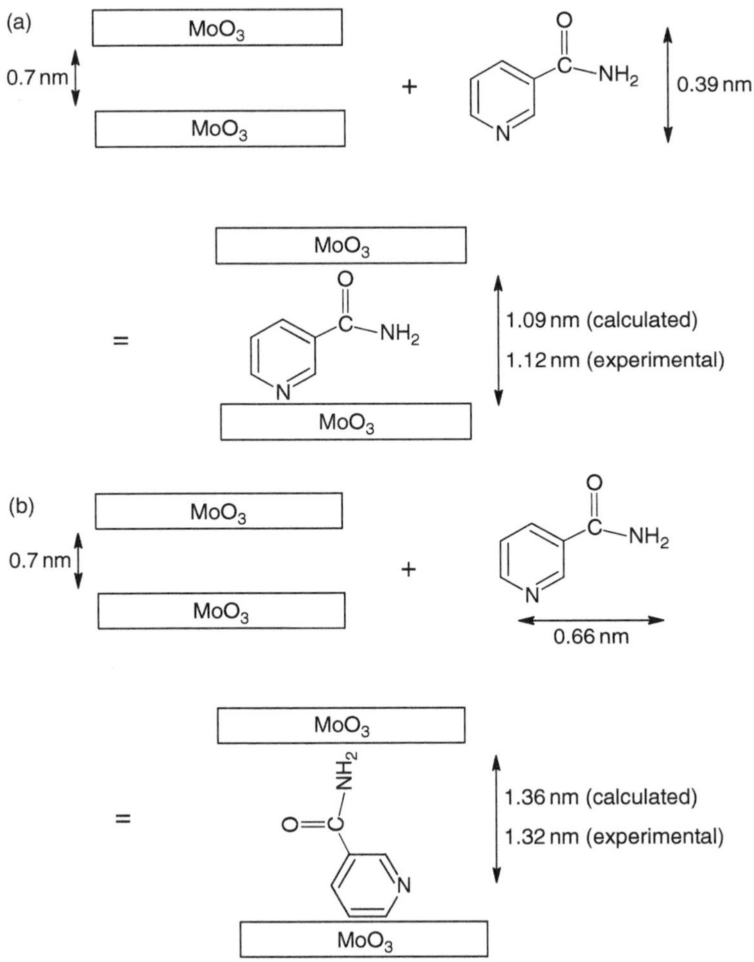

Figure 6.9 Schematic representation of the coordination features of nic molecules into MoO_3: Mo1, Mo2, and Mo3 (a) and Mo3 only (b).

For MoO_3–hmta intercalation compounds (Mo4, Mo5, and Mo6), the X-ray results are as follows: A diffraction peak at 8.7° ($d = 1.01$ nm) is present in the X-ray patterns of the three compounds. However, the intensity of this peak decreases in the sequence Mo6 < Mo5 < Mo4. For the Mo4 compound this peak is almost absent. Furthermore, this is not a diffraction peak due to free hmta. So, it is possible to suppose that at lower hmta/MoO_3 ratios, an intercalation phase, not present at higher concentration is formed. Based on the experimental interlayer distance for MoO_3 (0.70 nm) and the molecular length calculated for hmta (0.40 nm), the intercalation feature shown in Fig. 6.10 can be proposed.

In the hmta diffraction pattern, shown in Fig. 6.11, the most intense peak is at 18.1°. As can be observed in Figs. 6.7e–6.7g, this peak is absent for Mo4, is present for Mo5 and is very intense for Mo6.

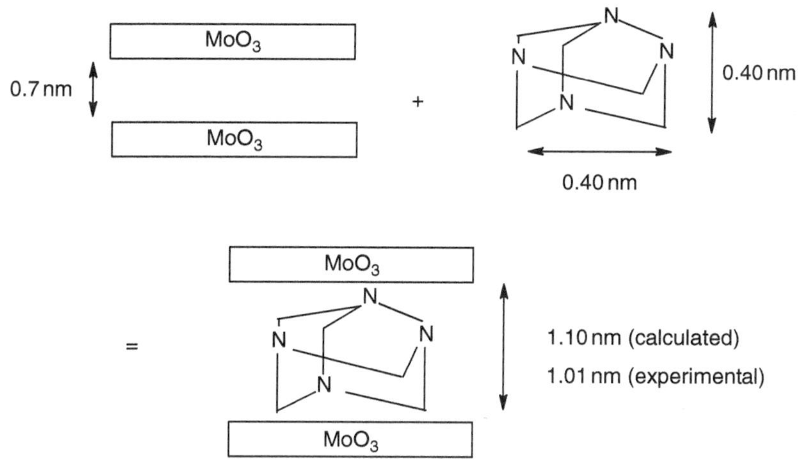

Figure 6.10 Schematic representation of the coordination feature of hmta molecules into MoO_3 (Mo1, Mo2, and Mo3 compounds).

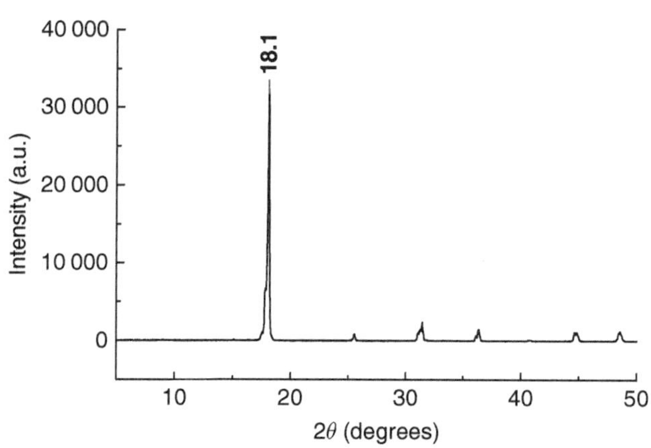

Figure 6.11 XRD pattern for hmta.

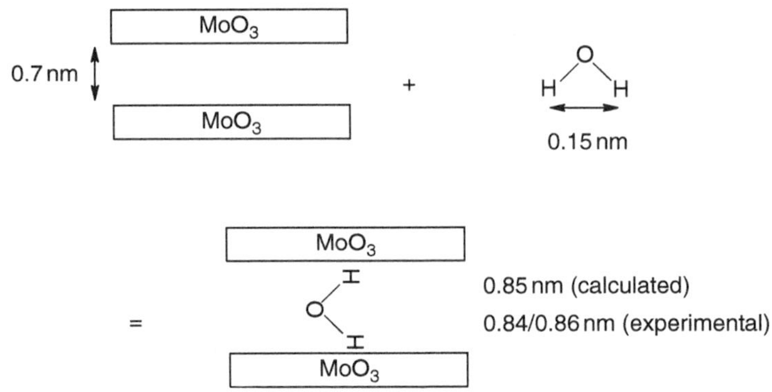

Figure 6.12 Schematic representation of the coordination features of water molecules into MoO_3.

Hence, based on the X-ray and elemental analysis data, it can be concluded that for hmta, the maximum intercalation amount in MoO_3 is in the range 2.04–2.63 mmol of hmta per gram of hybrid matrix. For all three MoO_3–hmta compounds, diffraction peaks at 10.2° ($d = 0.86$ nm) and 10.5° ($d = 0.84$ nm) are observed, suggesting the presence of two distinct intercalation phases. However, the minimum interlayer distance for and MoO_3–hmta intercalation compound should be 1.10 nm, as shown in Fig. 6.6. Hence, the 10.2° and 10.5° are not due to a MoO_3–hmta intercalation, but could be attributed to water intercalated ones, as shown in Fig. 6.12.

So, based on X-ray and quantum chemical calculations data, it can be inferred that at higher hmta/MoO_3 ratios, the hmta molecules in the lamellar space are replaced by water molecules, and so, the hmta molecules are only adsorbed on the surface. So, the replacement of hmta by water molecules into the interlayer space increase in the order Mo6 > Mo5 > Mo4. However, the total amount of intercalated water molecules is, in all cases, a minor one, as verified by TG [20], where the experimental mass losses are in good agreement with CHN elemental analysis results.

Comparing nic (Mo1, Mo2, and Mo3) with hmta (Mo4, Mo5, and Mo6) results indicate that MoO_3 exhibits a higher affinity toward nic than hmta, since for nic an increase of the organic moiety/inorganic substrate ratio increases the total amount of intercalated molecules, whereas for hmta the opposite behavior is observed. For nic, the maximum intercalation amount is above 2.63 mmol of nic per gram of hybrid matrix.

4.2 Adsorption of caffeine, dimethylglyoxime, and rodamin-B on lamellar molybdenum oxide

All reagents were analytical grade (Aldrich) and were used without further purification. The MoO_3 compounds with caff, dmg, and rod-B were synthesized by suspension of MoO_3 into aqueous solutions (40 cm^3 of deionized water) of the organic species at room temperature (28°C) under magnetic strring for 3 min [21]. The suspensions were aged for 5 days, and then, the remaining solvent was slowly evaporated at 30°C for another 2 days. The stoichiometric ratios used were as follows:

g. MoO_3 (1.0 g) + dmg (0.1 g) = Mo(dmg)1
h. MoO_3 (1.0 g) + dmg (0.2 g) = Mo(dmg)2
i. MoO_3 (0.6 g) + caff (0.06 g) = Mo(caff)1
j. MoO_3 (0.6 g) + caff (0.13 g) = Mo(caff)2
k. MoO_3 (1.0 g) + rod-B (0.05 g) = Mo(rod-B)

Hereafter, the six hybrid matrices will be named as MoDMG1, MoDMG2, Mocaff1, Mocaff2, and MoRODB from (a) to (e), respectively.

For the "extra" adsorption experiments performed with rod-B, a mass of 1.0 g of MoO_3 and 25.0 cm^3 each rod-B solution was used (Table 6.2).

The XRD patterns were obtained with a Shimadzu XD-3A instrument, using Cu Kα radiation (35 kV, 25 mA). Infrared spectra were recorded in a Gengis II FTIR spectrometer in the 4000–400 cm^{-1} range, with a resolution of 4 cm^{-1}. All infrared spectra were obtained from powders in KBr disks. Thermogravimetric curves were obtained using a Shimadzu TG- 50H apparatus under N_2 atmosphere with a heating rate of 15°C min^{-1}. The SEM micrographs were obtained using a JEOL electron microscope, model JSM T-300, with an accelerating voltage of 15 kV. The UV–Vis experiments were performed by using a FEMTO 600 spectrometer.

The spectrometer XRD patterns for the hybrid matrices are shown in Fig. 6.13. In the MoO_3 diffraction pattern (not shown), peaks at 14.8° and 29.9° are associated with the {001} and {002} diffraction planes. Taking into account the {001} diffraction peak position, the interlayer distance can be calculated as 0.69 nm. As can be verified by inspection of the XRD patterns, there is no intercalation (the interlayer distance is unchanged after reaction) in the hybrid matrices.

The morphology of the oxide grains was affected by the reaction processes, and, as illustrative examples, the SEM micrographs for the pure oxide and MoDMG2 are shown in Fig. 6.14. Due to the very intense infrared bands form MoO_3 the infrared spectra for the hybrid matrices (not shown) are not conclusive or indicative of the reaction processes.

Table 6.2 Data for the adsorption of rod-B on MoO_3 (1.0 g of the oxide was suspended in 25.0 cm^3 of rod-B solution)

RodB concentration (mol dm^{-3})	Initial absorbance	Absorbance after 48 h
0.6×10^{-3}	1.143	0.026
1.0×10^{-4}	0.745	0.018
0.5×10^{-4}	0.456	0.000
1.5×10^{-5}	0.147	0.056[a]
1.0×10^{-5}	0.097	0.000

[a] This "abnormal" absorbance value (taking into account the observed tendency down the table) is attributed to the presence of suspended particulate material.

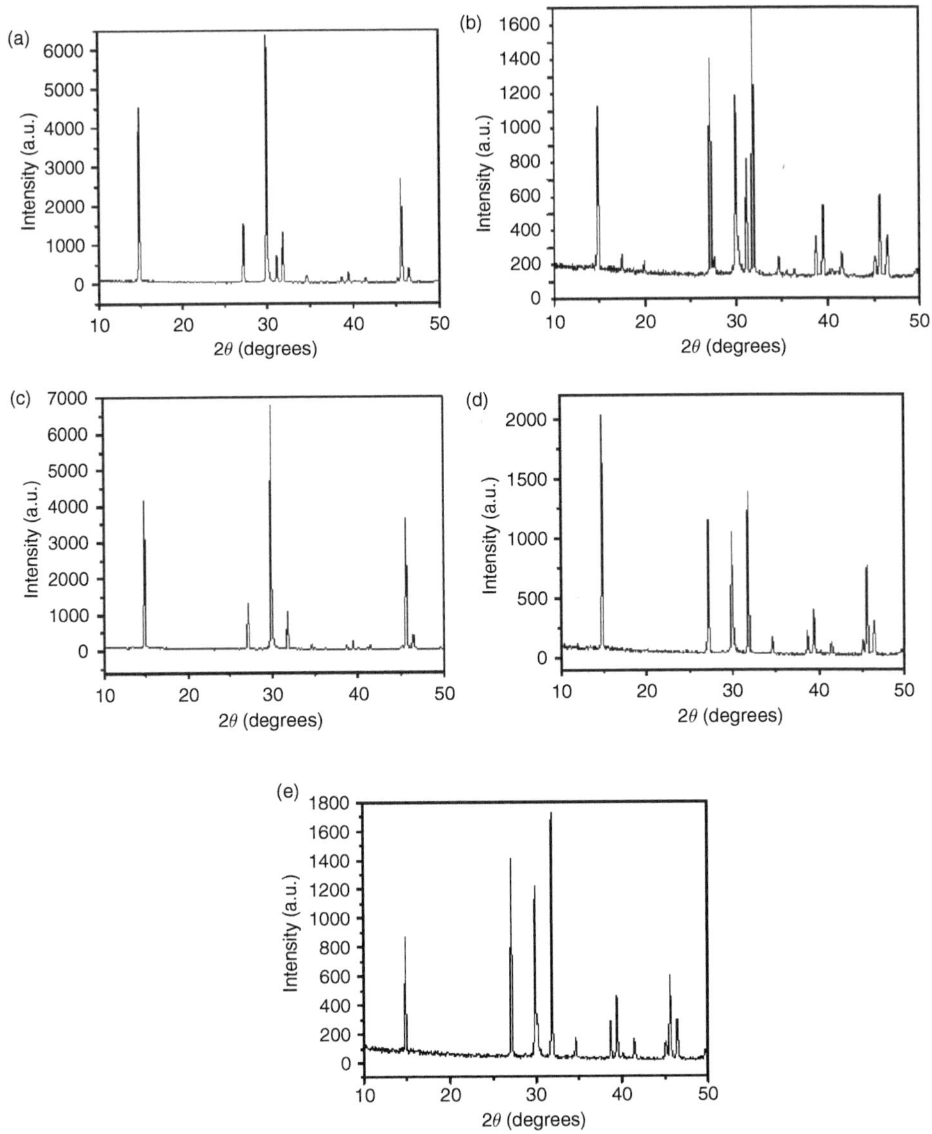

Figure 6.13 XRD patterns for (a) MoDMG1, (b) MoDMG2, (c) Mocaff1, (d) Mocaff2, and (e) MoRODB.

The thermogravimetric data are summarized in Table 6.3. The thermogravimetric curves for MoDMG1 and MoDMG2 are shown in Fig. 6.15. The mass losses are in the 50–400°C range, since the organic moiety is released in that temperature range, and that about 600°C the oxide sublimates.

The observed mass losses show that MoO_3 exhibits appreciable affinity toward dmg, but has no affinity (at least for these oxide/caff proportions) toward caff, since,

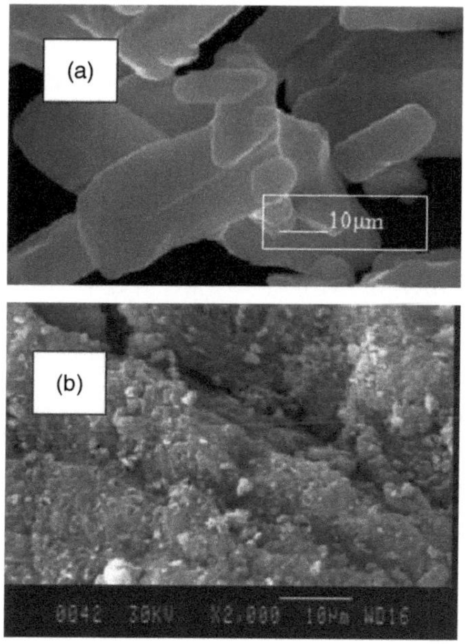

Figure 6.14 SEM micrographs for (a) MoO_3 and (b) MoDMG2.

Table 6.3 Thermogravimetric data summary for MoO_3 hybrid matrices

Matrix	Mass loss (%)
MoDMG1	8.4
MoDMG2	21.3
Mocaff1	0.0
Mocaff2	0.0
MoRODB	0.0

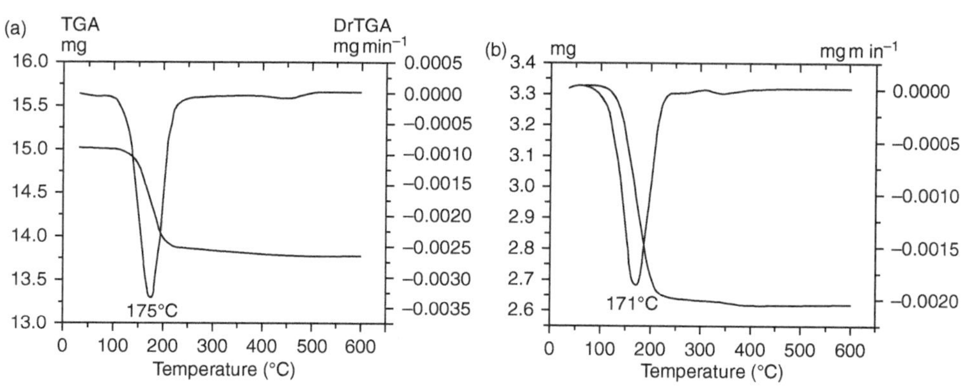

Figure 6.15 Thermogravimetric and derivative curves for (a) MoDMG1 and (b) MoDMG2.

for the latter, there was no mass loss, indicating that no caff molecules have been linked to the oxide surface. For the MoO_3–DMG matrices, based on the mass loss percentages verified for the synthesized matrices, it can be concluded that an increase of 100% in the used mass of DMG results in a much larger degree of adsorption.

For MoRODB, despite the null mass loss observed, by inspection of the matrix it is possible to verify that the pale green MoO_3 pure oxide turns pink after the adsorption process, showing that rod-B was indeed adsorbed. The observed null mass loss must be attributed to the low MoO_3/rod-B proportion used.

To evaluate the affinity of MoO_3 toward rod-B, more detailed experiments were performed, and the total amount of adsorbed rod-B evaluated by UV–Vis spectroscopy. The results are summarized in Table 6.2. As can be verified, the oxide exhibits, indeed, a very high affinity toward rod-B.

5. Sol–Gel Modified Molybdenum Oxide

To enlarge the possible applications of molybdenum oxide (as catalyst, for example) the modification of its surface was performed, by using organosilanes [22]. The modified molybdenum oxide surfaces were obtained by reaction with nitrogen containing organosilanes of the type $(C_2H_5O)_3Si—R$, R = $—(CH_3)_2NH_2$; $—(CH_3)_2NH(CH_2)_2NH_2$; and $—(CH_3)_2NH(CH_2)_2NH(CH_2)_2 NH_2$. Furthermore, a $MoO_3—SiO_2$ matrix was also obtained by reaction with TEOS. The main purposes were to prepare modified MoO_3 samples with higher surface areas than MoO_3 itself; to study the possible interactions of aminated silanes with MoO_3, in an attempt to produce hybrid samples with the Lewis acidity of the inorganic substrate, as well as the metal sequestering ability of the organic moiety.

TEOS and nitrogen-containing organosilicon-modified molybdenum oxide matrices were obtained by using the following synthetic procedure: MoO_3 (Baker) was dried under vacuum at 120°C for 8 h. Then, the dried oxide was exposed to an atmosphere saturated with water vapor for 2 min. Afterward, 2 g of the previously treated MoO_3 sample was suspended in 150 cm^3 of dry xylene. To this suspension, 5.0 mmol of $(C_2H_5O)_4Si$ or $(C_2H_5O)_3Si—R$ with R = $—(CH_3)_2NH_2$, $—(CH_3)_2NH(CH_2)_2NH_2$, or $—(CH_3)_2NH(CH_2)_2NH(CH_2)_2NH_2$ were added with vigorous stirring under a nitrogen atmosphere. The resulting suspensions were stirred for 5 days and then aged for another 12 days in air, after which they were filtered and washed with dry xylene, acetone, and ethanol. The solid products were ground for 30 min, and dried under vacuum for 24 h at room temperature. The four resulting modified matrices will be named as

$$MoO_3\ (0.3\ g) + TEOS\ (5.0\ cm^3) = MoSil$$

$$MoO_3\ (0.3\ g) + (C_2H_5O)_3Si(CH_3)_2NH_2\ (10.0\ cm^3) = MoN1$$

$$MoO_3\ (0.3\ g) + (C_2H_5O)_3Si(CH_3)_2NH(CH_2)_2NH_2\ (10.0\ cm^3) = MoN2$$

$$MoO_3(0.3g) + (C_2H_5O)_3Si(CH_3)_2NH(CH_2)_2NH(CH_2)_2NH_2(10.0cm^3) = MoN3$$

Thermogravimetric analyses were carried out in a Shimadzu TGA-50 apparatus under argon atmosphere with a heating rate of $8.3 \times 10^{-2}\,K\,s^{-1}$. The XRD patterns were obtained in a Shimadzu XD-3A instrument, using Cu K_{α} radiation (35 kV, 25 mA). Carbon, nitrogen, and hydrogen elemental analysis were performed in a Perkin-Elmer microelemental analyser. The infrared spectra were obtained in a Bomem apparatus in the range 4000–400 cm^{-1} with a resolution of 4 cm^{-1}, by dispersing the samples in KBr disks. The SEM micrographs were taken with a JEOL model JSM T-300 microscope with an accelerating voltage of 15 kV. The BET surface area measurements were performed in a FlowSorb 2300 apparatus from Micromeritics Instrument Co. ^{29}Si NMR CP-MAS spectra of solid samples were performed by using an AC 300/P Bruker instrument at room temperature, with the experimental conditions: pulse repetition time of 0.115 s, contact time of 3 ms, and 17 857 accumulations.

Carbon, hydrogen, and nitrogen elemental analysis results for the aminated silanes–molybdenum oxide hybrids samples are summarized in Table 6.4. The BET surface area results are summarized in Table 6.5.

As can be observed, the addition of TEOS to MoO_3, produced a mixed molybdenum–silicon oxide, resulting in a matrix of higher surface area when compared with pure MoO_3. On the other hand, the addition of nitrogen-containing silanes enhanced (or decreased, as observed for MoN2 sample) the oxide surface area in a minor extent.

The MoN matrices exhibited similar infrared spectra and the one for MoN1 sample is shown as example in Fig. 6.16.

The peaks at 996, 859, and 562 cm^{-1} are due to the inorganic substrate whereas those at 1612, 1509, 1225, and 1134 cm^{-1} can be attributed to the adsorbed organosilane. Based on infrared data, it can be inferred that the for *N*-silanes, the nitrogen atom is interacting with the Lewis acid sites of the oxide surface, as

Table 6.4 Carbon, hydrogen, and nitrogen elemental analysis results (%) for the aminated silanes–molybdenum oxide-modified matrices, with the corresponding silane content

Sample	C	H	N	Silane content ($mmol\,g^{-1}$)
MoN1	6.23	1.85	2.22	1.70
MoN2	12.95	3.23	5.59	2.00
MoN3	8.04	2.19	3.54	0.90

Table 6.5 BET surface area for pure and silicon-modified molybdenum oxide

Sample	BETsurface area ($m^2\,g^{-1}$)
MoO_3	0.61 ± 0.01
MoSil	78.12 ± 1.70
MoN1	1.85 ± 0.22
MoN2	0.59 ± 0.01
MoN3	2.53 ± 0.11

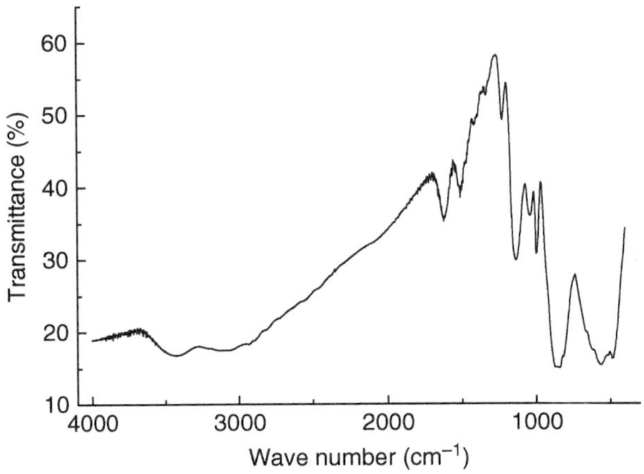

Figure 6.16 Infrared spectrum for MoN1 matrix.

observed for polyaniline (negative shifts for the NH vibration bands). Furthermore, a progressive change in the colors of the samples was observed with the increase in nitrogen content in the silane molecules. So, MoN1 is light gray whereas MoN2 is gray and MoN3 is deep gray. MoO_3 and MoSil are both light green compounds.

The XDD patterns for MoO_3 and the modified matrices are shown in Fig. 6.17. The measured 2θ values for the 001 diffraction plane were 12.64°, 13.20°, 12.82°, 12.84°, and 12.90° for MoO_3, MoSil, MoN1, MoN2, and MoN3 matrices, respectively. Such small variations of the interlayer distance, can not be associated to the insert of organic species inside the molybdenum oxide interlayer space, and could be attributed to variations in the degree of hydration of the oxide.

The SEM images for the pure oxide, MoSil, and MoN1 matrices are shown in Fig. 6.15. As can be observed in the SEM micrographs of MoO_3 and MoSil shown

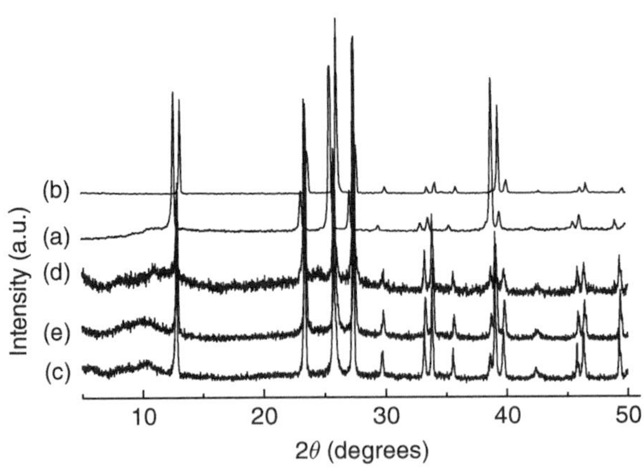

Figure 6.17 XRD patterns of MoO_3 (a), MoSil (b), MoN1 (c), MoN2 (d), and MoN3 (e).

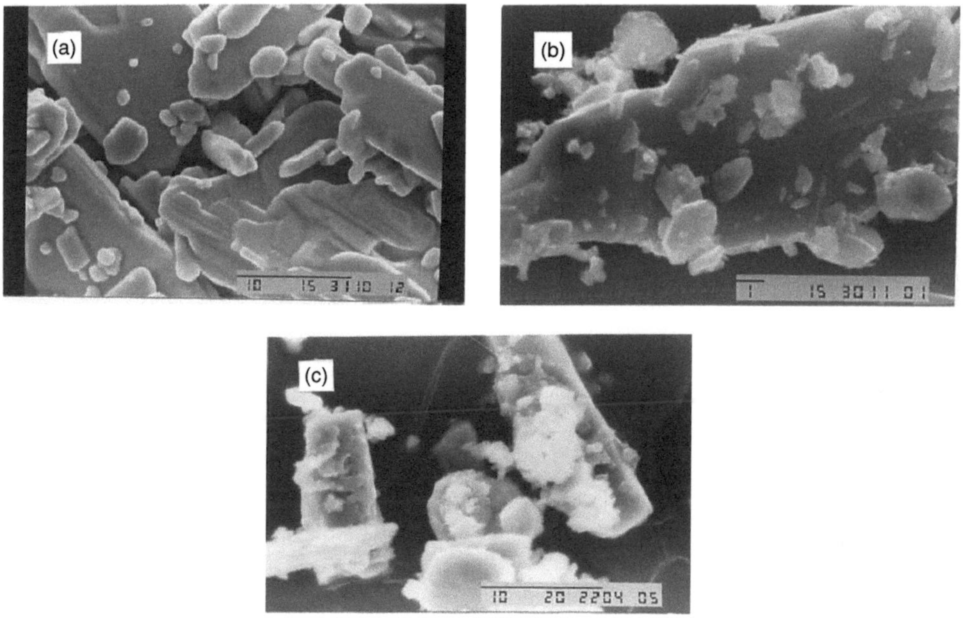

Figure 6.18 SEM micrographs for MoO_3, 3500× (a); and MoSil, 7500× (b); and MoN1, 3500× (c). The scale bar is in μm.

in Figs. 6.18a and 6.18b, respectively, the addition of TEOS to molybdenum oxide resulted in the production of "islands" of SiO_2 on the MoO_3 grain surfaces which can be considered as responsible for the increase in the surface area of the modified sample.

As can be observed in Fig. 6.18c for MoN1 sample, the nitrogen-containing silane is adsorbed on the molybdenum oxide grains surface constituting some kind of "coverage." So, it is a reasonable assumption that for the amine–silane-modified samples, the surface area suffered a minor modification in comparison with pure MoO_3.

The TG and derivative curves for MoO_3 and the synthesized hybrid matrices are shown in Fig. 6.19.

Molybdenum oxide exhibits a single mass loss step in the TG curve due to the sublimation of the oxide, starting at 730°C. The MoSil matrix exhibits a similar TG curve profile, but the thermal degradation takes place at 704°C. The MoN3 matrix shows the release of 23% mass due to the organic moiety followed by the sublimation of the molybdenum substrate. On the other hand, MoN1 and MoN2 matrices exhibited mass loss percentages of 20% and 30%, respectively. The three nitrogen–silane matrices show mass losses beginning at low temperatures, which could be attributed to the removal of water, alcohol, and unreacted alkoxy species. Comparing the thermogravimetric curves for MoN1 and MoN2, can be verified that the major mass loss occurs from 400°C to 700°C. So, it can be inferred that only near 400°C the silane–molybdenum oxide bonds are disrupted to a major extent.

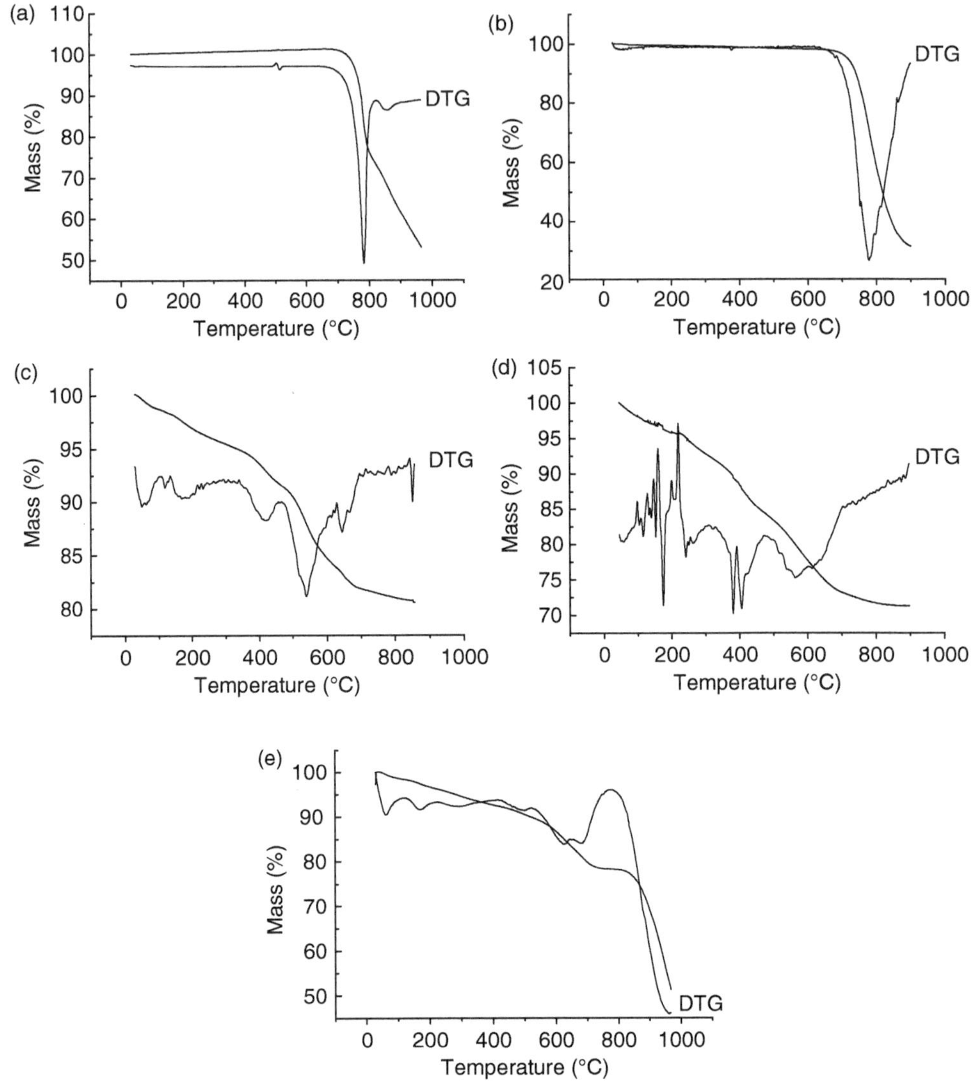

Figure 6.19 Thermogravimetric and derivative curves for MoO_3 (a), MoSil (b), MoN1 (c), MoN2 (d), and MoN3 (e).

Taking into account this thermal stability, as well as the applications of organosilane-modified inorganic matrices for chromatographic separations, very interesting possibilities could arise.

For silica gel samples, peaks at −90.6, −99.8, and −109.3 ppm in the ^{29}Si CP-MAS NMR spectrum can be observed, due to silicon atoms bound to free geminal silanols, free silanols, and siloxane groups, respectively [23]. These values are in good agreement with those obtained for the MoSil matrix, that is, −91.7, −101.36, and −110.63 ppm. So, it can be concluded that the MoSil matrix is composed of MoO_3 and amorphous silica gel, producing a MoO_3–SiO_2 mixed oxide with higher surface area, which is an important achievement for catalytic purposes [24].

Qualitative tests show that the inorganic–organic hybrid samples are able to coordinate transition metal cations such as copper (II) and cobalt (II).

The synthesis or MoO_3 inorganic–inorganic or inorganic–organic hybrids can be successfully performed through sol–gel process by using silicon alkoxides as guest species. Furthermore, MoO_3–SiO_2 samples with higher (in comparison with MoO_3) surface areas can be prepared.

In such inorganic–organic hybrids, there are interactions between the two oxides, with SiO_2 "islands" grown on the MoO_3 surface, and so both kinds of acidic sites, Brønsted (SiO_2) and Lewis (MoO_3) are available. Furthermore, the prepared inorganic–organic hybrids are able to coordinate transition metal cations.

6. Graphite Oxide

Graphite oxide (GO) can be defined/identified is an oxygen-rich carbogenic material that is typically derived by strong oxidation of crystalline graphite and contains oxygen in the concentration range 30–40% (w/w).

This solid exhibits an extended lamellar structure with randomly distributed aromatic and aliphatic regions, as well as a high amount of hydroxyl/carboxyl functional groups embedded in its layers. Hence, GO is endowed with swelling, intercalation and ion exchange properties [25–30]. This layered solid decomposes at relatively low temperature (<200°C), releasing CO_2 and H_2O.

As a recent example of the investigations on the chemistry of GO can be mentioned the study of the intercalation compounds of polyethylene oxide, polyvinylpyrrolidone, methyl cellulose, poly[oxymethylene (oxyethylene)], and poly[oligo (ethylene glycol oxalate)] into GO that has been recently reported [31]. It was verified that the forces of attraction between the polymers and GO are primarily dipole–dipole and/or hydrogen bonding arising from the polar groups in the polymers and the functional groups (COC, COOH, COH, and C=O) in the GO. Furthermore, it was verified that the addition of a small amount of dilute NaOH resulted in deprotonation of the acidic functionalities which caused exfoliation of the layered structure.

It has been shown that [32] GO is able to remove ammonia by two processes: intercalation and reactive adsorption. It was verified that the total amount of adsorbed ammonia is significantly high, compared with adsorption on activated carbon.

In an interesting interplay between GO and organosilanes (which are so extensively used to modify amorphous or nanostructured oxide surfaces), there are reports [33], of the preparation of intercalation compounds in which amino groups are inserted into the interlayer space of GO, using 3-aminopropylethoxysilanes. For such compounds, it was shown that the amino groups of

3-aminopropyltriethoxysilane were bonded to hydroxyl groups of GO. It was also found that the temperature in which the intercalation occurs affect the interlayer distance, with larger distances been observed to the products obtained at higher temperatures.

REFERENCES

1. L.M. Nunes, A.G. de Souza, R.F. de Farias, J. Alloys Compd. 319 (2001) 94.
2. C. Airoldi, L.M. Nunes, R.F. de Farias, Mater. Res. Bull. 35 (2000) 2081.
3. C.R. Alentejano, I.V. Aoki, Electrochim. Acta 49 (2004) 2779.
4. D. Carson, M. Forissier, J.C. Vedrine, J. Chem. Soc., Faraday Trans. 1 (1984) 1017.
5. R.F. de Farias, An. Assoc. Bras. Quím. 51 (2002) 107.
6. G.F. de Sá, O.L. Malta, C. de M. Donegá, A.M. Simas, R.L. Longo, P.A. Santa-Cruz, E.F. da Silva Jr., Coord. Chem. Ver. 196 (2000) 165.
7. R.F. de Farias, S. Alves Jr., M.F. Belian, G.F. de Sà, Opt. Mater. 18 (2002) 431.
8. R.F. de Farias, S. Alves Jr., M.F. Belian, M.R.S. Vieira, J.M. de Souza, G.G. Pedrosa, G.F. de Sá, J. Serb. Chem. Soc. 69 (2004) 387.
9. R.F. de Farias, S. Alves Jr., M.F. Belian, G.F. de Sá, Bull. Chem. Technol. Macedonia 23 (2004) 7.
10. J.M. de Souza, G.F. de Sá, W.M. de Azevedo, S. Alves Jr., R.F. de Farias, Opt. Mater. 27 (2005) 1187.
11. R.F. de Farias, C. Airoldi, M.F. Belian, S. Alvres Jr., J. Alloys Comp. 419 (2006) 50–53.
12. L. Kihlborg, The Crystal Chemistry of Molybdenum Oxides, Advances in Chemistry Series, vol. 39, American Chemical Society, Washington, 1963.
13. W.T. Elwell, D.F. Wood, Analytical Chemistry of Molybdenum and Tungsten, Pergamon Press, New York, 1971.
14. B.C. Gates, Catalytic Chemistry, Wiley, New York, 1992.
15. R.F. de Farias, J.M. de Souza, J.V. de Melo, C. Airoldi, J. Colloid Interface Sci. 212 (1999) 123.
16. R.F. de Farias, Effects of adsorption on the redox process of oxide surfaces, in: Arthur Hubbard (Ed.), Encyclopedia of Surface and Colloid Science, Marcel Dekker, New York, 2002.
17. R.F. de Farias, Int. J. Inorg. Mater. 3 (2001) 303.
18. R.F. de Farias, Int. J. Inorg. Mater. 3 (2001) 931.
19. R.F. de Farias, J. Phys. Chem. Solids 64 (2003) 2199.
20. R.F. de Farias, Mater. Chem. Phys. 90 (2005) 302.
21. R.F. de Farias, C.R. Franco, M.S. Refat, H.A. Hashem, Mens Agitat, 3(1) (2008) 41–44.
22. R.F. de Farias, J. Phys. Chem. Solids 64 (2003) 1241.
23. G.E. Maciel, D.W. Sindorf, J. Am. Chem. Soc. 102 (1980) 7606.
24. B.C. Gates, Catalytic Chemistry, Wiley, New York, 1992.
25. I. Dékány, R. Krüger-Grasser, A. Weiss, Colloid Polym. Sci. 276 (1998) 570.
26. A.B. Bourlinos, D. Gournis, D. Petridis, T. Szabó, A. Szeri, I. Dékány, Langmuir 19 (2003) 60505.
27. A. Lerf, H. He, M. Forster, J. Klinowski, J. Phys. Chem. B 102 (1998) 4477.
28. N.I. Kovtyukhova, P.J. Ollivier, B.R. Martin, T.E. Mallouk, S.A. Chizhik, E.V. Buzaneva, et al., Chem. Mater. 11 (1999) 771.
29. T. Szabó, A. Szeri, I. Dékány, Carbon 43 (2005) 87.
30. T. Szabó, O. Berkesi, P. Forgó, K. Josepovits, Y. Sanakis, D. Petridis, et al., Chem. Mater. 18 (2006) 2740.
31. R. Bissessur, S.F. Scully, Solid State Ionics 178 (2007) 877.
32. M. Seredych, T.J. Bandosz, Carbon 45 (2007) 2126.
33. Y. Matsuo, Y. Nishino, T. Fukutsuka, Y. Sugie, Carbon 45 (2007) 1384.

CHAPTER SEVEN

Chemistry on Modified Phosphates

Contents

1. Introduction

Lamellar α- and γ-titanium and zirconium hydrogen phosphates, which exhibit different proton exchange capacities, can be used as both cation exchange matrices and intercalation hosts producing intercalation compounds when reacted with organic species [1,2]. On the other hand, vanadyl phosphate, which exhibits catalytic applications, can be modified by the intercalation of organic species, and such a synthetic approach can be used to modify the phosphate surface morphology, affecting its catalytic action. In this chapter, calcium and barium phosphates and phosphonates will also be considered.

2. Titanium and Zirconium Phosphates

Lamellar α-titanium hydrogen phosphate (α-$Ti(HPO_4)_2.H_2O$) can be used to obtain, through proton ion exchange processes, other phosphates, such as Ca^{2+},

Interface Science and Technology, Volume 17
ISSN: 1573-4285, doi: 10.1016/S1573-4285 (08) 00407-9

Ba^{2+}, and La^{3+} ones, or can be modified with organic species producing intercalation compounds [1,2].

Zirconium phosphate, besides the traditional intercalation processes, has been used to produce pillared compounds, such as those involving mixed Fe–Cr oxides [3]. Furthermore, it has been shown [4] that the protonic conductivity of α-zirconium phosphate can be enhanced by the formation of a composite with alumina or silica, an example of synergic cooperation between the oxides and phosphates which are the main classes of compounds highlighted in the book.

3. VANADYL PHOSPHATE

3.1 Synthesis and characterization of a $VOPO_4$–NH_3 hybrid

Oxovanadium phosphate $VOPO_4.2H_2O$ is used as a catalyst, or catalyst precursor. It is a layered compound in which six oxygen atoms around the vanadium form an almost regular octahedron. The weaker interlayer binding makes $VOPO_4.H_2O$ and its analogs attractive as potential hosts for coordination–intercalation reactions [5–11].

Intercalation reactions of neutral host lattices usually occur via electron transfer between the guest species and the host lattice. Furthermore, vanadium (V) compounds are strong oxidants, and vanadium phosphate catalysts are able to promote, for example, oxidation of alcohols to ketones [12].

The synthesis and characterization of $VOPO_4$ intercalation compounds with aliphatic amines from C_1 to C_{10} have been reported [13,14]. However, the synthetic approach for the intercalation of such classes of molecules in this work is quite different from those used here. Furthermore (and most important), we think that the use of diamines could provide interesting results, from a structural point of view, since they exhibit two coordination sites (N atoms), and so are able to interact (coordinate) with two different vanadium atoms of opposite layers.

The aim of this section is to report the synthesis and characterization of a $VOPO_4$–NH_3 hybrid prepared at room temperature in aqueous solution. The main purpose was to investigate the possible effects of a such reaction on the nano- and microstructure of the oxovanadium lamellar matrix, since its structure exerts substantial influence on its possible catalytic applications [14].

$VOPO_4.2H_2O$ was synthesized as previously described [9], by refluxing a mixture of V_2O_5 (25 g), H_3PO_4 (85%, 223 g), and H_2O (577 cm^3) at 130°C for 16 h. The yellow product was filtered off, washed with acetone, and then dried under room temperature.

The $VOPO_4$–NH3 hybrid was synthesized by dissolution of 1.0 g of $VOPO_4.2H_2O$ in 60 cm^3 of H_2O. To this solution, 1.0 cm^3 of NH_4OH (28% NH_3 in water) were slowly added at room temperature. The resulting solution was evaporated in a fume hood for 5 days, and the red-brown crystals were dried under vacuum at room temperature for 12 h.

Hydrogen and nitrogen elemental analyses were performed in a Perkin-Elmer apparatus. Infrared spectra were obtained in a Bomem apparatus in the range 4000–400 cm^{-1} with a resolution of 4 cm^{-1} by dispersing the samples in KBr disks.

Thermogravimetric curves were obtained using a Shimadzu TGA 50 apparatus, under Ar atmosphere with a heating rate of 10°C min^{-1}. The XRD patterns were obtained in a Shimadzu XD-3A instrument, using Cu K_α radiation (35 kV, 25 mA). SEM micrographs were taken with a Jeol microscope, model JSM T-300, at an accelerating voltage of 15 kV.

The hydrogen and nitrogen elemental analysis results gave the values: H = 4.53 and N = 13.26%. In the NH_3 molecule, the N%/H% ratio is 82.4/17.6 = 4.68. For the synthesized hybrid, the same ratio is 13.26/4.53 = 2.93. So, it can be concluded that in the synthesized hybrid there are hydrogen atoms that originate not from the NH_3 molecule, but from physisorbed water molecules. Based on the nitrogen percentage, the formula $VOPO_4\ (NH_3)_2$ (calculated nitrogen content: 13.66%) was proposed for the synthesized hybrid.

The XRD patterns for α-$VOPO_4.2H_2O$ and the $VOPO_4$–NH_3 hybrid are shown in Fig. 7.1. The diffraction pattern obtained for $VOPO_4.2H_2O$ (Fig. 7.1a) is in agreement with previously reported results [13–15]. The oxovanadium phosphate exhibits diffraction peaks at 11.8°, 18.5°, and 23.9°, due to the 001, 101, and 002 diffraction planes [14], with associated *d* values of 0.75, 0.48, and 0.37 nm, respectively. The XRD pattern obtained for the synthesized hybrid clearly shows that it is a crystalline matrix. Unfortunately, unambiguous attributions could not be made for the observed diffraction peaks. However, only the diffraction peak at 23.9° observed in the oxovanadium phosphate is also observed in the hybrid. So, the results shown in Fig. 7.1 suggest that the layered structure of $VOPO_4.2H_2O$ was decomposed by the reaction, with the formation of a salt of V–P–O species [11].

The thermogravimetric and derivative curves are shown in Fig. 7.2. As can be observed, the vanadium phosphate matrix (Fig. 7.3a) exhibits two distinct mass loss steps, from 60°C to 110°C, associated with the dehydration process: $VOPO_4.2H_2O \rightarrow VOPO_4.H_2O + H_2O$ and $VOPO_4.H_2O \rightarrow VOPO_4 + H_2O$, respectively. This thermogravimetric curves are in good agreement with previously reported results [16]. However, the temparature ranges observed for the mass losses are quite different. This fact can be attributed to differing heating rates and atmospheres having been used.

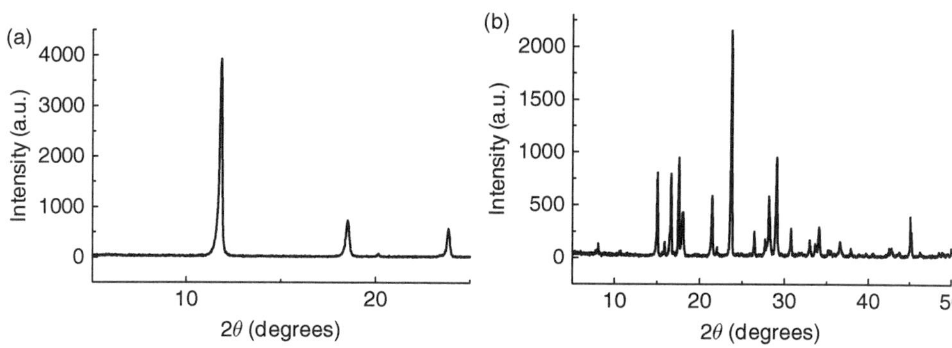

Figure 7.1 XRD patterns for $VOPO_4.2H_2O$ (a) and $VOPO_4$–NH_3 hybrid (b).

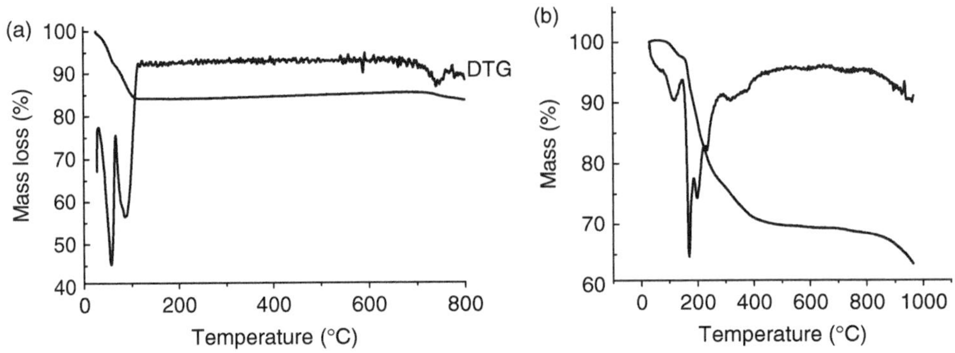

Figure 7.2 Thermogravimetric curves for $VOPO_4.2H_2O$ (a) and $VOPO_4$–NH_3 hybrid (b).

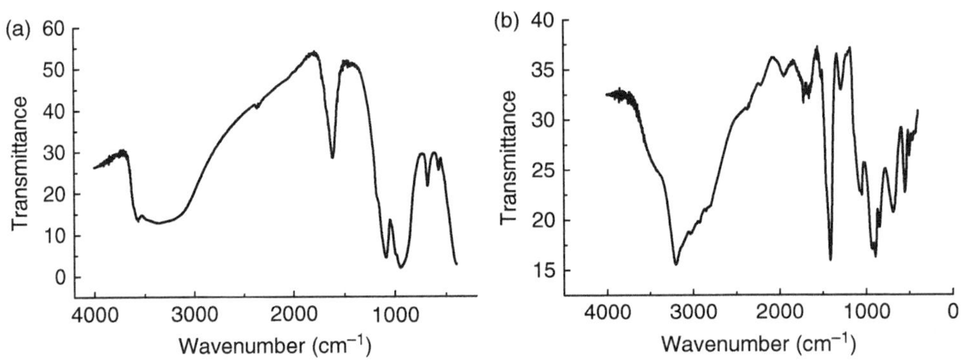

Figure 7.3 Infrared spectra for $VOPO_4.2H_2O$ (a) and $VOPO_4$–NH_3 hybrid (b).

Two types of water molecules are present in $VOPO_4.2H_2O$: one water molecule is coordinated to vanadium, in a *trans* position to the V = O group [17]. The other is situated between the layers, and both water molecules hold the layers together by hydrogen bonding to the oxide framework. Based on these considerations, the first mass loss can be attributed to the release of interlayer water molecules, and the second mass loss to the coordinated water.

The thermogravimetric curve for the $VOPO_4$–NH_3 hybrid (Fig. 7.3b) exhibits a 3% mass loss in the temperature range 95–160°C, due to the release of adsorbed water molecules. From 160°C to 450°C, a 27% mass loss step is observed. Since the NH_3 represents 17% of the hybrid mass, it appears that water molecules constitute 10% of this second mass loss. Taking into account that physisorbed water molecules are release at lower temperatures, it can be supposed that these water molecules are bonded to the hybrid through stronger intermolecular forces, such as hydrogen bonding.

The infrared spectra are shown in Fig. 7.4. For $VOPO_4.2H_2O$, the main bands are those located at 3583, 1606, 1093, 946, and 683 cm^{-1}, due to the ν(OH), ν(HOH), ν_{as}(P—O), ν(V—OH), and the ν(V—OH) or (P—OH) vibration modes, respectively [16,18].

(a)

(b)

Figure 7.4 SEM micrographs for $VOPO_4.2H_2O$, 750× (a) and $VOPO_4 NH_3$ hybrid 500× (b).

For the synthesized hybrid, the most prominent bands are those at 1650, 1414, 941, and 677 cm^{-1}. The bands at 1650, 941, and 677 cm^{-1} in the hybrid, could be related with those located at 1606, 946, and 683 cm^{-1} in the oxovanadium phospate matrix. The most intense infrared band for the hybrid, at 1414 cm^{-1} could be attributed to the δ N—H vibration mode.

The SEM micrographs are shown in Fig. 7.4. $VOPO_4.2H_2O$ exhibits a typical plate-like morphology. The SEM micrograph for the $VOPO_4.NH_3$ hybrid shows that the microstructure of $VOPO_4.2H_2O$ is changed after reaction, with a total exfoliation of the oxovanadium matrix.

The experimental results show that the intercalation of diaminopropane into lamellar $VOPO_4.H_2O$ (at least in the stoichiometric ratio used in this work), leads to the formation an equimolar (1:1) mixture of two compounds with formulas $VOPO_4.[H_2N—(CH_2)_3NH_2]$ and $VOPO_4.[H_2N—(CH_2)_3NH_2]_{0.5}$, which exhibits different coordination features. Furthermore, such intercalation promotes the exfoliation of the oxovanadium matrix, as confirmed by SEM images.

3.2 More vanadyl phosphate compounds

From the point of view of synthetic chemistry, is important to mention the intercalation compound $VOPO_4.eu.H_2O$ (eu = ethyleneurea) which is the first vanadyl phosphate compound prepared by both routes, the phosphate and the guest organic species in the solid state, without the use of any kind of solvent, that is, the compound was prepared by a solid-state route [19]. It was verified that the modifications on both, the nano- and microstructure of the phosphate, generally observed for solution reaction preparations are also observed for solid-state reactions, showing that this kind of phosphate surface modification can be successfully used to obtain new catalytic compounds. A comparative study of the effects of solid state and solution reaction intercalation on the nano- and microstructure of vanadyl phosphate was also performed, with dodecylamine and nic as guest molecules [20].

A very interesting modification of vanadyl phospahre surface can be performed by reaction with conducting polymers such as polyaniline [21].

4. Barium and Calcium Phosphate

Barium and calcium phosphates have been used to produce chemically modified electrodes, such as by intercalation of meldola blue, producing a stable and electrochemically active electrode for hydrazine [22].

Aromatic and aliphatic amines [23,24] can also be intercalated into barium phenylphosphonates; it has been verified that the total amount of intercalated amines is related to the basicity of such compounds, with exothermic enthalpy values, in all cases.

Bioactive films of apatite can be prepared by sol–gel method [25], and deposited on biomedical metallic implants, enhancing their bio-activity and so, reducing the risks of biological rejection.

Another example of oxide–phosphate synergic interaction is the immobilization of calcium phosphate nanoclusters into sol–gel TiO_2 [26].

5. Adsorption of Glycine on "Animal" (Bone) Phosphate

5.1 Introduction

The so-called "black stone," a "coal" of animal origins, has been used in some communities of Maranhão State (Brazil) to adsorb the poison from snakes in wounded persons. This kind of material is also used for the same purpose in other countries such as India, as well as in Africa.

Results related to the investigation of the adsorption of glycine on black stone are reported here. Glycine was chosen, taking into account that it exhibits coordination sites similar to those present in the poison of *Bothrops atrox*.

5.2 Experimental

All reagents were of analytical grade (Merck) and were used without further purification: glycine, ninhydrin, fructose, $Na_2HPO_4.12H_2O$, KH_2PO_4, KIO_3, Na_2HPO_4, and KH_2PO_4.

Figure 7.5 Some of the prepared black stone samples.

The black stone samples were prepared as follows:

1. An ox femur was cut into 2 cm pieces.
2. The grease was removed from the bone, and it was boiled in water.
3. The bone was then cooled to room temperature and the remaining grease was removed; then, the bone was dried in sunlight for 12 h.
4. The bone pieces were then heated until they became incandescent and cooled under inert atmosphere; After this, the black stone samples were boiled in water for 5 min and then dried at room temperature. Some of the black stone samples are shown in Fig. 7.5. The adsorption experiments were performed with powder samples of the black stones: the stones were ground in a mortar. A grain size of 250 mesh was chosen. To verify the reproducibility of the black stone preparations, two samples were prepared and investigated. The detailed description of the solutions preparations, etc., can be found elsewhere.

5.3 Results and discussion

The obtained results are summarized in Tables 7.1–7.4 and Figs. 7.6–7.11.

Taking into account the experimental conditions (33°C and a grain size of 250 mesh), it was verified that the time necessary to achieve an equilibrium state (maximum amount of adsorbed glycine) is 15 h. For this reason, the time interval

Table 7.1 Adsorption of glycine on black stone (sample 1, pH 7.4), where C_i is the initial concentration, C_{eq} is the final (equilibrium) concentration, and Γ is the adsorption capacity

Solution	C_i (mg cm^{-3})	C_{eq} (mg cm^{-3})	Γ (mg g^{-1})
1	9.6×10^{-4}	2.3×10^{-4}	7.3×10^{-2}
2	2.9×10^{-3}	3.4×10^{-4}	2.5×10^{-1}
3	4.8×10^{-3}	9.3×10^{-4}	3.9×10^{-1}
4	7.2×10^{-3}	1.0×10^{-3}	6.2×10^{-1}
5	9.6×10^{-3}	2.0×10^{-3}	7.5×10^{-1}
6	2.9×10^{-2}	6.7×10^{-3}	2.2
7	4.8×10^{-2}	1.4×10^{-2}	3.4
8	7.2×10^{-2}	2.4×10^{-2}	4.8

Table 7.2 Adsorption of glycine on black stone (sample 2, pH 7.4), where C_i is the initial concentration, C_{eq} is the final (equilibrium) concentration, and Γ is the adsorption capacity

Solution	C_i (mg cm^{-3})	C_{eq} (mg cm^{-3})	Γ (mg g^{-1})
1	9.6×10^{-4}	2.5×10^{-4}	7.1×10^{-2}
2	2.9×10^{-3}	6.2×10^{-4}	2.2×10^{-1}
3	4.8×10^{-3}	9.1×10^{-4}	4.0×10^{-1}
4	7.2×10^{-3}	1.8×10^{-3}	5.4×10^{-1}
5	9.6×10^{-3}	2.1×10^{-3}	7.5×10^{-1}
6	2.9×10^{-2}	7.9×10^{-3}	2.1
7	4.8×10^{-2}	1.5×10^{-2}	3.3
8	7.2×10^{-2}	2.4×10^{-2}	4.8

Table 7.3 Adsorption of glycine on black stone (sample 1, pH 6.9), where C_i is the initial concentration, C_{eq} is the final (equilibrium) concentration, and Γ is the adsorption capacity

Solution	C_i (mg cm^{-3})	C_{eq} (mg cm^{-3})	Γ (mg g^{-1})
1	9.6×10^{-4}	2.2×10^{-4}	7.4×10^{-2}
2	2.9×10^{-3}	3.2×10^{-4}	2.5×10^{-1}
3	4.8×10^{-3}	6.1×10^{-4}	4.1×10^{-1}
4	7.2×10^{-3}	1.1×10^{-3}	6.1×10^{-1}
5	9.6×10^{-3}	1.3×10^{-3}	8.3×10^{-1}
6	2.9×10^{-2}	3.4×10^{-3}	2.5
7	4.8×10^{-2}	5.2×10^{-3}	4.2
8	7.2×10^{-2}	1.0×10^{-2}	6.1

Table 7.4 Adsorption of glycine on black stone (sample 2, pH 6.9) C_i is the initial concentration, C_{eq} is the final (equilibrium) concentration, and Γ is the adsorption capacity

Solution	C_i (mg cm^{-3})	C_{eq} (mg cm^{-3})	Γ (mg g^{-1})
1	9.6×10^{-4}	6.1×10^{-5}	3.5×10^{-2}
2	2.88×10^{-3}	2.2×10^{-4}	2.5×10^{-1}
3	4.8×10^{-3}	5.9×10^{-4}	4.2×10^{-1}
4	7.2×10^{-3}	9.8×10^{-4}	6.2×10^{-1}
5	9.6×10^{-3}	1.5×10^{-3}	8.1×10^{-1}
6	2.88×10^{-2}	3.6×10^{-3}	2.5
7	4.8×10^{-2}	4.5×10^{-3}	4.3

of 24 h was chosen to facilitate the spectrophotometric determination of glycine (the black stone glycine mixture were filtered off and the spectrophotometric measurement performed on the supernatant).

As can be seen from Figs. 7.6 and 7.7, there is good reproducibility in the results for both black stone samples (considering a same pH value). So, adsorption experiments appear to be useful for testing the quality of black stone samples. Furthermore, it was found that in the concentration range used, saturation was not attained. However, the investigated concentration range encompasses the

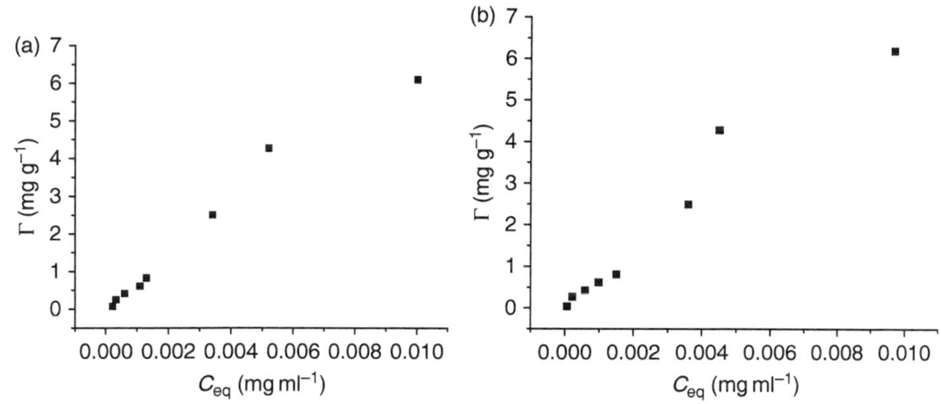

Figure 7.6 Adsorption isotherms for glycine on black stone at pH 6.9 (not buffered solution): (a) black stone sample 1 and (b) sample 2.

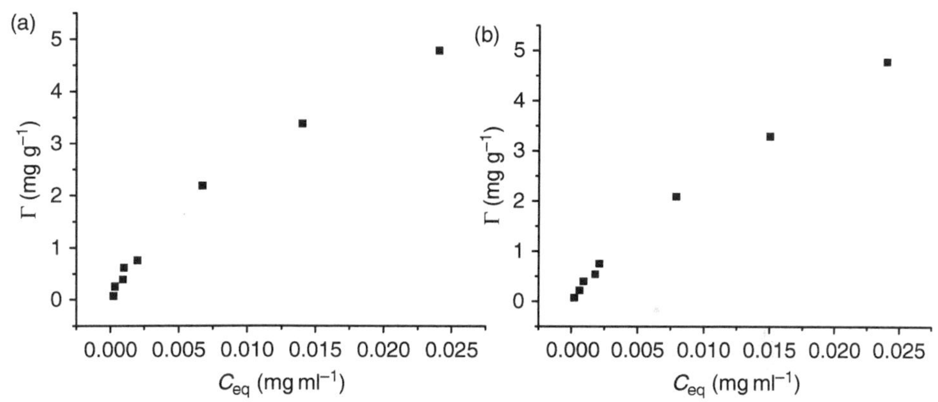

Figure 7.7 Adsorption isotherms for glycine on black stone, pH 7.4 (buffered solution): (a) black stone sample 1 and (b) sample 2.

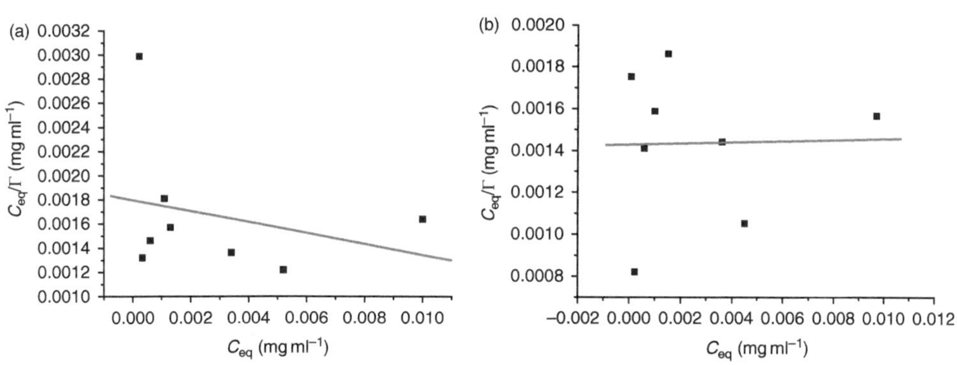

Figure 7.8 Application of Langmuir isotherm (linearized form) to the adsorption isotherms of glycine on black stone at pH 6.9: (a) sample 1 and (b) sample 2.

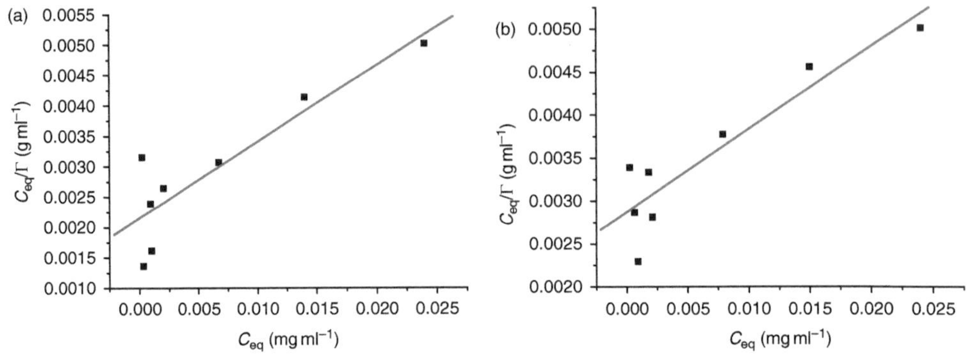

Figure 7.9 Application of Langmuir iostherm (linearized form) to the adsorption isotherms of glycine on black stone at pH 7.4: (a) sample 1 and (b) sample 2.

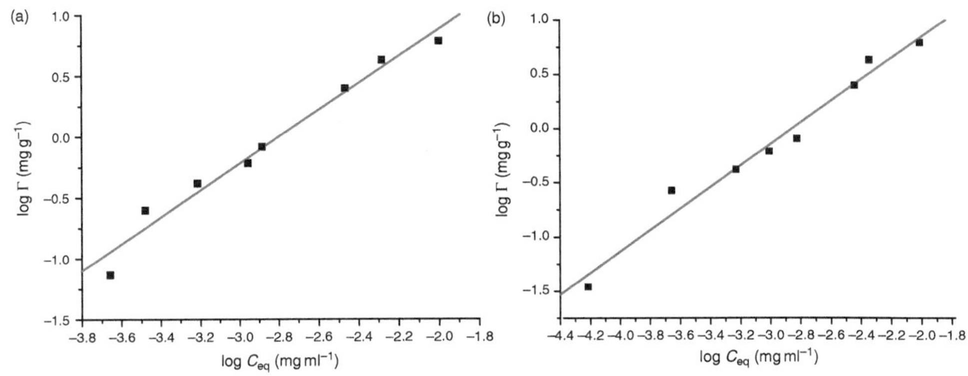

Figure 7.10 Application of Freundlich isotherm (linearized form) to the adsorption isotherms of glycine on black stone at pH 6.9: (a) sample 1 and (b) sample 2.

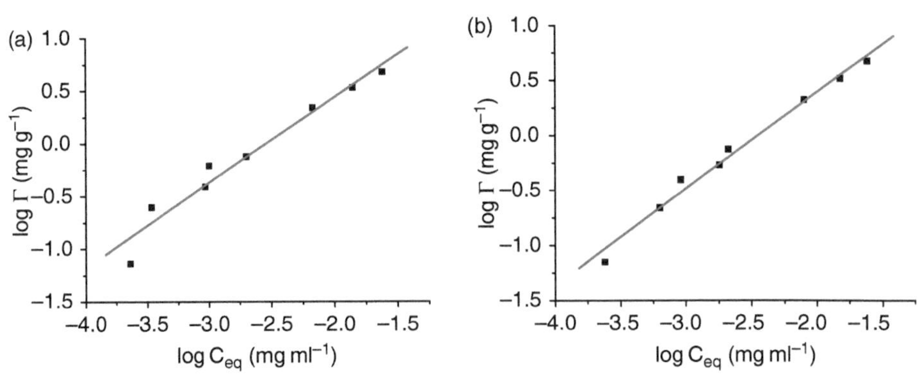

Figure 7.11 Application of Freundlich isotherm (linearized form) to the adsorption isotherms of glycine on black stone at pH 7.4: (a) sample 1 and (b) sample 2.

concentration range of poison in snake accidents. The results show that, under the investigated concentrations range, the adsorption capacity of black stone toward glycine is of 6.2 ± 0.05 and 4.8 ± 0.01 mg g^{-1} for unbuffered and buffered media, respectively. The higher adsorption capacity in buffered medium shows that black stone is a good substrate for adsorption of biological species at physiological pH.

The adsorption isotherms were interpreted using Langmuir and Freundlich adsorption models [27,28]. For the unbuffered system, the correlations between (C_{eq}/Γ) and C_{eq} are very poor, using the Langmuir model. A better correlation was observed for the buffered (pH 7.4) system.

As can be verified by inspection of Figs. 7.10 and 7.11, for buffered and unbuffered systems, better correlation is obtained with the Freundlich adsorption model, suggesting that the interaction of glycine with black stone is mainly a physical adsorption and not a chemical (with formation of chemical bonds) adsorption.

6. A Final Example

As a final example in this chapter, a study involving the sorption–desorption modeling of uranium (VI) onto lanthanum monophosphate, $LaPO_4$ [29], can be mentioned.

REFERENCES

1. R.F. de Farias, L.M. Nunes, C. Airoldi, J. Therm Anal. Calorim. 60 (2000) 517–521.
2. L.M. Nunes, C. Airoldi, R.F. de Farias, J. Solid State Chem. 145 (1999) 649–654.
3. F.J. Perez-Reina, P. Olivera-Pastor, E. Rodriguez-Castellon, A. Jimenez-Lopez, J. Solid State Chem. 122 (1996) 231.
4. X. Glipa, J.-M. Leloup, D.J. Jones, J. Rozière, Solid State Ionics 97 (1997) 227.
5. L. Benes, K. Melánová, V. Zima, J. Kalousová, J. Votinský, Inorg. Chem. 36 (1997) 2850.
6. A. De Stefanis, S. Foglia, A.G. Tomlinson, J. Mater. Chem. 5 (1995) 475.
7. H. Nakajima, G-E. Matsubayashi, J. Mater. Chem. 5 (1995) 105.
8. A. De Stefanis, A.G. Tomlinson, J. Mater. Chem. 5 (1994) 319.
9. J.W. Johnson, A.J. Jacobson, J.F. Brody, S.M. Rich, Inorg. Chem. 21 (1982) 3820.
10. Y. Yatabe, G.-E. Matsubayashi, J. Mater. Chem. 6 (1996) 1849.
11. T. Nakato, Y. Furumi, N. Terao, T. Okuhara, J. Mater. Chem. 10 (2000) 737.
12. I.J. Ellison, G.J. Hutchings, M.T Sananes, J-C. Volta, J. Chem. Soc. Chem. Commun. (1994) 1093.
13. L. Benes, R. Hyklová, Inorg. Chim. Acta. 71 (1990) 177.
14. T. Nakato, Y. Furumi, T. Okuhara, Chem. Lett. (1998) 611.
15. Wavefunction Inc., Irvine, CA, USA.
16. E. Bordes, P. Courtine, G. Pannetier, Ann. Chem. 8 (1973) 105.
17. C. Ŕkha, M.T. Vandenboore, L. Livage, R. Prost, E. Huard, J. Solid State Chem. 63 (1986) 202.
18. R.M. Silverstein, G.C. Bassler, T.C. Morril, Spectrometric Identification of Organic Compounds, Wiley, New York, 1991.
19. R.F. de Farias, C. Airoldi, J. Solid State Chem. 166 (2002) 277.
20. M.O. Machado, R.F. de Farias, C. Airoldi, J. Phys. Chem. Solids 65 (2004) 1697.
21. R.F. de Farias, C. Airoldi, Solid State Sci. 5 (2003) 611.
22. A.M. Lazarin, C. Airoldi, Sens. Actuators B 107 (2005) 446.
23. A.M. Lazarin, C. Airoldi, J. Inclusion Phenom. Macrocyclic Chem. 51 (2005) 33.

24. A.M. Lazarin, C. Airoldi, J. Chem. Thermodyn., 37 (2005) 243.
25. W. Weng, S. Zhang, K. Cheng, H. Qu, P. Du, G. Shen, J. Yuan, G. Han, Surf. Coat. Technol. 167 (2003) 292.
26. M. Shirkhanzadeh, S. Sims, J. Mater. Sci.: Mater. Med. 8 (1997) 595.
27. K.S.W. Sing, D.H. Everett, R.A.W. Haul, L. Moscou, R.A. Pierotti, J. Rouquérol, T. Siemieniewska, Pure Appl. Chem. 57 (1985) 603.
28. P.W. Atkins, Physical Chemistry, Oxford University Press, Oxford, 1994.
29. E. Ordoñez-Regil, R. Drot, E. Simoni, J. Colloid Interface Sci. 263 (2003) 391.

CHAPTER EIGHT

Chemistry on Modified Clay Surfaces

Contents

1. Introduction

Clays are a very important group of inorganic compounds due to their many possible practical applications, such as supports for catalysts, as lubricants, and for thermal insulation. Many clays such as vermiculite, montmorillonite, and steatite exhibit a lamellar nanostructure, which make them suitable hosts for the preparation of intercalation compounds. As recently [1] demonstrated for soapstone, a compact form of steatite (talcum powder), even very small variations in chemical composition can exert significant effects on physical and chemical properties.

2. Vermiculite Surface Modified with Carnauba Wax for Oil Cleaning

Taking into account their usefulness as adsorbants, clay minerals can be successfully hydrophobized with cationic surfactants [2]. Such hydrophobized matrices can be used for the interlayer adsorption of many organic species such as 1-pentanol.

Interface Science and Technology, Volume 17
ISSN: 1573-4285, doi: 10.1016/S1573-4285 (08) 00408-0

Vermiculite is a mica-like lamellar mineral that rapidly expands on heating to produce a lightweight material. This expanded (exfoliated) product is used as lightweight aggregate in concrete, plaster, and premixes. Horticultural and fertilizer products are also a large end user of vermiculite.

It has been verified that the degree of hydration of the vermiculite clay matrix can exert remarkable effects on the magnetic properties of its intercalation compounds. Furthermore, vermiculite has been recently used for studies involving the adsorption of oil-contaminated water samples [1].

Vermiculite samples were supplied by the source União Brasileira de Mineração, from Santa Luzia city, state of Paraiba, Brazil. For such vermiculite samples, the main components are (as oxide percentages): Si (44.7), Al (9.1), Fe (5.9), Mg (16.2), Na (0.6), Ca (5.1), K (2.7), and Ti (0.7); sulfur content: 0.2% (w/w); salt content: 862 mg NaCl dm^{-3}; paraffin: 13.6%.

The adsorption of crude oil was performed in dehydrated (expanded) and hydrophobized vermiculite samples, denoted here as EV (expanded vermiculite) and HV (hydrophobized vermiculite), respectively. Vermiculite samples with five different grain sizes were used: 100-150, 150-200, 200-250, 250-325, and 325-400 mesh. The crude oil had the following properties: density, 34.2 °API (at 15.6°C) and viscosity, 8.4 cP (at 37.8°C);

The hydrated vermiculite samples were expanded (dehydration process) by heating them at 800°C for 30 min. The hydrophobization process was performed by heating the EV samples with carnauba (*Copernicia cerifera*) wax to produce a 10% hydrophobized sample.

The samples were characterized as to BET surface area, by using a Micromeritics apparatus. The TG and DTA curves were obtained under nitrogen atmosphere with a heating rate of 10°C min^{-1}, by using a TGA-7 and a DTA-2000 apparatus, respectively, both from Perkin-Elmer. The XRD patterns were obtained with a Phillips instrument, using Cu K_α radiation. The infrared spectra were obtained in KBr pellets, in the 4000–400 cm^{-1} range, using a Perkin-Elmer spectrometer. The UV–Vis studies were performed on a Bomem spectrophotometer.

The adsorption studies used two distinct approaches: (a) by using columns filled with the vermiculite matrices, and promoting the flow of the crude oil through it and (b) by dispersion of the vermiculite samples in an oil-water emulsion (50 ppm of oil) with constant magnetic stirring. The water–oil emulsion was prepared by dispersion of the desired amount of oil on water under vigorous mechanic stirring for 1 h.

The XRD patterns for the hydrated, anhydrous (expanded), and HV matrices are shown in Fig. 8.1.

Based on the 2θ values for the {100} diffraction peak, the interlayer distance for the hydrated and anhydrous samples can be calculated as 1.34 and 0.95 nm, respectively. The observed decrease of the interlayer distance for the expanded sample can be associated with the release of water molecules observed under heating [3]. On the other hand, the hydrophobized matrix exhibits the same XRD pattern of the expanded one. Based on this fact, it can be concluded that the hydrophobizant molecules are adsorbed on the surface of the vermiculite grains and not into the interlayer space, that is, there is not the formation of an vermiculite–hydrophobizant intercalation compound.

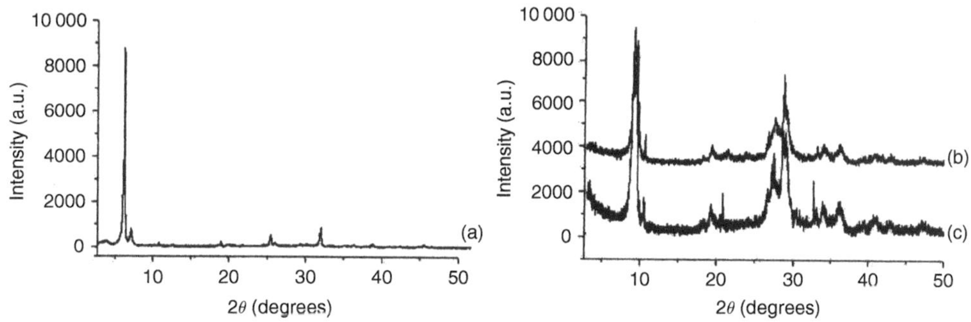

Figure 8.1 XRD patterns for hydrated (a), anhydrous (b), and hydrophobized (c) vermiculite samples.

The obtained infrared spectra (not shown) confirm the presence of the hydrophobizant agent on the clay matrix.

The thermogravimetric and derivative curves for the hydrated, expanded (anhydrous), hydrophobized, and hydrophobized with adsorbed crude oil vermiculite samples are shown in Fig. 8.2. For the hydrated samples, the first and second mass loss steps are associated with the release of physisorbed/intercalated water molecules and the water molecules due to the condensation of hydroxyl groups, respectively. As expected, for the anhydrous sample, no mass loss step is observed. On the other hand, for the hydrophobized sample, a mass loss of 10%, due to the release of the organic moiety can be observed, in agreement with the hydrophobizant/clay mineral ratio. As expected, the hydrophobized sample with adsorbed crude oils exhibits the larger mass loss percentage.

The thermogravimetric curve for the crude oil sample is shown in Fig. 8.3. As can be observed from Fig. 8.3, the crude oil exhibits four mass loss steps. The fist one, from 25°C to 220°C is due to the release of the most volatile compounds in the

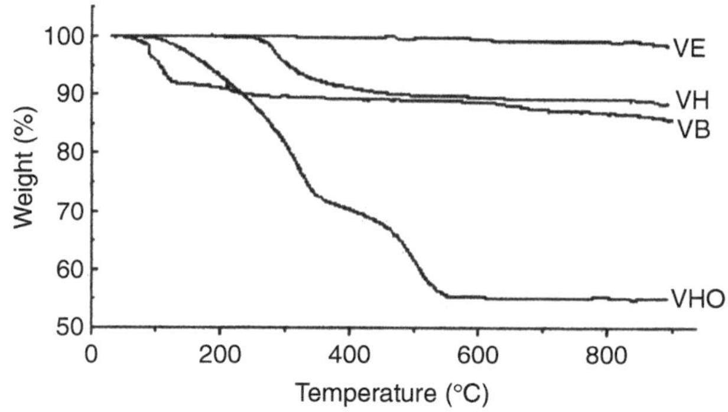

Figure 8.2 Thermogravimetric curves for hydrated (VB), anhydrous (VE), hydrophobized (VH), and hydrophobized with adsorbed crude oil (VHO) vermiculite samples.

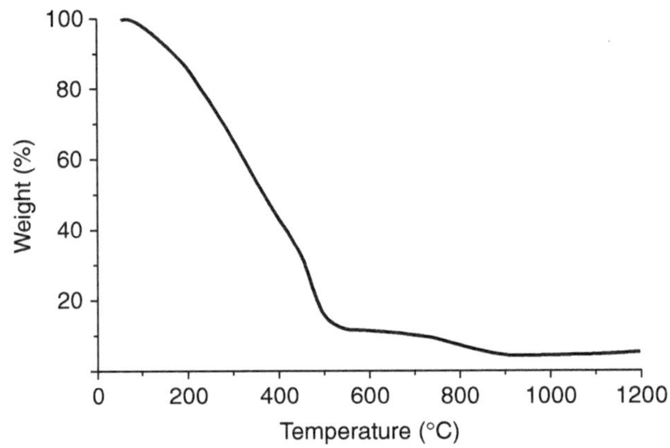

Figure 8.3 Thermogravimetric curve for crude oil.

oil. The second (230–350°C), third (350–510°C), and fourth (600–950°C) mass loss steps are due to the release of the heavier compounds in the oil sample.

As can be observed in Fig. 8.2, the HV sample with adsorbed crude oil exhibits a TG curve showing the release of adsorbed oil, as well as hydrophobizant agent molecules in the temperature range considered. For all samples, each mass loss step is associated with an equivalent endothermic peak in the DTA curves, which confirms the endothermic nature of the desorption processes.

The results for the adsorption of crude oil on columns, and the BET surface area for the studied samples are summarized in Table 8.1. The adsorption factor (AF) is defined here as: (oil mass)/(vermiculite mass). For both, expanded and HV samples, the adsorption studies were performed at two different mass values: 2.0 and 4.0 g. However, it was found that the AF values were not affected by the used vermiculite mass. Adsorption experiments were also performed for hydrated (not expanded) vermiculite samples, and it was observed that the AF for these sample is very low (0.3). All adsorption experiments were performed in triplicate, and it was verified that the results are reproducible.

Table 8.1 Results for adsorption of crude oil on expanded and hydrophobized vermiculite samples

Sample	Grain size (mesh)	Surface area ($m^2\,g^{-1}$)	AF
VE	150–200	3.14±0.01	0.7±0.1
VE	200–250	4.05±0.03	0.7±0.1
VE	250–325	3.65±0.02	0.8±0.1
VE	325–400	7.30±0.10	0.8±0.1
VH	100–150	–	0.8±0.1
VH	150–200	0.94±0.04	0.9±0.1
VH	200–250	0.97±0.06	0.9±0.1
VH	250–325	0.63±0.06	1.0±0.1
VH	325–100	1.97±0.05	1.1±0.1

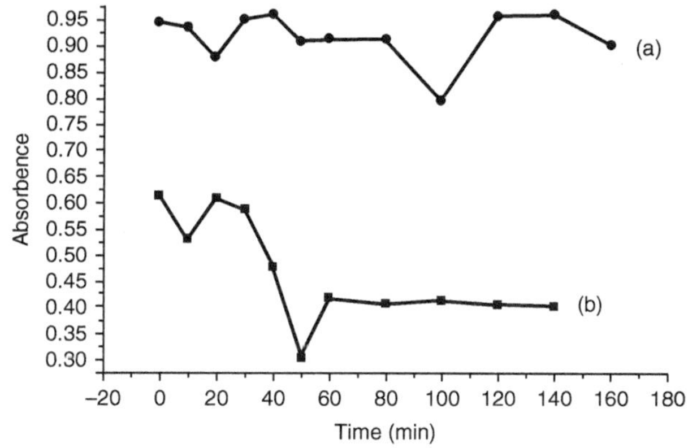

Figure 8.4 UV–Vis spectroscopic results for the adsorption of crude oil from a water–oil emulsion on anhydrous (a) and hydrophobized (b) vermiculite samples.

As can be seen from Table 8.1, the hydrophobized samples exhibit very high AFs, compared with the expanded ones. This fact can be attributed to the presence of the organic hydrophobic moiety, which has a high affinity toward other hydrophobic compounds/mixtures such as crude oil.

For the adsorption studies performed on the water–oil emulsion, UV-Vis spectroscopy was used to compare the AF for the expanded and HV samples. The results are summarized in Fig. 8.4.

As can be seen, the hydrophobized sample exhibits a large adsorption capacity, since the absorbance of oil on the emulsion is lower. Furthermore, as seen from the data in Fig. 8.4 that, after 60 min, an equilibrium state is achieved, with the saturation of the solid matrix.

As a general rule it is is observed that the AF increases with an increase of the BET surface area.

Based on the experimental data, it can be concluded that hydrated vermiculite exhibits a very low adsorption capacity for crude oil. On the other hand, anhydrous (expanded) and hydrophobized matrices show high adsorption capacities. The 10% hydrophobized matrix shows a 50% increase in adsorption capacity, in comparison with the expanded one. For adsorption process performed in a water–oil emulsion (50 ppm of oil), the saturation of the solid hydrophobized matrix is achieved after 60 min.

3. Water as a Surface Modifier: The Hydrophobicity of Vermiculite

Since water molecules can adsorb on the clay surface, water can be considered as a modifier. To investigate the hydration process of vermiculite, a calorimetric study was performed with the main results as shown in this section. The vermiculite sample used was the same as described in the previous section.

Hydrated and dehydrated (expanded) vermiculite samples in the 325-400 mesh range were used. The expanded sample was calcined at 800°C for 30 min.

The X-ray powder diffractograms were obtained with a Shimadzu XD-3A instrument (Cu K_α radiation; 35 kV; 25 mA). The SEM micrographs were obtained on a JEOL microscope model JSM T-300 with an accelerating voltage of 15 kV. The BET surface area measurements were performed on Micromeritics (FlowSorb 2300) equipment. TG and DTA curves were obtained on TGA-7 and DTA-2000 instruments, both from Perkin-Elmer. Nitrogen atmosphere and a heating rate of 10°C min^{-1} were used. The infrared spectra were obtained using a Bomem spectrometer in KBr disks in the 4000–400 cm^{-1} range with a 4 cm^{-1} resolution. The calorimetric study was performed on a LKB 2250 calorimeter using the so-called ampoule-breaking procedure. Vermiculite samples between 5.0 and 60.0 mg were used, and the ampoules were broken in a reaction vessel with 100 cm^3 of double-distiled water, and the enthalpic effect was measured for 5.0 min.

The infrared spectra, as well as the TG and DTA curves (not shown), are in agreement with the presence (hydrated samples) or absence (expanded samples) of water in the vermiculite samples.

The thermal effect associated with breaking an empty ampoule is very low (200 mJ). So, the thermal effect occurring when ampoules containing vermiculite are broken must be associated only with the interaction of vermiculite with water molecules. Since vermiculite is not water soluble, any kind of dissolution can also be ruled out.

For both, hydrated and dehydrated vermiculite samples, the wetting enthalpy values are the same: $\Delta H^{\theta}{}_{wett} = 5.30 \pm 0.20\,J\,g^{-1}$. This shows that at least for the considered reaction time (5.0 min), the dehydrated vermuclite does not behave as a hydrophilic matrix. This hypothesis is reinforced by the X-ray powder diffraction patterns for contact times (between dehydrated vermiculite and water) of 24, 48, and 72 h, as shown in Fig. 8.5.

As can be seen in Fig. 8.1a, the hydrated vermiculite sample exhibits its {001} diffraction peak at 6.12°, corresponding to an interlayer distance of 1.44 nm. On the other hand, the dehydrated sample, Fig. 8.1b exhibits the {001} diffraction peak 9.40°, corresponding to an interlayer distance of 0.92 nm. However, in such a diffractogram, the 001 diffraction peak of the hydrated sample could still be observed, showing that the dehydration process was not complete.

The dehydration process implies also a reduction in the surface area: $7.30 \pm 0.10\,m^2\,g^{-1}$ for the dehydrated sample and $8.8 \pm 0.10\,m^2\,g^{-1}$ for the hydrated sample (a reduction of 20.5% in the superficial area).

The diffraction patterns of the dehydrated samples dispersed in water for 24, 48, and 72 h are shown in Figs. 8.5c–8.5e, respectively. As can be verified, the intensity of the {001} diffraction peak of the hydrated samples enhances in intensity for largest contact times. However, even after 72 h, less than half of the sample was rehydrated. So, both data sets, calorimetric and XRD, show that the dehydrated vermiculite does not behave as a hydrophilic compound, with its "total surface," that is, the external and internal space of the lamella, exhibiting a very low affinity toward water molecules. This indicates that besides being unfavorable from a thermodynamic point of view (endothermic), the rehydration process, that is, adsorption of water molecules on the surface, is kinetically slow.

As can be seen by comparing the SEM micrographs shown in Fig. 8.6, the dehydration process affects the microstructure of vermiculite. After dehydration, the

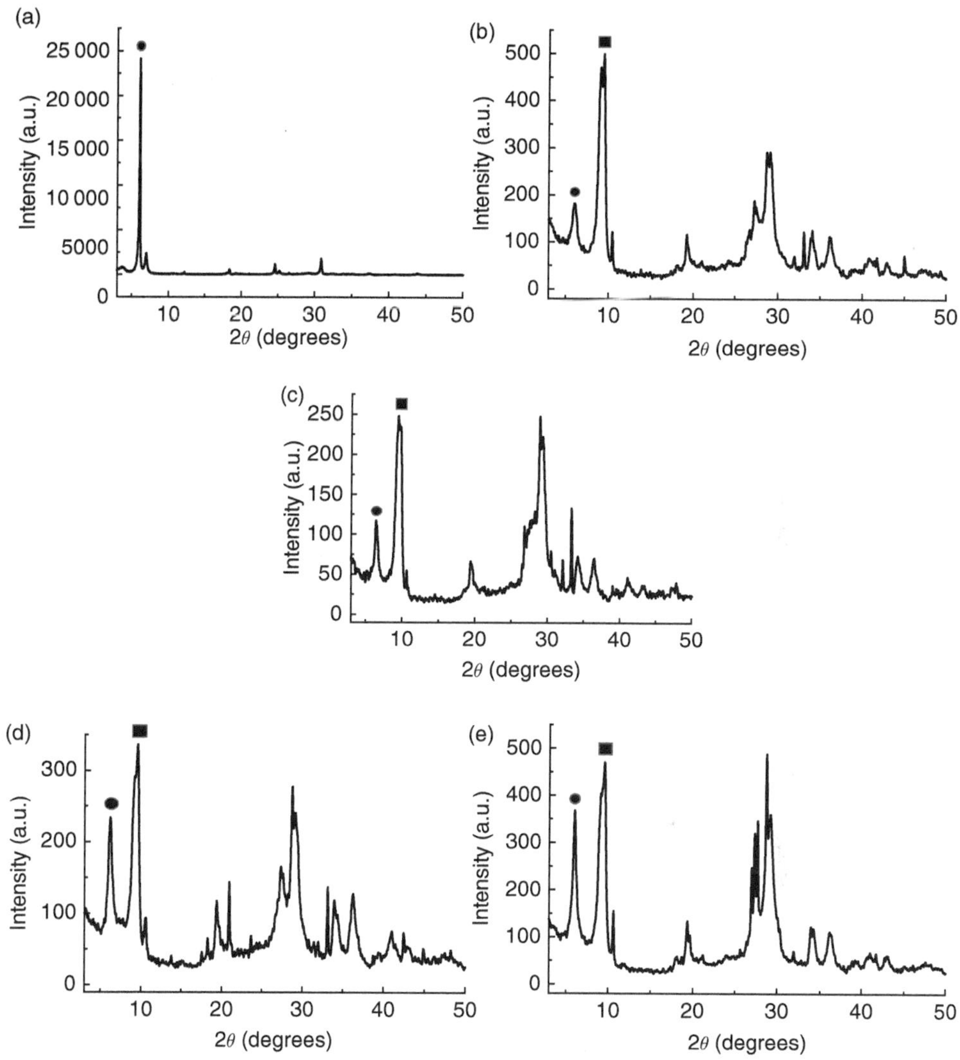

Figure 8.5 X-ray diffraction patterns for: hydrated vermiculite (a), dehydrated vermiculite (b), and dehydrated vermiculite in 24 h (c), 48 h (d), and 72 h (e). The symbols (●) and (■) are associated with the 001 diffraction plane of vermiculite in the hydrated and dehydrated samples, respectively.

interlayer distance and surface area are reduced, in comparison with the hydrated sample, and it is supposed that both effects are responsible for the slow rehydration of vermiculite, since a shorter interlayer distance signifies stonger interlayer interactions, making the intercalation of water molecules less favorable, and a reduced surface area signifies a minor amount of acidic sites with which the donor atom (oxygen) from water can interact.

Figure 8.6 SEM micrographs for hydrated vermiculite, 1500× (a) and dehydrated vermiculite, 750× (b).

4. The Effect of Surface Chemical Composition on the Acid Corrosion of *Pedra Sabão* (Soapstone)

The so-called *pedra sabão* ("soapstone") was extensively used in the nineteenth century in the production of many artistic artifacts in Brazil, such as the statues of the prophets in Congonhas City, in Minas Gerais State, an example of the baroque art of South America. The statues were made by a Brazilian artist, Antônio Francisco Lisboa, called *o aleijadinho*. Nowadays, the progressive corrosion of the statues provoked mainly by air pollution, is a very serious problem to be solved.

The name "soapstone" comes from the fact that the stone is very soft. From a formal point of view, soapstone is a compact form of talcum powder (steatite), the "fundamental" chemical formula of which is $Mg_3Si_4O_{16}(OH)_2$. This basic silicate is the softest one on the scale of Mohs. Nowadays, soapstone is used to produce artistic artifacts such as statues as well as utensils such as pottery (Fig. 8.7). Care must be taken in its manipulation, due to its possible inhalation toxicity [3].

Steatite is found most often in the United States, Japan, and Brazil. In Brazil it is found mainly in the States of Minas Gerais, Bahia, Paraná, Goiás, and São Paulo. Its colors vary from white to shades of yellow, beige, green, and gray, and from light to almost black, depending on the chemical composition of the associated minerals.

A basic study of the chemical composition and structure of two soapstone samples to improve scientific knowledge of this marvelous mineral is reported here. An investigation about the effects of a sulfuric acid aqueous solution on the physical integrity of the stone is also performed to provide some clues as to the main causes of corrosion (acid rain), and so to the possible approach to avoiding it. Since the acid attack begins at the surface, it is the surface chemical composition and structure that controls the acid corrosion.

To perform the desired investigations, two soapstone samples were used, both from Ouro Preto city, Minas Gerais State, Brazil. From a macroscopic point of view, the only difference between the two samples is their color: one is light gray (the stone itself, since the powder is white) whereas the other is red-brown. The two samples will be denoted hereafter as SS 1 and SS 2, respectively [4].

Figure 8.7 Utensils made of soapstone.

To simulate the effects of acid rain on soapstone, a 0.1 mol dm^{-3} sulfuric acid solution was prepared, and cubic samples (5 mm × 5 mm × 5 mm) were immersed in 150 cm^{3} of the acid solution for 40 days.

Infrared spectra were obtained from soapstone powders in a Bomem spectrometer in the range 4000–400 cm^{-1} with a resolution of 4 cm^{-1} by dispersing the samples in KBr disks. Thermogravimetric curves were obtained using a Shimadzu TGA 50 apparatus, under Ar atmosphere with a heating rate of 10°C min^{-1}. The XRD patterns were obtained in a Shimadzu XD-3A instrument, using Cu K_{α} radiation (35 kV, 25 mA). SEM micrographs were taken with a JEOL microscope, model JSM T-300, with an accelerating voltage of 15 kV ^{29}Si NMR CP-MAS spectra of solid samples were performed by using a AC 300/P Bruker instrument at room temperature, with the experimental conditions: pulse repetition time of 0.115 s, contact time of 3 ms, and 17,857 accumulations.

The infrared spectrum obtained for SS 1 is shown in Fig. 8.8. The band at 3677 cm^{-1} is associated with the vibrations of isolated silanols (Si—OH). It is worth

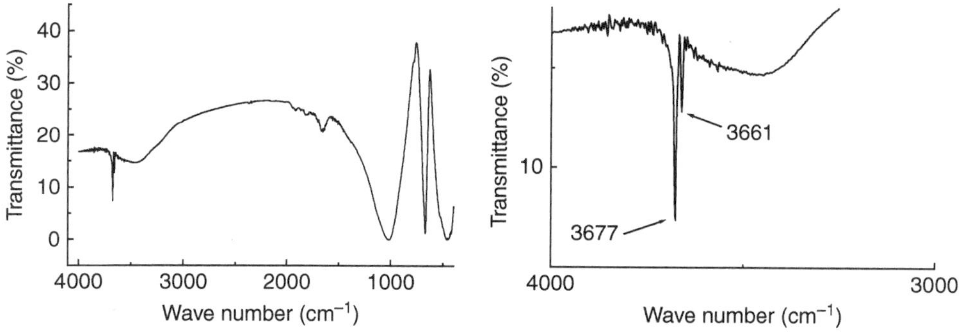

Figure 8.8 Infrared spectra for SS 1.

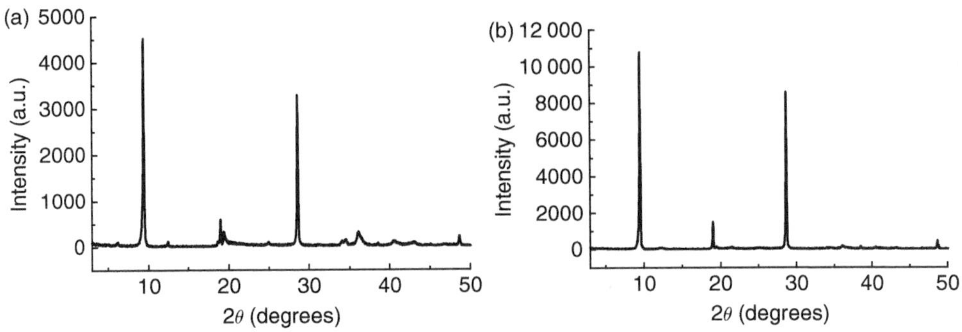

Figure 8.9 XRD patterns for SS 1 (a) and SS 2 (b).

noting that such a band is very difficult to observe in silica samples, requiring a careful drying process. This result strongly suggests that the silanol groups on soapstone are mainly isolated ones. The band at 3661 cm^{-1} could be associated with the geminal silanol groups ($Si(OH)_2$). The bands at 1918, 1810, and 1023 cm^{-1} are associated with the vibrations of the silica backbone, and the band at 1662 cm^{-1} with the ν —H—OH vibration mode from adsorbed water molecules. A very similar IR spectrum was obtained for the SS 2 sample.

The XRD patterns are shown in Fig. 8.9. As can be seen, the X-ray patterns for both samples are virtually identical, with main diffraction peaks at 9.5°, 19.0°, and 28.5°, associated with diffractions from the {001}, {002}, and {003 planes}, respectively. Such a diffraction profile is compatible with a lamellar nanostructure. Based on the 2θ value for the {001} plane, the interlayer distance can be calculated as 0.93 nm.

An EDX analysis was performed on both samples, and only three elements were detected: Si, Mg, and Fe. The reliability of such analysis was previously verified by using a Pd (80%)/Au (20%) standard disk, and the observed ratio was as expected. The quantitative results are as follows: SS 1 (Si = 66.0%, Mg = 29.1% and Fe = 4.9%); SS 2 (Si = 63.6%, Mg = 31.1% and Fe = 5.3%). That is, SS 2 exhibits larger amounts of Mg and Fe in comparison with SS 1. The red-brown color of SS 2 can be attributed to its higher iron content.

The presence of sulfur was not detected, and so the hypothesis that Mg and/or Fe could be present in the rock as sulfates is ruled out. The thermogravimetric analysis performed for both samples in the range 25–1000°C show that no mass loss is observed (a single plateau is observed in the TG curve). Since $MgCO_3$ suffers thermal degradation at 540°C [5], the hypothesis that magnesium could be present as carbonate can be discarded. So, based on the obtained experimental results it can be concluded that silicon, magnesium and iron are present as their respective oxides, that is, soapstone is an iron–magnesium silicate, or a magnesium silicate contaminated with iron, in this case. So, the results are not in complete agreement with the "fundamental" chemical composition provided in the introduction.

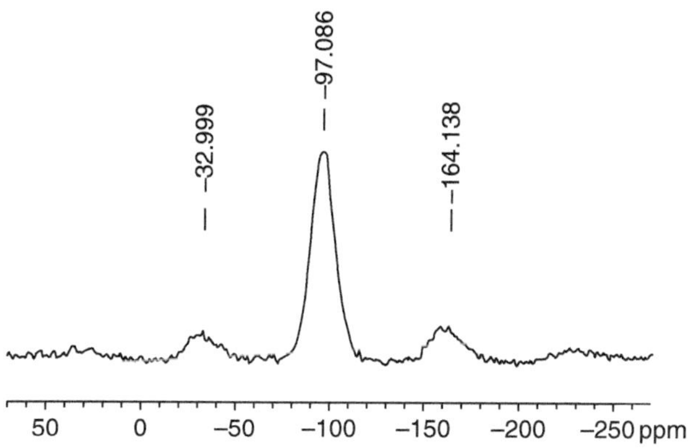

Figure 8.10 ^{29}Si CP-MAS NMR spectra for SS 1.

The ^{29}Si NMR CP-MAS spectrum for SS 1 is shown in Fig. 8.10. Peaks at –33.0, –97.1 and –164.1 ppm are observed. For SS 2 sample, a spectrum with similar profile was obtained. However, the peaks are shift to higher field values: –29.4, –94.8, and –154.5 ppm, respectively. This fact reinforced the evidence provided by EDX analysis, that the chemical compositions of both samples are not the same.

The observed shifts can be associated with the higher amount of iron and magnesium in the SS 2 sample. The peak at –97.1 ppm for SS 1 and at –94.8 ppm for SS 2, can be attributed to the isolated silanols, where the silicon atom is bonded to a single —OH group and three siloxane groups, that is, (OH)Si*—(OSi≡)$_3$ [6]. So, these data confirm the FTIR results. It is worth noting that this peak is the most intense in both NMR spectra. So, the majority of silicon atoms present in soapstone are bonded to OH groups. The peaks at –33.0 and –164.1 ppm (SS 1) or –29.4 and –154.5 ppm (SS 2) are resonance peaks related to the main peak. No peaks related with siloxane groups Si* (OSi≡)$_4$, normally observed around –109 ppm [5] are found in the spectra. So, it can be concluded that such groups are absent from both samples.

The SEM micrographs are shown in Figs. 8.11 and 8.12. In Figs. 8.11c and 8.11d it can be seen that soapstone exhibits a plate-like microstructure. However, the SS 2 sample exhibits larger grain sizes in comparison with SS 1. By comparison of Figs. 8.5a and 8.5b with Figs. 8.5c and 8.5d, it can be verified that the immersion in the sulfuric acid solution has not affected the microstructure of the SS 1 sample. On the other hand, by comparison of Figs. 8.12a and 8.12b with Figs. 8.6c, 8.6d, and 8.6e, it can be verified that the microstructure of the SS 2 sample is strongly affected by immersion in the sulfuric acid solution. An exfoliation of the matrix is readily apparent after the acid treatment. Thus, for SS 2 sample exposure to an acid aqueous environment can provoke the rupture of the interlayer forces, and then, the physical degradation of the stone.

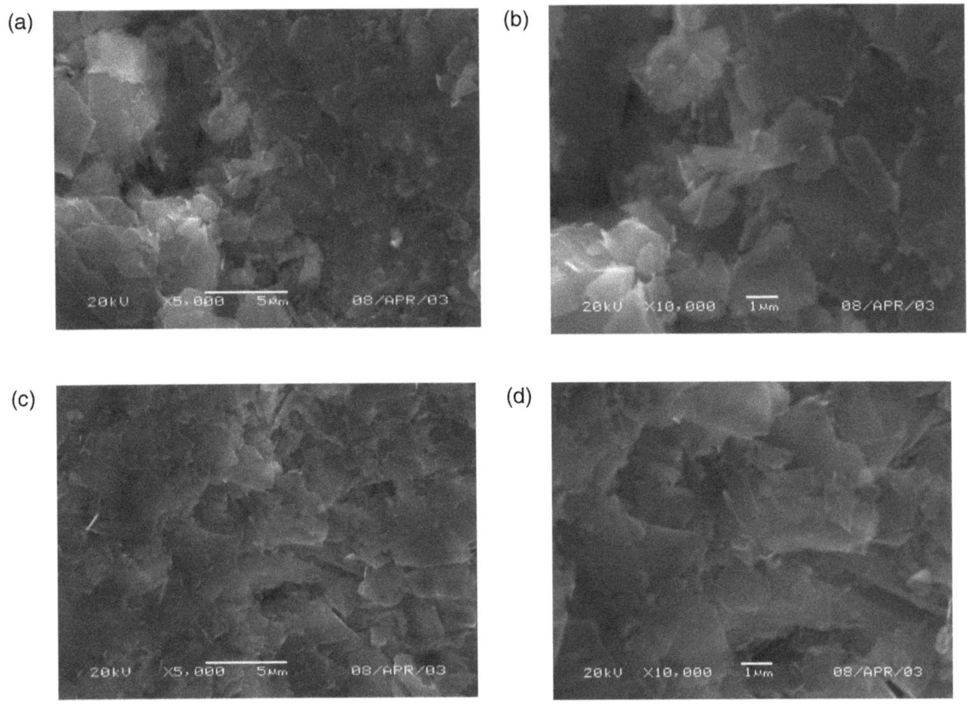

Figure 8.11 SEM micrographs for SS 1 before (a) and (b) and after (c) and (d) immersion in a 0.1 mol dm^{-3} sulfuric acid solution. The magnifications used and scale bar are in the micrographs.

Based on the EDX and SEM data, it can be inferred that the higher magnesium and/or iron content of SS 2 is related to its slightly increased resistance to an acid attack. Despite the fact that solubility analysis of the crystalline compounds was not performed here, the results are in agreement with those previously reported [7] for a study involving the investigation of cookware made of soapstone.

In this study, 3% and 5% acetic acid solutions were used as a food stimulant, and it was verified that the mineral migration followed the solubility of the crystalline components of the rock, with Fe exhibiting a higher migration than Mg.

Based on the experimental results it can be concluded that:

1. In soapstone samples, a minor resistance to acid attack is associated with a larger magnesium and/or iron content in the stone.
2. Soapstone exhibits a lamellar nanostructure, which makes it able to promote intercalation reactions. Furthermore, they exhibit isolated OH groups, as indicated by FTIR data, and the major number of silicon atoms present in soapstone are bonded to OH groups, as proved by ^{29}Si CP-MAS NMR analysis.

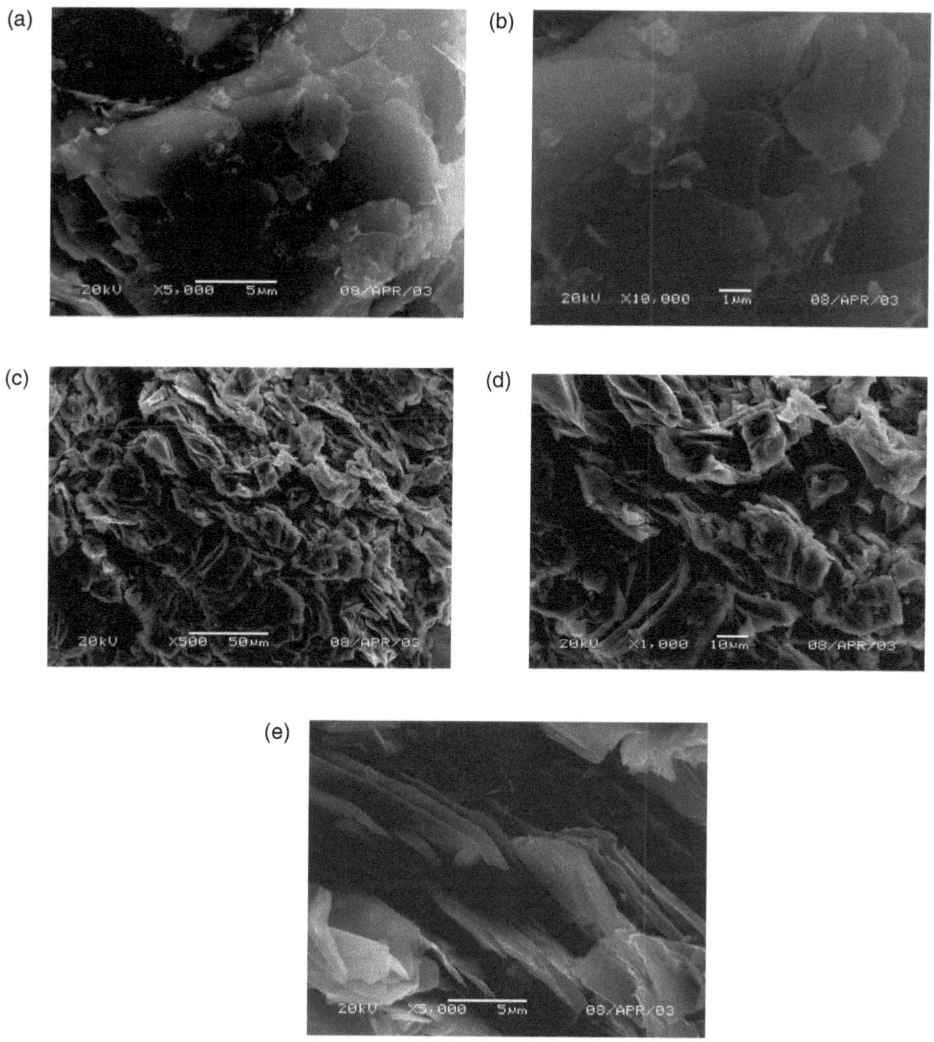

Figure 8.12 SEM micrographs for SS 2 before (a) and (b) and after (c), (d), (e) immersion in a $0.1\,\mathrm{mol\,dm^{-3}}$ sulfuric acid solution. The magnifications used and scale bar are in the micrographs.

5. Thermal Degradation of Vegetable Oils on Clays (Biodiesel Production)

5.1 Introduction

Despite the fact that it was used on a large scale as an emergency energy source in World War II, biosiesel has been seriously considered as a "clean" fuel, able to reduce the production of the so called "green house effect" gases only in the last

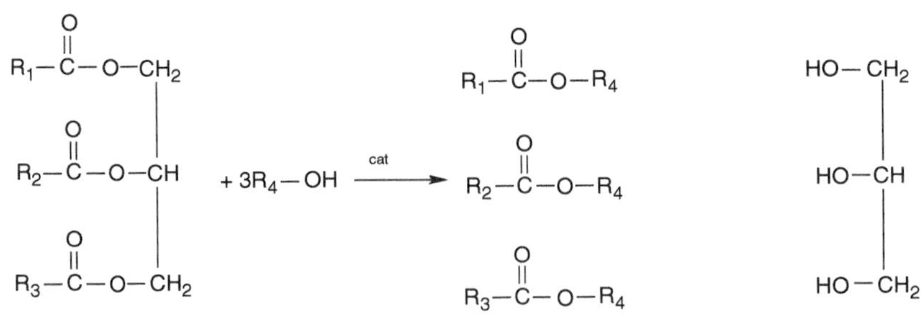

Figure 8.13 Summary of the chemical preparation of biodiesel by trans-esterification route.

three decades. In Brazil, were there is a large and economical production of ethanol; the trans-esterification route was chosen by the government as the official route to the preparation of biodiesel. The law No. 11.097 of 01/13/2005 officially, introduces, biodiesel into the Brazilian energetic matrix.

In Fig. 8.13 is shown a summary of the chemical reaction involved in the production of biodiesel by trans-esterification route. The triglyceride came from the vegetable oil and the catalyst is generally homogeneous (sodium or potassium hydroxide dissolved in alcohol). The specific vegetable oil and alcohol to be used is a choice to be made by any country. In the United States, for example, soja oil and methanol are used. In Brazil, ethanol is used, and a series of vegetable oils are used, rather than a single one, since in Brazil the biodiesel program is very closely related with social programs to enhance the production and uses of regional vegetable species. Through this route, the biodiesel is a long chain ester, and glycerine is obtained as a by-product.

The so-called cracking route is based on the thermal degradation of vegetable oils on some kind of inorganic catalyst, for example, silicates. In this route, the main products are not esters, as in the trans-esterification route, but hydrocarbons very similar to those present in the diesel oil obtained from petroleum. Hence, to investigate the effects of clays and similar compounds on the thermal degradation of vegetable oils is relevant. Furthermore, the investigation of heterogeneous catalysts for the preparation of biodiesel by trans-esterification has been carried out by some research groups in Brazil. Some results of such investigations, specifically those involving clays as inorganic substrates are presented here (Figs. 14 and 15).

In the thermogravimetric curve shown in Fig. 8.16 (all curves were obtained under nitrogen atmosphere, with a flux of 50 cm^3 min^{-1} and heating rate of 10°C min^{-1}), the first mass loss step is associated with the release of water molecules from montmorillonite K-10. The other two mass loss steps are due to the thermal degradation of soy oil. So, comparing Figs. 8.15 and 8.17, it can be verified that the presence of the clay lowered (from 400°C to 200°C) the temperature of the thermal degradation of the oil. For comparison, Fig. 8.16 shows the thermogravimetric curve for biodiesel (methyl esters from soy oil). As can be verified, the first fraction of the soy oil to suffer pyrolysis on the clay surface does so in the same temperature range as exhibited by the biodiesel.

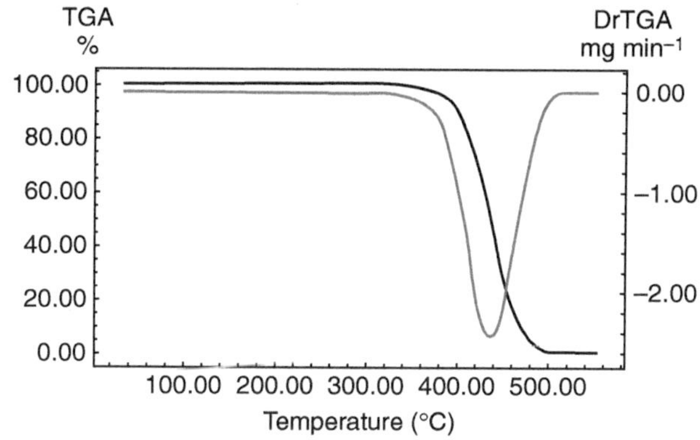

Figure 8.14 Thermogravimetric and derivative curves for soy oil.

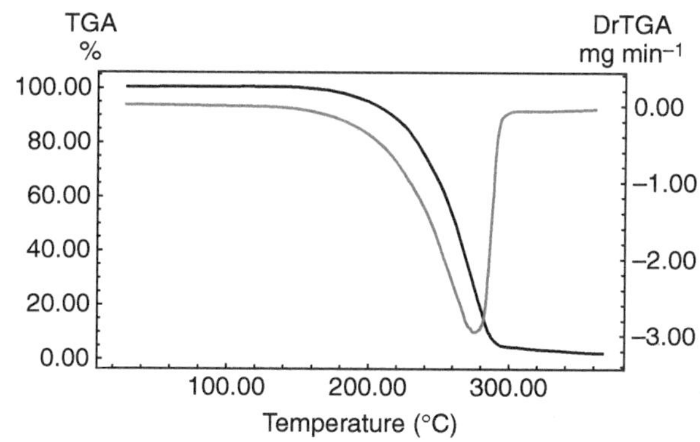

Figure 8.15 Thermogravimetric and derivative curves for biodiesel (methyl esters from soy oil).

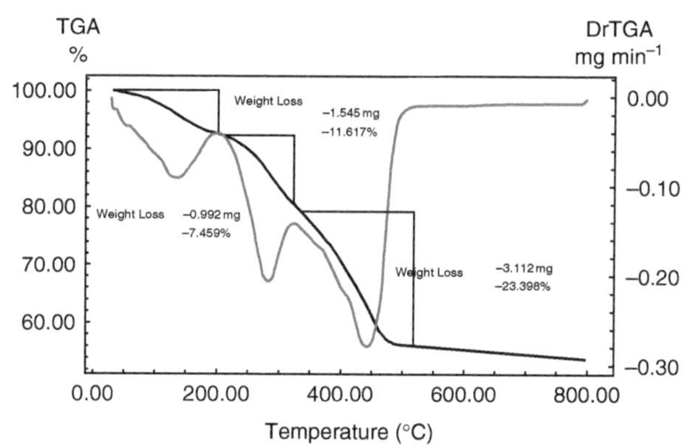

Figure 8.16 Thermogravimetric and derivative curves for the degradation of soy oil on montmorillonite K-10.

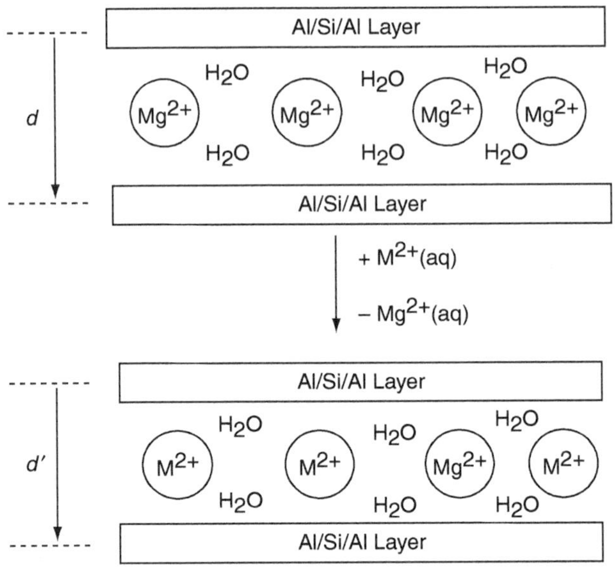

Figure 8.17 Proposed general mechanism for ion-exchange involving the initial saturated magnesium vermiculite with divalent cations in aqueous solution. *d* is the interlamellar distance.

6. Miscellaneous Examples

It has been shown that pH exerts a significant effect on oil wastewater treatment by using expanded and hydrophobized (carnauba wax) vermiculite [8]; pH 9 is the optimum value. Furthermore, it was verified that for standard mineral oil, canola oil, Kutwell oil, as well as for refinery effluents, EV exhibits a higher adsorption capacity in comparison with the hydrophobized samples. In this study it was shown that the hydrophobization process does not necessarily enhance oil removal capability, but that such phenomena will be influenced by the clay/hydrophobizant ratio to enhance oil removal by the presence of an hydrophobic compound without provoking a decrease of oil removal capability by clogging the clay pores.

The ion-exchange capacity of vermiculite was investigated in doubly distilled water solution to exchange magnesium inside the lamella with the copper, nickel, cobalt, and lead at the solid/liquid interface [9]. It was verified that the extension of each exchange reaction was dependent on time, pH, and cation concentration. The maximum time presented the following order $Pb^{2+} < Ni^{2+} < Cu^{2+} < Co^{2+}$ (12, 24, 48, and 72 h, respectively). From the point of view of cation exchange capacity, the sequence is $Pb^{2+} < Co^{2+} < Cu^{2+} < Ni^{2+}$ (0.59, 0.76, 0.84, and 0.93 mmol g^{-1}, respectively). This capacity is significantly increased under alkaline conditions.

REFERENCES

1. U.G. da Silva Jr., M.A. de F. Melo, A.F. da Silva, R.F. de Farias, J. Colloid Interface Sci. 260 (2003) 302.
2. J.R. Hindman, Vermiculite, in: Industrial Minerals and Rocks, Society for Mining, Metallurgy and Exploration Inc., Littleton, 1994.
3. Al. Mclaughl, Practitioner 204 (1970) 337.
4. R.F. de Farias, J. Serbian Chem. Soc. 70 (2005) 261.
5. J.D. Lee, Concise Inorganic Chemistry, Chapman & Hall, London, 1991.
6. C.R. Silva, C. Airoldi, J. Colloid Interface Sci. 195 (1997) 381.
7. K.D. Quintaes, J. Amaya-Farfan, M.A. Morgano, D.M.B. Mantovani, Food Add. Cont. 19 (2002) 134.
8. E. Mysore, T. Viraraghavan, Y.-C. Jin, Water Res. 39 (2005) 2643.
9. M.G. da Fonseca, M.M. de Oliveira, L.N.H. Arakaki, J.G.P. Espinola, C. Airoldi, J. Colloid Interface Sci. 285 (2005) 50.

CHAPTER NINE

Zeolites

Contents

1. Introduction

The chemistry of zeolites began in 1756 with the discovery of the mineral stibilite $NaCa_2Al_5Si_{13}O_{36}.14H_2O$ by Crö nstedt. He named this class of compounds zeolites, from the greek *zeo* and *lithos* ("boiling stones"). Synthetic zeolites are a class of highly porous materials that have found widespread use since their advent in late 1940s and early 1950s. The very special structural feature of these crystalline microporous solids is the presence of pores and cavities of molecular dimensions (0.3–1.0 nm). This fact makes zeolites suitable for applications in the realms of catalysis, separation, purification, ion exchange, radioactive waste clean-up, etc. More novel applications for zeolites are expected in electrochemistry, photochemistry, pharmaceutical engineering, membrane science, and technology (including nanotechnology).

In a certain sense, zeolites are the "sum" of the themes of this book, since zeolites can be both things: modified oxides and phosphates. Zeolites are crystalline aluminosilicates, but they can be modified to form the analogous phosphates, such as SAPO (silicoaluminophosphate). A SAPO molecular sieve is considered to be an AlPO (crystalline aluminophosphate molecular sieve) analog when some silicon atoms are located in its framework at the position of Al or P. SAPO exhibits a very interesting framework, since when a tetravalent Si atom substitutes a pentavalent P atom, a negative charge arises. When a proton is exchanged as a countercation for this negatively charged site, a Brønsted acid site is formed.

Zeolites are a well-known class of compounds due to their many catalytic applications, mainly in the petroleum industry. Zeolite β, for example, which

Interface Science and Technology, Volume 17
ISSN: 1573-4285, doi: 10.1016/S1573-4285 (08) 00409-2

possesses an open-framework structure of three-dimensional 12-membered ring channels, has received considerable attention for its prospects in petroleum industry and fine-chemical synthesis.

2. Some Examples and Applications

There are some natural counterparts of synthetic zeolites. A good example is lutinaite, a zeolite from Antarctica, that is the natural counterpart of ZSM-5 [1]. Mutinaite was the third new zeolite from Ferrar dolerites at Mt. Adamson (Northern Victoria Land, Antarctica), occurring as subspherical aggregates of tiny radiating lath-like fibers or as aggregates of transparent, colorless to pale-milky, tiny tabular crystals; it has vitreous luster, white streaks, and good {100} cleavage.

To be used as modified catalysts or ion exchangers, the incorporation of d or f block elements into zeolite structures has been extensively investigated (the ion exchange of cations in crystalline aluminosilicates and related materials is a standard method used for modifying their catalytic and sorption properties. For example, the solid-state interaction of lanthanum(III) chloride with zeolites under anhydrous conditions was investigated [2].

By using organofunctionalized silicon alkoxides, it was shown [3] that hydroxy–sodalite nanocrystals (sodalite is a small-pore zeolite whose framework consists of a six-membered ring aperture with a pore size of 0.28 nm) having organic functional groups ($\equiv$Si—$(CH_3)(CH_2)_3NH_2$ or $\equiv$Si—CH_3) can be synthesized by the direct transformation of organofunctionalized silicalite nanocrystals.

As a typical example of the remarkable ability of zeolites as catalysts, can be mentioned the selective single step synthesis of substituted octahydrophenanthridines and annelated pyridines was achieved using simple cyclic ketones, aliphatic aldehydes and ammonia over zeolites as catalysts under high pressure (autogeneous pressure) conditions [4]. High yields and high conversions were obtained for the first time, and the catalyst was inexpensive and environmentally friendly. Furthermore, it could be reused many times without significant loss of activity.

Regarding new methods for synthesizing "traditional" zeolites, can be mentioned the use of electrophoretic deposition as the seeding method for formation of zeolite ZSM-5 layers on alumina [5]. With such method, alumina disks were fully covered with ZSM-5 crystal seeds. Furthermore, the resulting seeded supports were submitted to hydrothermal treatment yielding uniform dense zeolite membranes. It was also verified that the thickness and density of the membranes could be modified by varying the Si/Al ratio of zeolite ZSM-5 particles, concentration of seeding suspension and seeding time. A selectivity of the produced ZSM-5 membrane for i-propanol permeation over water was observed.

The SAPO-34 structure is three-dimensional, with a pore diameter of 0.38 nm, and H–SAPO-34 has mild-to-moderate acidity. The separation of gases, such as CO_2 and CH_4, can be achieved by using H–SAPO-34 zeolite membranes modified with Li^+, Na^+, K^+, NH_4^+, and Cu^{2+} as ion exchangers [6]. It was verified that the presence of the cations on the zeolite increases the CO_2–CH_4 separation selectivities up to 60%.

Taking into account the importance of the acidic sites on the properties and uses of zeolites, investigations focusing on the elucidation of their structure and compositions are very important. In this connection, by making a detailed examination of the infrared data concerning ZSM-5 zeolites [7], specifically the infrared spectral region corresponding to the vibrations of the silanols, four different kinds of silanols were identified: two inner species, located in the channels, and two species corresponding to external sites, at the boundaries of the crystallites (Fig. 9.1).

To control the basicity of zeolites, such as Si–MCM-41, some strategies have been used, such as the functionalization of the surface with compounds that contain terminal amines. The influence of surfactant chain length on basic catalytic properties of Si–MCM-41 has been investigated [8] and it was verified that the material with higher catalytic activity (the catalytic properties of as-synthesized MCM-41 materials were explored in the base catalyzed Knoevenagel condensation of butyraldehyde with ethyl cyanoacetate) is the one with higher ordering degree, prepared with longer carbon chain surfactants and with lower cosolvent concentration in the gel synthesis.

The MCM-41 material has attracted a lot of attention, and it is the most widely studied mesoporous material. It possesses a hexagonal array of unidirectional tubular pores, very high surface area and porosity, a narrow pore size distribution, and pore size adjustable from ~2 to 10 nm. In this connection, a study of thermal and mechanical stability of MCM-41 materials containing different titanium contents was performed [9] and it was verified that at all temperatures the structural changes are less pronounced for Ti–MCM-41, and the complete collapse of the ordered

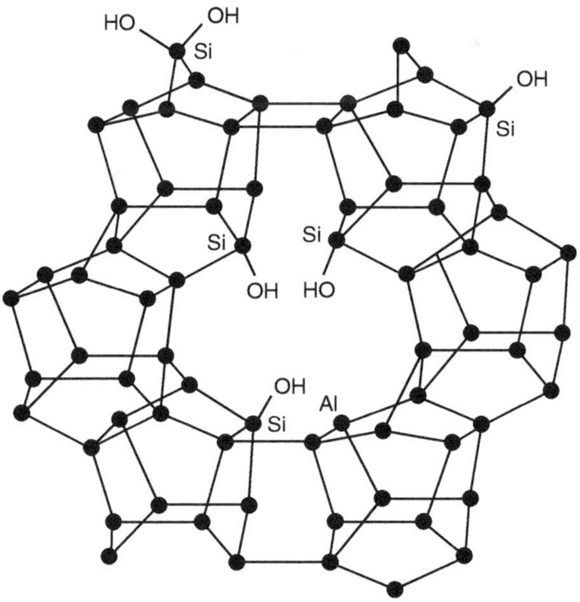

Figure 9.1 Structural model of a ZSM-5 zeolite indicating the configuration of the different silanols.

mesopore structure occurs at 1373 K while for the pure silica it occurs at 1273 K. Furthermore, tetracoordinated titanium incorporated in the walls remains stable up to 1073 K and seems to be responsible for the enhancement of thermal stability. Concerning mechanical stability in air, it was found that for all freshly calcined pure silica- and titanium-containing samples tested, the MCM-41 structure is still observed after the application of a unidirectional external pressure of 518 MPa and the collapse of the pore structure occurs at about 814 MPa.

REFERENCES

1. E. Galli, G. Vezzalini, S. Quartieri, Zeolites 19 (1997) 318.
2. B. Suliiowski, J. Find, H.G. Karge, D. Herein, Zeolites 19 (1997) 395.
3. D. Li, J. Yao, H. Wang, N. Hao, D. Zhao, K.R. Ratinac, S.P. Ringer, Micropor. Mesopor. Mater. 106 (2007) 262.
4. K.V.V.K. Mohan, N. Narender, S.J. Kulkarni, Micropor. Mesopor. Mater. 106 (2007) 229.
5. M. Abdollahi, S.N. Ashrazadeh, A. Malekpour, Micropor. Mesopor. Mater. 106 (2007) 192.
6. M. Hong, S. Li, H.F. Funke, J.L. Falconer, R.D. Noble, Micropor. Mesopor. Mater. 106 (2007) 140.
7. P. Hoffmann, J.A. Lobo, Micropor. Mesopor. Mater. 106 (2007) 122.
8. L. Martins, D. Cardoso, Micropor. Mesopor. Mater. 106 (2007) 8.
9. C. Galacho, M.M.L.R. Carrott, P.J.M. Carrott, Micropor. Mesopor. Mater. 108 (2007) 283.

Further Reading

1. R.F. de Farias (Org.), O.A. de Oliveira, C.W.B. Bezerra, A.B. de Oliveira, C. Airoldi, E.C.P. Maia, C.P. Demicheli, M.R. Sales, N.S. Gonçalves, S. de P. Machado, Química de Coordenação – fundamentos e atualidades, Editora Átomo, Campinas, 2005.
2. R.F. de Farias, Práticas de Química Inorgânica, Editora Átomo, Campinas, 2004.
3. A.O. de Souza, R.F. de Farias, Elementos de Química Quântica, Editora Átomo, Campinas, 2007.
4. R.F. de Farias, Introdução à Química Forense, Editora Átomo, Campinas, 2007.
5. C.N.R. Rao, J. Gopalakrishnan, New Directions in Solid State Chemistry, Cambridage University Press, Cambridge, 1997.
6. B.C. Gates, Catalytic Chemistry, Wiley, New York, 1992.
7. V. Gentil, Corrosão, Editora Guanabara, Rio de Janeiro, 1987.
8. W.G. Krauledat, Notação e Nomenclatura de Química Inorgânica, Edgard Blücher, São Paulo, 1970.
9. G.-A. Mazzocchin, L.M. Moretto, Ácidos e Bases, EDUCS, Caxias do Sul, 1996.
10. J.J.C.T. Dias, Espectroscopia Molecular, Fundação calouste Gulbenkian, Lisboa, 1986.
11. S.A. de Souza, Composição Química dos Aços, Edagard Blücher, São Paulo, 1989.
12. M.A. de Brito, Química Inorgânica: Compostos de Coordenação, EDIFURB, Blumenau, 2002.
13. Z. Marczenko, M. Balcerzak, Separation, Preconcentration and Spectrophotometriy in Inorganic Analysis, Elsevier, Amsterdam, 2000.
14. A. Leprevost, Química Analítica dos Minerais, Livros Técnicos e Científicos, Rio de Janeiro, 1975.
15. D.M. Bishop, Group Theory and Chemistry, Dover, New York, 1973.
16. S.J. Formosinho, A.J.C. Varandas, Estrutura e Reactividade Molecular, Fundação Calouste Gulbenkian, Lisboa, 1986.
17. A. Guinier, A Estrutura da Matéria, EDUSP, São Paulo, 1996.
18. E.V. Benvenutti, Química Inorgânica, Editora da UFRGS, Porto Alegre, 2003.
19. A.L. Barbosa, Dicionário de Química, AB Editora, Gioânia, 2000.
20. Oxford Chemical Dictionary, Oxford University Press, Oxford, 1985.
21. O. Sala, Fundamentos da Espectroscopia Raman e no Infravermelho, Editora da UNSEP, São Paulo, 1995.
22. D.L. Perry, S.L. Phillips, Handbook of Inorganic Compounds, CRC Press, Boca Raton, 1995.
23. J.M.T. Adad, Controle Químico de Qualidade, Guanabara Dois, Rio de Janeiro, 1982.
24. J.J.C.T. Dias, Química Quântica, Fundação Calouste Gulbenkian, Lisboa, 1982.
25. T. Moeller, Inorganic Chemistry: An Advanced Textbook, Wiley, New York, 1952.
26. R.B. Reslop, K. Jones, Química Inorgânica, Fundação Calouste Gulbenkian, Lisboa, 1988.
27. J.O. Edwards, Inorganic Reaction Mechanisms, W.A. Benjamin, New York, 1965.
28. F. Basolo, R. Johnson, Química de los Compuestos de Coordinación, Editorial Reverté, Barcelona, 1976.
29. G. Rayner-Canham, Descriptive Inorganic Chemistry, W.H. Freeman, New York, 1996.
30. J.E. Huheey, E.A. Keiter, R.L. Keiter, Inorganic Chemistry, Harper Collins, New York, 1993.
31. D.F. Shriver, P.W. Atkins, C.H. Langford, Inorganic Chemistry, Oxford University Press, Oxford, 1992.
32. R.J.P. Williams, J.J.R.F. da Silva, The Natural Selection of the Chemical Elements, Oxford University Press, Oxford, 1996.
33. F.A. Cotton, G. Wilkinson, Advanced Inorganic Chemistry, Wiley, New York, 1988.

Interface Science and Technology, Volume 17
ISSN: 1573-4285, doi: 10.1016/S1573-4285 (08) 00410-9

34. I.S. Butler, J.F. Harrod, Química Inorgánica, Addsion-Wesley, Wilmington, 1992.
35. G.J. Moody, J.D.R. Thomas, Dipole Moments in Inorganic Chemistry, Edward Arnold, London, 1971.
36. R. Saez, P.A. Caro, Rare Earths, Editorial Complutence, Madrid, 1998.
37. A.L. Companion, Ligação Química, Edgard Blücher, São Paulo, 1975.
38. W.A. Bueno, Ligação de Hidrogênio, McGraw-Hill, São Paulo, 1978.
39. M.J. Winter, d-Block Chemistry, Oxford University Press, Oxford, 1999.
40. D.E. Fenton, Biocoordination Chemistry, Oxford University Press, Oxford, 1997.
41. D.M.P. Mingos, Essential Trends in Inorganic Chemistry, Oxford University Press, Oxford, 1998.
42. J.D. Lee, Química Inorgânica, Edgard Blücher, São Paulo, 1980.
43. N.N. Greenwood, A. Earnshaw, Chemistry of the Elements, Butterworth-Heinemann, Cambridge, 1984.
44. F.A. Cotton, G. Wilkinson, P.L. Gauss, Basic Inorganic Chemistry, Wiley, New York, 1995.
45. J.J.R.F. da Silva, R.J.P. Williams, The Biological Chemistry of the Elements, Oxford University Press, Oxford, 1994.
46. P.C. Wilkins, R.G. Wilkins, Inorganic Chemistry in Biology, Oxford University Press, Oxford, 1997.
47. D.S. Vaitsman, J.C. Afonso, P.B. Dutra, Para que Servem os Elementos Químicos, Editora Interciência, Rio de Janeiro, 2001.
48. J.J.R.F. da Silva, Introdução à Química da Vida, Universidade Nova de Lisboa, Lisboa, 1985.
49. F.A. de Azevedo, A.A. da M. Chasin, Metais – gerenciamento da toxicidade, Ateneu, São Paulo, 2003.
50. H. Tomlinson, Utensílios de Alumínio e Doenças, Ícone, São Paulo, 1993.
51. M. Karplus, R.N. porter, Atoms & Molecules, Benjamin/Cummings, Menlo Park, 1970.
52. D. Cruz-Garritz, J.A. Camizo, A. Garritz, Estrutura Atômica, Addison-Wesley, Wilmington, 1987.
53. D.C. Harris, M.D. Bertolucci, Symmetry and Spectroscopy, Dover, New York, 1978.
54. A. Fazzio, K. Watari, Introdução à Teoria de Grupos, Editora da UFSM, Santa Maria, 1998.
55. A.B. do Nascimento, Problemas Selecionados de Química, Edição do autor, João Pessoa, 1987.
56. G.S. Girolami, T.B. rauchfuss, R.J. Angelici, Synthesis and Techniques in Inorganic Chemistry, University Science Boks, Sausalito, 1999.
57. K.D. Karlin, Progress in Inorganic Chemistry, Wiley, New York, 2003.
58. M.T. Weller, Inorganic Chemistry of Materials, Oxford University Press, Oxford, 1996.
59. D.M. Smyth, The Defect Chemistry of Metal Oxides, Oxford University Press, Oxford, 2000.
60. G. Attard, C. Barnes, Surfaces, Oxford University Press, Oxford, 1998.
61. C.F. Gomes, Argilas, Fundação Calouste Gulbenkian, Lisboa, 1988.
62. I.N. Levine, Quantum Chemistry, Prentice-Hall, Englewood Cliffs, 1991.
63. T.P. Softley, Atomic Spectra, Oxford University Press, Oxford, 1994.
64. W.G. Richards, P.R. Scott, Energy Levels in Atoms and Molecules, Oxford University Press, Oxford, 1994.
65. M.J. Winter, Chemical Bonding, Oxford University Press, Oxford, 1995.
66. H.W. Jaffe, Crystal Chemistry and Refractivity, Dover, New York, 1996.
67. A.K. Varshneya, Fundamentals of Inorganic Glasses, Academic Press, San Diego, 1994.
68. P.J. van der Put, The Inorganic Chemistry of Materials, Plenum Press, New York, 1998.
69. J.L. Figueiredo, F.R. Ribeiro, Catálise Heterogênea, Fundação calouste Gulbenkian, Lisboa, 1989.
70. W.D. Callister Jr., Materials Science and Engineering, Wiley, New York, 1997.
71. L.H. Van Vlack, Princípios de Ciências dos Materiais, Edgard Blücher, São Paulo, 1988.
72. G. Faure, Principles and Applications of Inorganic Geochemistry, Macmillan Publishing, New York, 1992.
73. L. Pauling, The Nature of the Chemical Bond, Cornell University Press, Ithaca, 1960.
74. R.J. Gillespie, P.L.A. Popelier, Chemical Bondind and Molecular Geometry, Oxford University Press, New York, 2001.
75. R.G. Pearson, Chemical Hardness, Wiley-VCH, Weinheim, 1997.
76. W.B. Jensen, The Lewis Acid-Base Concepts, Wiley, New York, 1980.

77. V. Gutmann, The Donor-Acceptor Approach to Molecular Interactions, Plenum Press, New York, 1978.
78. S.J. Ashcroft, C.T. Mortimer, Thermochemistry of Transition Metal Complexes, Academic Press, London, 1970.
79. L. Pauling, General Chemistry, Dover, New York, 1988.
80. R. McWeeny, Spins in Chemistry, Dover, New York, 2004.
81. A.J. Jacobson, Intercalation reaction of layered compounds, in: A.K. Cheetham, P. Day (Eds.), Solid State Chemistry Compounds, Clarendon Press, Oxford, 1992, p. 182.
82. A. Lerf, Intercalation compounds in layered host lattices: Supramolecular chemistry in nanodimension, in: H.S. Nalwa (Ed.), Handbook of Nanostructured Materials and Nanotechnology, vol. 5, Academic Press, New York, 2000, p. 5.
83. M. Ogawa, K. Kuroda, Chem. Rev. 95 (1995) 399–438.
84. A. Clearfield, Inorganic Ion Exchange Material, CRC Press Inc, Boca Raton, 1982, p. 1.
85. G. Alberti, T. Bein (Eds.), Comprehensive Supramolecular Chemistry, first ed., vol. 7, Pergamon Press, New York, 1996.
86. A. Cabreza, M.A.G. Aranha, M.M. Lara, S. Bruque, J. Sanz, Acta Crystallogr. B 52 (1996) 982–988.
87. A. Nangia, G.R. Desiraju, Acta Crystallogr. A 54 (1998) 934–944.
88. P. Capková, H. Schenk, J. Inclusion Phenom. 47 (2003) 1–10.
89. P. Capková, M. Pospísil, Z. Weiss, J. Mol. Model. 9 (2003) 195–205.
90. A. Clearfield, Prog. Inorg. Chem. 47 (1998) 371–510.
91. A. Clearfield, Chem. Rev. 88 (1988) 125–148.
92. F. Wypych, W.H. Schreiner, N. Mattoso, D.A. Mosca, R. Marangoni, C.A.S. Bento, J. Mater. Chem. 13 (2003) 304–307.
93. A. Espina, F. Menéndez, E. Jaimez, S.A. Khainakov, C. Trobajo, J.R. García, J. Rodrigues, Chem. Mater. 10 (1998) 2490–2496.
94. F. Fredoueil, D. Massiot, P. Janvier, F. Gingl, M.B. Doeuff, M. Evain, A. Clearfield, B. Bujoli, Inorg. Chem. 38 (1999) 1831–1833.
95. K. Goubitz, P. Capková, K. Melánová, W. Molleman, H. Schenk, Acta Crystallogr. B 57 (2001) 178–183.
96. G.B. Hix, A. Turner, B.M. Kariuki, M. Tremayne, E.J. MacLean, J. Mater. Chem. 12 (2002) 3220–3227.
97. N. Miyamoto, K. Kuroda, M. Ogawa, J. Mater. Chem. 14 (2004) 165–170.
98. M. Trchová, P. Capková, P. Matejka, K. Melánová, L. Benes, J. Solid State Chem. 145 (1999) 1–9.
99. M.G. Fonseca, C. Airoldi, J. Mater. Chem. 10 (2000) 1457–1463.
100. A.T. Kan, J.E. Oddo, M.B. Tomson, Langmuir 10 (1994) 1450–1455.
101. L.N.H. Arakaki, L.M. Nunes, J.A. Simoni, C. Airoldi, J. Colloid Interface Sci. 228 (2000) 46–51.
102. C.B.A. Lima, C. Airoldi, Thermochim. Acta 400 (2003) 51–59.
103. L.M. Nunes, C. Airoldi, Chem. Mater. 11 (1999) 2069–2075.
104. C. Airoldi, S. Roca, J. Mater. Chem. 6 (1996) 1963–1966.
105. L.M. Nunes, C. Airoldi, Mater. Res. Bull. 34 (1999) 2121–2130.
106. L.M. Nunes, C. Airoldi, J. Solid State Chem. 154 (2000) 557–563.
107. D.M. Poojary, B. Zhang, A. Cabeza, M.A.G. Aranda, S. Bruque, A. Clearfield, J. Mater. Chem. 6 (1996) 639–644.
108. M. Suarez, J.R. Garcia, J. Rodrigues, Mater. Chem. Phys. 8 (1983) 451–458.
109. D.M. Poojary, D. Grohol, A. Clearfield, Angew. Chem. Int. Ed. 34 (1995) 1508–1510.
110. C. Airoldi, S. Roca, Sol. Chem. 22 (1993) 707–713.
111. C. Airoldi, E.F.C. Alcântara, J. Chem. Thermodyn. 27 (1995) 623–632.
112. C.B.A. Lima, C. Airoldi, Solid State Sci. 4 (2002) 1321–1329.
113. M.G. Fonseca, C. Airoldi, J. Colloid Interface Sci. 240 (2001) 533–538.
114. J.L. Bideau, C. Payen, P. Palvadeau, B. Bujoli, Inorg. Chem. 33 (1994) 4885–4890.
115. C. Airoldi, L.N.H. Arakaki, Polyhedron 20 (2001) 929–936.
116. A.M. Lazarin, C. Airoldi, Anal. Chim. Acta 523 (2004) 89–95.

117. V.S.O. Ruiz, C. Airoldi, J. Therm. Anal. Calorim. 71 (2003) 459–465.
118. N. Chiron, R. Guilet, E. Deydier, Water Res. 37 (2003) 3079–3086.
119. O.A.C. Monteiro, C. Airoldi, J. Colloid Interface Sci. 212 (1999) 212–219.
120. M. Jaraniec, R. Madey, Physical Adsorption on Heterogeneous Solids, Elsevier, New York, 1988.
121. C. Airoldi, S.F. Oliveira, Struct. Chem. 2 (1991) 41–46.
122. M.O. Machado et al., J. Chem. Thermodyn. 38 (2006) 130–135.
123. D.A. Wardle, A. Ghani, Soil Biol. Biochem. 27 (1995) 1601.
124. J.W. Raich, C.S. Potters, Global Biogeochem. Cycles 9 (1995) 23.
125. T. George, R. Magbanua, D.P. Garrity, B.S. Tubãna, J. Quiton, Agron. J. 94 (2002) 981.
126. G.A. O'Coonor, H.A. Elliot, N.T. Basta, R.K. Bastian, G.M. Pierzynski, R.C. Sims, J.E. Smith Jr., J. Environ. Qual. 34 (2005) 7.
127. W.M. Stewart, D.W. Dibb, A.E. Johnston, T.J. Smyth, Agron. J. 97 (2005) 1.
128. S.A.M. Critter, S.S. Freitas, C. Airoldi, Thermochim. Acta 394 (2002) 145.
129. S.A.M. Critter, S.S. Freitas, C. Airoldi, Thermochim. Acta 394 (2002) 133.
130. N. Barros, S. Feijóo, S. Fernández, Thermochim. Acta 406 (2003) 161.
131. A.G.S. Prado, C. Airoldi, Thermochim. Acta. 394 (2002) 155.
132. N. Barros, S. Feijóo, Biophys. Chem. 104 (2003) 561.
133. S. Saggar, A. Parshotam, C. Hedley, G. Salt, Soil Biol. Biochem. 31 (1999) 2025.
134. J.N. Ladd, J. Monroizer, M. Amato, Soil Biol. Biochem. 24 (1992) 359.
135. N. Barros, S. Feijóo, S. Fernández, J.A. Simoni, C. Airoldi, Entropie 224 (2000) 75.
136. N. Barros, S. Feijóo, S. Fernández, J.A. Simoni, C. Airoldi, Thermochim. Acta 356 (2000) 1.
137. P. Tissot, J. Therm. Anal. Calorim. 57 (1999) 303.
138. Y. Laor, M. Raviv, M. Borisover, Thermochim. Acta 420 (2004) 119.
139. H. Debray, Ann. Chim. Phys. 61 (1961) 437.
140. J.F. McCullough, E. City, M.L. Salutsky, US Patent 3,141,732, July 21 (1964).
141. A.M. Erskine, G. Grim, S.C. Horning, Ind. Eng. Chem. 36 (1944) 456.
142. S.I. Vol'fkovich, R.E. Remen, Chem. Abstr. 50 (1956) 6243.
143. G.L. Bridger, M.L. Salutsky, R.W. Starostka, J. Agric. Food Chem. 10 (1962) 181.
144. V. Barrón, J. Torrent, J. Agric. Food Chem. 42 (1994) 105.
145. O.B. Michelsen, Anal. Chem. 29 (1957) 60.
146. S.A.M. Critter, S.S. Freitas, C. Airoldi, Thermochim. Acta 410 (2004) 35.
147. S.A.M. Critter, S.S. Freitas, C. Airoldi, Thermochim. Acta 417 (2004) 275.
148. E.D. Vance, P.C. Brookes, D.S. Jenkinson, Soil Biol. Biochem. 19 (1987) 697.
149. U. von Stockar, J.S. Liu, Biochim. Biophys. Acta 1412 (1999) 191.
150. G.P. Sparling, J. Soil Sci. 34 (1983) 381.
151. L. Nuñez, N. Barros, I. Barja, Thermochim. Acta 237 (1994) 73.
152. T. Kimura, K. Takahashi, J. Gen. Microbiol. 131 (1985) 3083.
153. U. von Stockar, L. Gustafsson, C. Larsson, I. Marison, P. Tissot, E. Gnaiger, Biochim. Biophys. Acta 1183 (1993) 221.
154. E.H. Battley, Thermochim. Acta 309 (1998) 17.
155. S.G. Carling, P. Day, D. Visser, Inorg. Chem. 34 (1995) 3917.
156. J. Salgado, M.M. Mato, A. Vázquez-Galiñanes, M.I. Paz-Andrade, T. Carballas, Thermochim. Acta 410 (2004) 141.
157. J.B. Dixon, S.B. Leed, Minerals in Soil Environments, Soil Science Society of America, Wiscosin, 1989.
158. R.C. Mackenzie, B.D. Mitchell, Differential Thermal Analysis, vol. 2, Academic Press, New York, 1972.
159. J. Salgado, M.I. González, J. Armada, M.I. Paz-Andrade, M. Carballas, T. Carballas, Thermochim. Acta 259 (1995) 165.
160. G. Giovannini, S. Luchessi, M. Giachetti, Soil Sci. 143 (1987) 220.
161. A.G.M. Vasandani, M. Raza Shah, J. Therm. Anal. Calorim. 41 (1994) 1053.
162. N. Barros, S. Feijóo, J.A. Simoni, S.A.M. Critter, C. Airoldi, J. Therm. Anal. Calorim. 63 (2001) 577.

163. U. von Stockar, I. Marison, Thermochim. Acta 229 (1993) 157.
164. E.H. Battley, Energetics of Microbial Growth, John Wiley & Sons, New York, 1987.
165. W.M. van Gulik, J.J. Heijnen, Biotechnol. Bioeng. 48 (1995) 681.
166. A. Bauer, A.L. Black, Soil Sci. Soc. Am. J. 45 (1981) 1166.
167. K. Eisele, D.S. Shimel, L.A. Kaprustka, W.J. Parton, Oecologia 79 (1989) 471.
168. J. Irvine, B.E. Law, Glob. Chang. Biol. 8 (2002) 1183.
169. P. van Bodegom, F. Stams, L. Mollema, S. Boeke, P. Leffelaar, Appl. Environ. Microbiol. 67 (2001) 3586.
170. D.M. Sylvia, J.J. Fuhrman, P.G. Hartel, D.A. Zuberer, Principles and Applications of Soil Microbiology, Prentice Hall, New York, 1998.
171. O. Kedem, S.R. Caplan, Trans. Faraday Soc. 61 (1965) 1897.
172. J.W. Stucky, Eur. J. Biochem. 109 (1980) 269.
173. A. Clearfield, Inorganic Ion Exchange Materials, CRC Press, Boca Raton, FL, 1982.
174. G. Alberti, U. Costantino, in: J.L. Atwood, J.E. Davies, D.D. Mac-Nicol (Eds.), Inclusion Compounds, Oxford University Press, New York, 1991.
175. A. Clearfield, U. Costantino, in: G. Alberti, T. Bein (Eds.), Comprehensive Supramolecular Chemistry, vol. 7, Pergamon, New York, 1996.
176. D. O'Hare, Inorganic Materials, John Wiley, England, 1992.
177. I. Dekany, L. Szirtes, J. Radioanal. Nucl. Chem. 190 (1995) 167.
178. K. Kakiguchi, Y. Baba, T. Yanagida, M. Danjo, M. Tsuhako, H. Nariai, S. Yamaguchi, I. Motooka, Phosphorus Res. Bull. 5 (1995) 131.
179. C. Airoldi, S.F. Oliveira, Struct. Chem. 2 (1991) 41.
180. C. Airoldi, S. Roca, J. Mater. Chem. 6 (1996) 1963.
181. M. Suarez, M.L. Rodríguez, R. Llavona, L.M. Barcina, A. Veja, J. Rodríguez, J. Chem. Soc., Dalton Trans. (1997) 2757.
182. C.B.A. Lima, C. Airoldi, Thermochim. Acta 400 (2003) 51.
183. C. Airoldi, L.M. Nunes, R.F. Farias, Mater. Res. Bull. 35 (2000) 2081.
184. H. Nakayama, T. Eguchi, N. Nakamura, M. Danjo, M. Tsuhako, H. Nariai, S. Yamaguchi, I. Motooka, Bull. Chem. Soc. Jpn. 70 (1997) 1053.
185. L.M. Nunes, C. Airoldi, Chem. Mater. 11 (1999) 2069.
186. L.M. Nunes, C. Airoldi, Mater. Res. Bull. 34 (1999) 2121.
187. L.M. Nunes, C. Airoldi, Thermochim. Acta 328 (1999) 297.
188. C.B.A. Lima, C. Airoldi, Int. J. Inorg. Chem. 3 (2001) 907.
189. P. Capkova, H. Schenk, J. Inclusion Phenom. 47 (2003) 1.
190. A. Espina, E. Jaimez, S.A. Khainakov, C. Trobajo, J.R. García, J. Rodríguez, Chem. Mater. 10 (1998) 2490.
191. A.R. Katritzky, C.W. Rees, Comprehensive Heterocyclic Chemistry, vol. 2, Pergamon Press, Oxford, 1984.
192. C. Airoldi, E.F.C. Alcântara, J. Chem. Thermodyn. 27 (1995) 623.
193. S. Roca, C. Airoldi, J. Chem. Soc., Dalton Trans. (1997) 2517.
194. E.F.S. Vieira, J.A. Simoni, C. Airoldi, J. Mater. Chem. 7 (1997) 2249.
195. C. Airoldi, L.M. Nunes, Langmuir 16 (2000) 1436.
196. M.G. Fonseca, C. Airoldi, J. Colloid Interface Sci. 240 (2001) 533.
197. M.G. Fonseca, C. Airoldi, J. Therm. Anal. Calorim. 64 (2001) 273.
198. S.F. Ashcroft, C.T. Mortimer, Thermochemistry of Transition Metal Complexes, Academic Press, London, 1970.
199. M.A.V. Ribeiro da Silva, M.A.R. Matos, M.J.S. Monte, J. Chem. Thermodyn. 22 (1990) 609.
200. C. Airoldi, A.P. Chagas, Coord. Chem. Rev. 119 (1992) 29.
201. C. Airoldi, L.N.H. Arakaki, J. Colloid Interface Sci. 249 (2002) 1.
202. P.K. Jal, S. Patel, B.K. Mishra, Talanta 62 (2004) 1005.
203. F.A. Pavan, I.S. Lima, E.V. Benvenutti, Y. Gushikem, C. Airoldi, J. Colloid Interface Sci. 275 (2004) 386.
204. A.G.S. Prado, L.A.A. Sales, R.M. Carvalho, J.C. Rubin, C. Airoldi, J. Non-Cryst. Solids 333 (2004) 61.

205. A. Goswani, A.K. Singh, Anal. Chim. Acta 454 (2002) 229.
206. A.G.S. Prado, L.H.N. Arakaki, C. Airoldi, Green Chem. 236 (2002) 161.
207. L.G.A. van de Water, W.L. Driessen, M.W. Glenny, J. Reedijk, M. Schroder, React. Funct. Polym. 51 (2002) 33.
208. A.G.S. Prado, C. Airoldi, Fresenius J. Anal. Chem. 371 (2001) 1028.
209. A.G.S. Prado, C. Airoldi, Anal. Chim. Acta 432 (2001) 201.
210. D.J. Macquarrie, Green Chem. 1 (1999) 195.
211. H.J. Im, E. Barnes, S. Dai, Z. Xue, Microporous Mesoporous Mater. 70 (2004) 57.
212. J. Shah, T.J. Pinavaia, Chem. Mater. 17 (2005) 947.
213. Y. Lin, S. Fiskum, W. Yantase, H. Wu, S.W. Mattigod, E. Vorpagel, G.E. Fryxell, K.N. Raymond, J. Xu, Environ. Sci. Technol. 39 (2005) 1332.
214. G.E. Fryxell, Y. Lin, S.K. Fiskum, J.C. Birnbaum, H. Wu, K. Kemner, S. Kelly, Environ. Sci. Technol. 39 (2005) 1324.
215. A. Stein, B.J. Melde, R.C. Schroden, Adv. Mater. 12 (2000) 1403.
216. C. Airoldi, L.N.H. Arakaki, Polyhedron 20 (2001) 929.
217. G.E. Fryxell, H. Wu, Y. Lin, W.J. Shaw, J.C. Birnbaum, J.C. Linehan, Z. Nie, K. Kemner, S. Kelly, J. Mater. Chem. 14 (2004) 3356.
218. T. Yokoi, H. Yoshitake, T. Tatsumi, J. Mater. Chem. 14 (2004) 951.
219. H. Yoshitake, E. Koiso, T. Tasumi, H. Horie, H. Yoshimura, Chem. Lett. (2004) 872.
220. P.T. Tanev, T.J. Pinavaia, Science 267 (1995) 865.
221. J.W. Kondo, Y. Takahara, D.L. Lu, K. Domen, Chem. Mater. 13 (2001) 1200.
222. P.M. Price, J.H. Clark, D.J. Macquarrie, J. Chem. Soc., Dalton Trans. (2000) 101.
223. A.M. Tshavhungwe, M. Layh, N.J. Coville, J. Sol-Gel Sci. Technol. 29 (2004) 167.
224. D.J. Macquarrie, D.B. Jackson, J.E.G. Mdoe, J.H. Clark, New J. Chem. 23 (1999) 539.
225. M.H. Valkenberg, W.F. Hölderich, Catal. Rev. 44 (2002) 321.
226. T. Prub, D.J. Macquarrie, J.H. Clark, J. Mol. Catal. A 211 (2004) 209.
227. G.E. Fryxell, J. Lin, T.A. Hauser, Z. Nie, K.F. Ferris, S. Mattigod, M. Gong, R.T. Hallen, Chem. Mater. 11 (1999) 2148.
228. J. Brown, R. Richer, L. Mercier, Microporous Mesoporous Mater. 37 (2000) 41.
229. V. Antochshuk, M. Jaroniec, Chem. Commun. (2002) 258.
230. C.C. Chen, E.J. Mckimmy, T.J. Pinavaia, K.F. Hayes, Environ. Sci. Tecnhol. 38 (2003) 4758.
231. S. Babel, T.A. Kurniawan, J. Hazard. Mater. B 97 (2003) 219.
232. J.A.A. Sales, C. Airoldi, J. Non-Cryst. Solids 330 (2003) 142.
233. A.G.S. Prado, C. Airoldi, J. Mater. Chem. 12 (2002) 3823.
234. J.A.A. Sales, C. Airoldi, Thermochim. Acta 427 (2005) 77.
235. J.A.A. Sales, A.G.S. Prado, C. Airoldi, Polyhedron 21 (2002) 2647.
236. M. Ogawa, N. Yamamoto, Langmuir 15 (1999) 2227.
237. R.J.P. Corriu, A. Mehdi, C. Reye, C. Thieuleux, Chem. Mater. 16 (2004) 159.
238. A.M. Liu, K. Hidajat, S. Kawi, D.Y. Zhao, Chem. Commun. (2000)1145.
239. P.T. Tanev, T.J. Pinnavaia, Chem. Mater. 8 (1996) 2068.
240. J. Kramer, W.L. Driessen, K.R. Koch, J. Reedijk, Hydrometallurgy 64 (2002) 59.
241. H. Yoshitake, T. Yokoi, T. Tatsumi, Bull. Chem. Soc. Jpn. 76 (2003) 847.
242. X. Wang, K.S.K. Lin, J.C.C. Chan, S. Cheng, Chem. Commun. (2004)2762.
243. I. Park, Z. Wang, T.J. Pinnavaia, Chem. Mater. 17 (2005) 383.
244. O.A.C. Monteiro Jr., C. Airoldi, J. Colloid Interface Sci. 282 (2005) 32.
245. A.M. Lazarin, C. Airoldi, J. Chem. Thermodyn. 37 (2005) 243.
246. J. Thornon, Pure Appl. Chem. 93 (2001) 1231.
247. J.A.A. Sales et al., Surf. Sci. 590 (2005) 51–62.
248. R. Buchmeiser, J. Chromatogr. A 918 (2001) 233.
249. J. Nawrocki, C. Dunlap, A. McCormick, P.W. Carr, J. Chromatogr. A 1028 (2004) 1.
250. J.J. Kirkland, J.W. Henderson, J.J. DeStefano, M.A. van Straten, H.A. Claessens, J. Chromatogr. A 762 (1997) 97.
251. U. Tründinger, G. Müller, K.K. Unger, J. Chromatogr. 535 (1990) 111.
252. J. Yu, Z. El Rassi, J. Chromatogr. 631 (1993) 91.
253. A. Kurganov, U. Tündinger, T. Isaeva, K. Unger, Chromatographia 42 (1996) 217.

254. K. Tani, Y. Suzuki, J. Liq. Chromatogr. Relat. Technol. 19 (1996) 3037.
255. J.J. Pesek, V.H. Tang, Chromatographia 649 (1994) 649.
256. A. Ellwanger, M.T. Matyska, K. Albert, J.J. Pesek, Chromatographia 49 (1999) 424.
257. J. Nawrocki, C. Dunlap, J. Li, J. Zhao, C.V. McNeff, A. McCormick, P.W. Carr, J. Chromatogr. A 1028 (2004) 31.
258. R.B. Silva, K.E. Collins, C.H. Collins, J. Chromatogr. A 869 (2000) 137.
259. R.B. Silva, Y. Gushikem, C.H. Collins, J. Sep. Sci. 869 (2001) 49.
260. D.A. Fonseca, K.E. Collins, C.H. Collins, J. Chromatogr. A 1030 (2004) 209.
261. C.R. Silva, C. Airoldi, K.E. Collins, C.H. Collins, LC-GC 22 (2004) 632.
262. C.R. Silva, C. Airoldi, K.E. Collins, C.H. Collins, J. Chromatogr. A 1073 (2005) 155.
263. J.J. Kirkland, J.W. Henderson, J.D. Martosella, B.A. Bidlingmeyer, J. Vasta-Russell, J.B. Adams Jr., LC-GC 17 (1999) 634.
264. J.E. O'Gara, D.P. Walsh, C.H. Phoebe Jr., B.A. Alden, E.S.P. Bouvier, P.C. Iraneta, M. Capparella, T.H. Walter, LC-GC 19 (2001) 632.
265. U.D. Neue, Y.F. Cheng, B.A. Alden, P.C. Iraneta, C.H. Phoebe, K. van Tran, Chromatographia 54 (2001) 169.
266. C.R. Silva, S. Bachmann, R.R. Schefer, K. Albert, I.C.S.F. Jardim, C. Airoldi, J. Chromatogr. A 948 (2002) 85.
267. C.R. Silva, I.C.S.F. Jardim, C. Airoldi, J. Chromatogr. A 987 (2003) 127.
268. J. Horak, W. Lindner, J. Chromatogr. A 1043 (2004) 177.
269. M. Przybyciel, R.E. Majors, LC-GC 20 (2002) 584.
270. R.D. Morrison, J.W. Dolan, LC-GC 18 (2000) 936.
271. C.R. Silva, C. Airoldi, Process of preparation of new alkoxysilanes with polar urea groups Brazilian Patent PI 9, 903,110–118 (1999).
272. U.D. Neue, HPLC Columns, Theory, Technology and Practice, Wiley-VCH, New York, 1997.
273. H. Engelhardt, M. Arangio, T. Lobert, LC-GC 15 (1997) 856.
274. K. Kimata, K. Iwaguchi, S. Onishi, K. Jinno, R. Eksteen, K. Hosoya, M. Araki, N. Tanaka, J. Chromatogr. Sci. 27 (1989) 721.
275. G.E. Berendsen, K.A. Pikaart, L. de Galan, J. Liq. Chromatogr. 3 (1980) 1437.
276. K. Albert, J. Sep. Sci. 215 (2003) 215.
277. M.D. Bruch, H.O. Fatunmbi, J. Chromatogr. A 1021 (2003) 61.
278. H. Engelhardt, R. Grüner, M. Scherer, Chromatographia 53 (2001) S154.
279. U.D. Neue, K. Van Tran, P.C. Iraneta, B.A. Alden, J. Sep. Sci. 26 (2003) 174.
280. M.R. Euerby, P. Petersson, J. Chromatogr. A 994 (2003) 13.
281. M.R. Euerby, P. Petersson, LC–GC Eur. 13 (2000) 665.
282. J. Horak, N.M. Maier, W. Lindner, J. Chromatogr. A 1045 (2004) 43.
283. J.J. Kirkland, M.A. van Straten, H.A. Claessens, J. Chromatogr. A 797 (1998) 111.
284. J.J. Kirkland, J.B. Adams Jr., M.A. van Straten, H.A. Claessens, Anal. Chem. 70 (1998) 4344.
285. M.J. Wirth, R.W.P. Fairbank, H.O. Fatumbi, Science 275 (1997) 44.
286. Y.F. Cheng, T.H. Walter, Z. Lu, P. Iraneta, B.A. Alden, C. Genderau, U.D. Neue, J.M. Grassi, J.L. Carmody, J.E. O'Gara, R.P. Fisk, LC-GC 18 (2000) 1162.
287. K.D. Wyndhan, J.E. O'Gara, T.H. Walter, K.H. Glose, N.L. Lawrence, B.A. Alden, G.S. Izzo, C.J. Hudalla, C.P. Iraneta, Anal. Chem. 75 (2004) 6781.
288. J.D. Thompson, P.W. Carr, Anal. Chem. 74 (2002) 4150.
289. K. Albert, E. Bayer, J. Chromatogr. 544 (1991) 345.
290. M. Pursch, L.C. Sander, K. Albert, Anal. Chem. 68 (1996) 4107.
291. S. Bachmann, L.F.C. Melo, R.B. Silva, T.A. Anazawa, I.C.S.F. Jardim, K.E. Collins, C.H. Collins, K. Albert, Chem. Mater. 13 (2001) 1874.
292. C. Hellriegel, U. Skogsberg, K. Albert, M. Lämmerhofer, N.M. Maier, W. Lindner, J. Am. Chem. Soc. 126 (2004) 3809.
293. W.B. Kim, S.H. Choi, J.S. Lee, J. Phys. Chem. B 104 (2000) 8670.
294. K. Asakura, J. Inukai, Y. Iwasawa, J. Phys. Chem. 96 (1992) 829.
295. U.D. Neue, E. Serowik, P. Iraneta, B.A. Alden, T.H. Walter, J. Chromatogr. A 849 (1999) 87.
296. H.A. Claessens, M.A. van Straten, J.J. Kirkland, J. Chromatogr. A 728 (1996) 259.

297. D.A. Fonseca, H.R. Gutierrez, K.E. Collins, C.H. Collins, J. Chromatogr. A 1030 (2004) 149.
298. J.J. Kirkland, M.A. van Straten, H.A. Claessens, J. Chromatogr. A 691 (1995) 3.
299. J.W. Moore, S. Ramamorthy, Heavy Metals in Natural Waters, Springer-Verlag, New York, 1994.
300. M.R.M.P. Aguiar, A.C. Novaes, A.W.S. Guarino, Quím. Nova 25 (2002) 1145.
301. J. Rubio, F. Tessele, Miner. Eng. 10 (1997) 671.
302. M.F. Briagatti, F. Corradini, G.C. Franchini, S. Mazzoni, I. Medici, L. Poppi, Appl. Clay Sci. 9 (1995) 383.
303. M.L. Schlegel, L. Charlet, A. Manceau, J. Colloid Interface Sci. 220 (1999) 392.
304. T. Mathialagan, T. Viraraghavan, Sep. Sci. Technol. 38 (2003) 57.
305. J.F. Blais, S. Shen, N. Meunier, R.D. Tyagi, Environ. Technol. 24 (2003) 205.
306. E. Alvarez-Ayuso, A. Garcia-Sanchez, Environ. Technol. 24 (2003) 615.
307. R.-S. Juang, S.-H. Lin, F.-C. Huang, C.-H. Cheng, J. Colloid Interface Sci. 274 (2004) 337.
308. P. Jacquier, J. Ly, C. Beaucaire, Appl. Clay Sci. 26 (2004) 163.
309. G.W. Brindely, G. Brown, Crystal Structures of Clay Minerals and Their X- Ray Identification, first ed., Mineralogical Society, London, 1980.
310. B. Velde, Introduction to Clay Minerals, Chapman & Hall, London, 1992.
311. J. Konta, Appl. Clay Sci. 10 (1995) 275.
312. M.J. Potter, Am. Ceram. Bull. 8 (1999) 145.
313. A. Liu, R.D. Gonzalez, J. Colloid Interface Sci. 218 (1999) 225.
314. G. Abate, J. Lichtig, J.C. Masini, Talanta 58 (2002) 433.
315. D.A. Skoog, D.M. West, F.J. Holler, Fundamentals of Analytical Chemistry, seventh ed., Saunders, Philadelphia, 1996.
316. D.W. O'Hare, in: D.W. Bruce, D. O'Hare (Eds.), Inorganic Materials, Wiley, New York, 1997.
317. V.C. Farmer, Spectrochim. Acta 20 (1964) 1149.
318. P.J. Launer, Am. Mineral. 51 (1967) 765.
319. C.H. Gilles, D. Smith, J. Colloid Interface Sci. 47 (1974) 755.
320. A.W. Adamson, Physical Chemistry of Surfaces, Wiley, New York, 1990.
321. G.A. Krestov, Thermodynamics of Solvation: Solution and Dissolution; Ions and Solvents; Structure and Energetics, first ed., Ellis Horwood, London, 1991.
322. Y. Marcus, Ion Solvation, Wiley, London, 1985.
323. A. Bem-Naim, Solvation Thermodynamics, Plenum, New York, 1987.
324. K. Wilson, J.H. Clark, Pure Appl. Chem. 72 (2000) 1313.
325. J.H. Clark, D.J. Macquarrie, Chem. Commun. 8 (1998) 853.
326. U. Deschler, P. Kleinschmit, P. Panster, Angew. Chem. Int. Ed. Engl. 25 (1986) 236.
327. E.S. Ribeiro, S.S. Rosatto, Y. Gushikem, L.T. Kubota, J. Solid State Electrochem. 7 (2003) 665.
328. C.R. Silva, I.C.S.F. Jardim, C. Airoldi, J. Chromatogr. A 987 (2003) 139.
329. C.R. Silva, I.C.S.F. Jardim, C. Airoldi, J. Sep. Sci. 24 (2001) 197.
330. C. Bresson, M.J. Menu, M. Dartiguenave, Y. Dartiguenave, J. Environ. Monit. 2 (2000) 240.
331. L.N.H. Arakaki, V.L.S. Augusto, J.G.P. Espinola, M.G. Fonseca, S.F. Oliveira, T. Arakaki, C. Airoldi, J. Environ. Monit. 5 (2003) 366.
332. J.A.A. Sales, A.G.S. Prado, C. Airoldi, J. Therm. Anal. Calorim. 70 (2002) 135.
333. J.P. Bernal, E. Rodriguez de San Miguel, J.C. Aguilar, G. Salazar, J. Gyves, Sep. Sci. Technol. 35 (2000) 1661.
334. P.D. Verweij, M.J. Haanepen, J.J. de Ridder, W.L. Driessen, J. Reedijk, Recl. Trav. Chim. Pays-Bas 111 (1992) 371.
335. S. Rio-Segade, B. Perez-Cid, C. Bendicho, Fresenius J. Anal. Chem. 351 (1995) 798.
336. L.N.H. Arakaki, C. Airoldi, Polyhedron 19 (2000) 367.
337. A.G.S. Prado, L.N.H. Arakaki, C. Airoldi, J. Chem. Soc., Dalton Trans. (2001) 2206.
338. L.N.H. Arakaki, C. Airoldi, Quim. Nova 22 (1999) 246.
339. N. Rauf, S.S. Tahir, J. Chem. Thermodyn. 32 (2000) 651.
340. M.G. Voronkov, N.N. Vlasova, Y.N. Pozhidaev, Appl. Organomet. Chem. 14 (2000) 287.
341. C.R. Silva, C. Airoldi, J. Colloid Interface Sci. 195 (1997) 381.
342. A.G.S. Prado, L.N.H. Arakaki, C. Airoldi, Green Chem. 4 (2002) 42.
343. M.G. Fonseca, C. Airoldi, Thermochim. Acta 359 (2000) 1.

344. C. Airoldi, M.R.M.C. Santos, J. Mater. Chem. 4 (1994) 1479.
345. M.G. Fonseca, J.A. Simoni, C. Airoldi, Thermochim. Acta 369 (2001) 17.
346. L.N.H. Arakaki, L.M. Nunes, C. Airoldi, J. Colloid Interface Sci. 228 (2000) 46.
347. L.D. White, C.P. Tripp, J. Colloid Interface Sci. 227 (2000) 237.
348. A.R. Cestari, C. Airoldi, J. Therm. Anal. Calorim. 44 (1995) 79.
349. J.G.P. Espínola, S.F. Oliveira, W.E.S. Lemus, A.G. Souza, C. Airoldi, J.C.A. Moreira, Colloids Surf. A 166 (2000) 45.
350. H. Irving, R.J.P. Williams, J. Chem. Soc. (1958)3192.
351. M.G. Fonseca, A.S. Oliveira, C. Airoldi, J. Colloid Interface Sci. 240 (2001) 533.
352. M.G. Fonseca, C. Airoldi, J. Colloid Interface Sci. 240 (2001) 229.
353. M.G. Fonseca, C. Airoldi, J. Chem. Soc., Dalton Trans. (1999)3687.
354. A. Clearfield, Prog. Inorg. Chem. 47 (1998) 47.
355. M. Ogawa, K. Kuroda, Chem. Rev. 95 (1995) 399.
356. B.G.B. Hix, S.J. Kitchin, K.D.M. Harris, J. Chem. Soc., Dalton Trans. (1998) 2315.
357. U. Costantino, M. Casciola, L. Massinelli, M. Nocchetti, R. Vivani, Solid State Ionics 97 (1997) 203.
358. Y. Murakami, H. Imai, J. Ceram. Soc. Jpn. 100 (1992) 439.
359. A.H. Mahmoudkhani, V. Langer, Solid State Sci. 3 (2001) 519.
360. C.B.A. Lima, C. Airoldi, Solid State Sci. 4 (2002) 1321.
361. G. Alberti, U. Costantino, J. Mol. Catal. 27 (1984) 235.
362. B. Shpeizer, D.M. Poojary, K. Ahn, C.F. Runyan, A. Clearfield, Science 266 (1994) 1357.
363. J.L. Miquel, L. Facchini, A.P. Legrard, C. Rey, J. Lemaitre, Colloids Surf. 45 (1990) 427.
364. E. Jaimez, A. Bortun, G.B. Hix, J.R. Garcia, J. Rodriguez, R.C.T. Slade, J. Chem. Soc., Dalton Trans. (1996) 2285.
365. A. Bortun, E. Jaimez, R. Leavona, J.R. García, J. Rodriguez, Mater. Res. Bull. 30 (1995) 413.
366. A.N. Christensen, E.K. Andersen, I.G.K. Andersen, G. Alberti, M. Nielsen, M.B. Lehmann, Acta Chem. Scand. 44 (1990) 865.
367. G. Alberti, E. Giontella, S.M. Mascarós, Inorg. Chem. 36 (1997) 2844.
368. A. Goni, J. Rius, M. Insausti, L.M. Lezama, J.L. Pizarro, M.I. Arriortua, T. Rojo, Chem. Mater. 8 (1996) 1052.
369. J. Fraissard, J.J Etienne, Bull. Soc. Fr., Minéral. Crystallogr. 90 (1967) 162.
370. U. Costantino, in: A. Clearfield (Ed.), Inorganic Ion Exchange Materials, CRC Press, Boca Raton, FL, 1982.
371. K.G. Bhattachatyya, Langmuir 8 (1992) 2284.
372. A.I. Vogel, A Text-Book of Quantitative Inorganic Analysis: Theory and Practice, fourth ed., Longmans, London, 1978, pp. 320, 756.
373. W.P. Rothwell, J.S. Waugh, J.P. Yesirowski, J. Amer. Chem. Soc. 10 (1980) 2637.
374. S. Carlino, M.J. Hudson, W.J. Locke, J. Mater. Chem. 7 (1997) 813.
375. C. Airoldi, S. Roca, J. Solution Chem. 22 (1993) 707.
376. G. Cao, T.E. Mallouk, Inorg. Chem. 30 (1991) 1434.
377. Y. Zhang, K.J. Scott, A. Clearfield, Chem. Mater. 5 (1993) 495.
378. D.J. Machlan, K.R. Morgan, J. Phys. Chem. Mater. 94 (1990) 7656.
379. A. Clearfield, Prog. Inorg. Chem. 47 (1998) 371.
380. K. Peeters, P. Grobet, E.F. Vansant, J. Mater. Chem. 6 (1996) 239.
381. R. Hoppe, G. Alberti, U. Contantino, C. Dionigi, G.S. Ekloff, R. Vivani, Langmuir 13 (1997) 7252.
382. G. Alberti, in: G. Alberti, T. Bein (Eds.), Comprehensive Supramolecular Chemistry, vol. 7, Pergamon, 1996 (Chapter 5), p. 151.
383. S. Bruque, M.A.G. Aranha, E.R. Losilla, P.O. Pastor, P.M. Torres, Inorg. Chem. 893 (1995) 34.
384. M. Suárez, M.L. Rodríguez, R. Llavona, L.M. Barcina, A. Veja, J. Rodrígues, J. Chem. Soc., Dalton Trans. (1997) 163.
385. T.E. Mallouk, J.A. Gavin, Acc. Chem. Res. 31 (1998) 209.
386. F. Fredoueil, D. Massiot, P. Janvier, F. Gingl, E.M.B. Doeuff, A. Clearfield, B. Bujoli, Inorg. Chem. 38 (1999) 1831.

387. G.B. Hix, V.J. Carter, D.S. Wragg, R.E. Morris, P.A. Wright, J. Mater. Chem. 9 (1999) 179.
388. J.C. Amicangelo, W.R. Leenstra, J. Am. Chem. Soc. 120 (1998) 6181.
389. G. Alberti, U. Costantino, S. Allulli, N. Tomassini, J. Inorg. Nucl. Chem. 40 (1978) 1113.
390. C.B.A. Lima, C. Airoldi, Int. J. Inorg. Mater. 3 (2001) 907.
391. A. Anillo, M.A.V. García, R. Llavona, M. Suárez, J. Rodríguez, Mater. Res. Bull. 34 (1999) 627.
392. K.J. Scott, Y. Zhang, R. Wang, A. Clearfield, Chem. Mater. 7 (1995) 1095.
393. P.A. Williams, M.J. Hudson, Recent Developments in Ion-Exchange, Elsevier, London, 1987.
394. Z. Liu, K. Ooi, H. Kanoh, W. Tang, X. Yang, T. Tomida, Chem. Mater. 13 (2001) 473.
395. J. Morizzi, M. Hobday, C. Rix, J. Mater. Chem. 10 (2000) 1693.
396. D. Cunningham, P.J.D. Hennelly, Inorg. Chim. Acta 37 (1979) 95.
397. L.M. Nunes, C. Airoldi, J. Solid State Chem. 154 (2000) 557.
398. C. Airoldi, D.S. Prandini, Thermochim. Acta 328 (1999) 25.
399. M.S. Whittingham, A. Jacobson, Intercalation Chemistry, Academic Press, New York, 1982.
400. T. Kanazawa, Inorganic Phosphate Materials, Elsevier, Amsterdam, 1989.
401. A. Clearfield, Chem. Rev. 88 (1988) 125.
402. C. Ruan, F. Yang, J. Xu, C. Lei, J. Deng, Electroanalysis 15 (1997) 1180.
403. Y. Ye, R. Baldwin, Anal. Chem. 88 (1988) 125.
404. G. Zaitseva, Y. Gushikem, E.S. Ribeiro, S.S. Rosatto, Electrochim. Acta 47 (2002) 1469.
405. Z. Lu, S. Dong, J. Chem. Soc., Faraday Trans. 84 (1988) 2979.
406. A.R. Guadalupe, K.E. Liu, H.D. Abruña, Electrochim. Acta 36 (1991) 881.
407. S. Clark, R.A. Song, E.F. Bowden, M.J. Tailov, J. Phys. Chem. 97 (1993) 6564.
408. C.P. Audrieux, P. Adebert, B. Divisa-Blohom, S. Linquette-Marllet, J. Electroanal. Chem. 353 (1993) 289.
409. R.W. Murray, A.G. Ewing, R.A. Durst, Anal. Chem. 59 (1987) 379.
410. A. Amine, J.M. Kauffmann, G.G. Guibault, Anal. Lett. 26 (1993) 1281.
411. J. Hu, A.P. Turner, Anal. Lett. 24 (1991) 15.
412. E.I.S. Andreotti, Y. Gushikem, L.T. Kubota, J. Brazil. Chem. Soc. 3 (1992) 21.
413. G. Bremle, B. Person, L. Gorton, Electroanalysis 3 (1991) 77.
414. C.A. Pessôa, Y. Gushikem, L.T. Kubota, Electroanalysis 10 (1997) 800.
415. A.M. Lazarin, C.A. Borgo, Y. Gushikem, Sens. Actuators B 87 (2002) 498.
416. C.B.A. Lima, C. Airoldi, Solid State Sci. 1321 (4) (2002) 1321.
417. H.-N. Kim, S.W. Keller, T.E. Mallouk, J. Schmitt, Chem. Mater. 9 (1997) 1414.
418. C.A. Pessoa, Y. Gushikem, L.T. Kubota, L. Gorton, J. Electroanal. Chem. 431 (1997) 23.
419. L.T. Kubota, L. Gorton, Electroanalysis 11 (1999) 719.
420. L.T. Kubota, F. Munteanu, A. Roddick-Lanzilotta, A.J. MeQuillian, L. Gorton, Quim. Anal. 19 (2000) 15.
421. L.T. Kubota, Y. Gushikem, J. Perez, J. Electroanal. Chem. 509 (1995) 1009.
422. A. Tortenson, L. Gorton, J. Electroanal. Chem. 130 (1981) 199.
423. J. Moiroux, P.J. Elving, Anal. Chem. 50 (1978) 1056.
424. J.M. Anderson, P.W. Flanagan, in: D.C. Coleman, J.M. Oades, G. Uehara (Eds.), Dynamics of Soil Organic Matter in Tropical Ecosystems, Niftal Project, Hawai, 1989, pp. 97–122.
425. A.J. Franzluebbers, R.L. Haney, F.M. Hons, D.A. Zuberer, Soil Biol. Biochem. 28 (1996) 1367.
426. H. Bossuyt, K. Denel, J. Six, S.D. Frey, R. Merckx, K. Paustian, Appl. Soil Ecol. 16 (2001) 195.
427. M.M. Kononova, T.Z. Nowakowski, A.C.D. Newman, Soil Organic Matter, Oxford, Pergamon Press, 1966, pp. 47–108.
428. B.P. Degens, L.A. Schipper, G.P. Sparling, M. Vojvodic-Vukovic, Soil Biol. Biochem. 32 (2000) 189.
429. A.C. Kennedy, Agric. Ecosyst. Environ. 74 (1999) 65.
430. G.T. Hill, N.A. Mitkowski, L. Aldrich-Wolfe, L.R. Emele, D.D. Jurkonie, A. Ficke, S. Maldonado-Ramirez, S.T. Lynch, E.B. Nelson, Appl. Soil Ecol. 15 (2000) 25.
431. A.E. Beezer, Biological Microcalorimetry, Academic Press, London, 1980, p. 311.
432. S.A.M. Critter, S.S. Freitas, C. Airoldi, Appl. Soil Ecol. 18 (2001) 217.

433. R.S. Criddle, A.J. Fontana, D.R. Rank, D. Paige, J.D. Hansen, R.W. Breidenbach, Anal. Biochem. 194 (1991) 413.
434. P. Johansson, I. Wadso, J. Biochem. Biophys. Methods 35 (1997) 103.
435. M. Raubuch, F. Beese, Soil Biol. Biochem. 31 (1999) 949.
436. I. Wadso, Thermochim. Acta 269/270 (1995) 337.
437. N. Barros, S. Feijoó, R. Balsa, Thermochim. Acta 296 (1997) 53.
438. S.A.M. Critter, J.A. Simoni, C. Airoldi, Thermochim. Acta 232 (1994) 145.
439. C. Airoldi, S.A.M. Critter, Thermochim. Acta 288 (1996) 73.
440. N. Barros, S. Feijoó, J.A. Simoni, A.G.S. Prado, F.D. Barboza, C. Airoldi, Thermochim. Acta 328 (1999) 99.
441. A.G.S. Prado, C. Airoldi, Thermochim. Acta 349 (2000) 17.
442. S.A.M. Critter, C. Airoldi, J. Environ. Qual. 30 (2001) 954.
443. L. Yu, X. Hu, R. Lin, H. Zhang, Z. Nan, F. Li, Thermochim. Acta 359 (2000) 95.
444. I. Wadso, N. Markova, Thermochim. Acta 360 (2000) 101.
445. Soil Surfey Staff, Soil Taxonomy: A Basic System of Soil Classification for Making and Interpreting Surveys, Washington, DC, 1975, p. 640.
446. E.K. Triegel, in: L.H. Keith (Ed.), Principles of Environmental Sampling, American Chemical Society, Washington, 1988.
447. C. Airoldi, S.A.M. Critter, Clays Clay Miner. 45 (1997) 125.
448. S.A.M. Critter, C. Airoldi, Geoderma 111 (2003) 57.
449. J. Tuite, Plant Pathological Methods, Burgess Publishing Company, Minneapolis, 1969, p. 239.
450. A.J. Salle, Laboratory Manual on Fundamental Principles of Bacteriology, McGraw Hill, New York, 1973, p. 201.
451. P. Bäckman, M. Bastos, L.E. Briggner, S. Hägg, D. Hallén, P. Lönnbro, S.O. Nilsson, G. Olofsson, A. Schön, J. Suurkuusk, C. Teixeira, I. Wadso, Pure Appl. Chem. 66 (1994) 375.
452. A.G.S. Prado, C. Airoldi, Pest Manage. Sci. 56 (2000) 419.
453. A.G.S. Prado, C. Airoldi, J. Colloid Interface Sci. 236 (2001) 161.
454. A.A. El Nasser, R.V. Parish, J. Chem. Soc., Dalton Trans. (1999)3463.
455. S. Nir, T. Undabeytia, D. Yaron-Marcovitch, Y. El-Nahhal, T. Polubesova, C. Serban, G. Rytwo, G. Lagaly, B. Rubin, Environ. Sci. Technol. 34 (2000) 1269.
456. P.A. Wight, M.E. Davis, Chem. Rev. 102 (2002) 3589.
457. H.E. Fisher, A.S. King, J.B. Miller, J.Y. Ying, J.B. Benzinger, Inorg. Chem. 30 (1991) 4403.
458. Y. Kurusu, J. Macromol. Sci. Chem. 27 (1990) 1389.
459. T.N. Shekhovtsova, S.V. Chernetskaya, I.F. Dolmonova, J. Anal. Chem. 48 (1993) 94.
460. G. Schwedt, D.O. Waldheim, K.D. Neumann, K. Stein, Fresenius J. Anal. Chem. 346 (1993) 659.
461. M.G. Fonseca, C. Airoldi, Thermochim. Acta 6258 (2000) 1.
462. C.R. Silva, I.C.S.F. Jardim, C. Airoldi, J. High Resolut. Chromatogr. 22 (1999) 103.
463. C.J. Brinker, G.W. Scherer, Sol–Gel Science – The Physics and Chemistry of Sol–Gel Processing, Academic Press, New York, 1990.
464. Y. Fukushima, M. Tami, Chem. Commun. 24 (1995).
465. Y. Fukushima, M. Tami, Bull. Chem. Soc. Jpn. 69 (1996) 3667.
466. S.L. Burkett, A. Press, S. Mann, Chem. Mater. 9 (1997) 1071.
467. L. Ukrainczyk, J. Phys. Chem. B 101 (1997) 531.
468. Y.-S. Hong, S.-J. Kim, Bull. Korean Chem. Soc. 18 (1997) 2.
469. W. Whilton, S.L. Burkett, S. Mann, J. Mater. Chem. 8 (1998) 1927.
470. M.G. Fonseca, C.R. Silva, C. Airoldi, Langmuir 15 (1999) 5048.
471. M.G. Fonseca, C.R. Silva, J. Barone, C. Airoldi, J. Mater. Chem. 3 (2000) 789.
472. M.G. Fonseca, C. Airoldi, J. Mater. Chem. 10 (2000) 1457.
473. T.P. Lishko, L.V. Glushchenko, Y.V. Kholin, Z.N. Zaitev, A. Bugaevskii, N.D. Donskaya, Russ. J. Phys. Chem. 65 (1991) 1584.
474. I. Taylor, A.G. Howard, Anal. Chim. Acta 271 (1993) 77.
475. D.M. Moore, R.C. Reynolds Jr., X-Ray Diffraction and Identification and Analysis of Clay Minerals, second ed., Oxford University Press, Oxford, 1997.

476. K. Nakamoto, Infrared Spectra of Inorganic and Coordination Compounds, fourth ed., Wiley, New York, 1986.
477. R.M. Silverstein, G.C. Basser, T.C. Morrill, Spectrometric Identification of Organic Compounds, second ed., Wiley, New York, 1991.
478. D.L. Pavia, G.M. Lampman, G.S. Kriz, Introduction to Spectroscopy, second ed., Saunders College Publishing, New York, 1996.
479. R.W.T. Wilkins, J. Ito, Am. Min. 1649 (1967) 52.
480. J. Xiao, G. Villemure, Clays Clay Miner. 48 (1998) 2.
481. D.W. Sinford, G.E. Maciel, J. Am. Chem. Soc. 105 (1983) 3767.
482. J.J. Jang, I.M. El-Nahhal, G.E. Maciel, J. Non-Cryst. Solids 204 (1996) 205.
483. J.J. Jang, I.M. El-Nahhal, I.E. Chuang, G.E. Maciel, J. Non-Cryst. Solids 209 (1997) 19.
484. B. Meunier, Chem. Rev. 92 (1992) 1411.
485. D. Mansuy, Coord. Chem. Rev. 125 (1993) 129.
486. C.C.Y. Wang, D.M. Ho, J.T. Groves, J. Am. Chem. Soc. 118 (1990) 12094.
487. J. Yang, R. Breslow, Angew. Chem. Int. Ed. 39 (2000) 2692.
488. J.T. Groves, J. Porphyrins Phthalocyanines 4 (2000) 350.
489. J.L. McLain, J. Lee, J.T. Groves, B. Meunier, Biomimetic Oxidation Catalysed by Transition Metal Complexes, Imperial College Press, London, 2000 (Chapter 3).
490. B. Meunier, A. Robert, G. Pratviel, J. Bernadou, The Porphyrin Handbook, vol. 4, Academic Press, San Diego, 1999, p. 119.
491. R.A. Sheldon, B. Meunier, Biomimetic Oxidations Catalysed by Transition Metal Complexes, Imperial College Press, London, 2000 (Chapter 14).
492. S. Nakagaki, A.M. Machado, F. Wypych, S.M. Drechsel, J. Colloid Interface Sci. 254 (2002) 158.
493. F. Bedioui, Coord. Chem. Rev. 144 (1995) 39.
494. L. Barloy, J.P. Lallier, P. Battioni, D. Mansuy, Y. Piffard, M. Tournous, J.B. Valim, W. Jones, New J. Chem. 16 (1992) 71.
495. M.A. Schiavon, Y. Iamamoto, O.R. Nacsimento, M.D. Assis, J. Mol. Catal. A: Chem. 174 (2001) 213.
496. C. Gilmartin, J.R.L. Smith, J. Chem. Soc., Perkin Trans. 2 (1995) 243.
497. M.D. Assis, J.R.L. Smith, J. Chem. Soc., Perkin Trans. 2 (1998) 2221.
498. F.G. Doro, J.R. Lindsay Smith, A.G. Ferreira, M.D. Assis, J. Mol. Catal. A: Chem. 164 (2000) 97.
499. S. Evans, J.R. Lindsay Smith, J. Chem. Soc., Perkin Trans. 2 (2001) 174.
500. S.S. Candy, T.J. Pinnavaia, Inorg. Chem. 17 (1978) 1501.
501. T.J. Pinnavaia, Science 220 (1983) 4595.
502. M.G. Fonseca, C. Airoldi, Mater. Res. Bull. 36 (2001) 277.
503. M.A. Martinez-Lorente, P. Battioni, W. Kleemiss, J.F. Bartoli, D. Mansuy, J. Mol. Catal. A: Chem. 113 (1996) 343.
504. L. Barloy, P. Battioni, D. Mansuy, J. Chem. Soc., Chem. Commun. (1990) 1365.
505. J.G. Sharefkin, H. Saltzman, Org. Synth. 5 (1963) 660.
506. J.G. Sharefkin, H. Saltzman, Org. Synth. 5 (1963) 658.
507. H. Lucas, E.R. Kennedy, J. Org. Chem. 3 (1955) 484.
508. A.D. Adler, F.R. Longo, F. Kampas, J. Kim, J. Inorg. Nucl. Chem. 32 (1970) 2443.
509. S. Evans, J.R. Lindsay Smith, J. Chem. Soc., Perkin Trans. 2 (2000) 1541.
510. F.P. Guengerich, Chem. Res. Toxicol. 14 (2001) 611.
511. J.T. Groves, Y.Z. Han, P.R. Montellano, Cytochrome Structure, Mechanism and Biochemistry, Plenum Press, New York, 1995 (Chapter 1).
512. P.R. Crook, C. Gilmartin, G.W. Gray, J.R. Lindsay Smith, J. Chem. Soc., Perkin Trans. 2 (1995) 1573.
513. R.M. Silverstein, G.C. Bassler, T.C. Morrel, Spectrometric Identification of Organic Compounds, fifth ed., Wiley, New York, 1991.
514. H. Fujii, Coord. Chem. Rev. 226 (2002) 51.
515. T. Murakami, K. Yamaguchi, Y. Watanabe, I. Morishima, Chem. Soc. Jpn. 71 (1998) 1343.
516. G.W. Brindely, G. Brown, Crystal Structure of Clays. Minerals and their X-ray Identification, Mineralogical Society, London, 1998.

517. M. Wesolowski, Thermochim. Acta 78 (1984) 395.
518. P.R. Cooke, J.R.L. Smith, Tetrahedron Lett. 33 (1992) 2737.
519. D. Mont, A. Pastorini, G. Mancini, S. Borocci, P. Tagliatesta, J. Mol. Catal. A: Chem. 179 (2002) 125.
520. S.S. Kurek, P. Michorczyk, A.M. Balisz, J. Mol. Catal. A: Chem. 38822 (2002) 1.
521. Y.Z. Li, J.Y. Liu, X.F. Li, W.B. Chang, A.J. Huang, J. Mol. Catal. A: Chem. 187 (2002) 163.
522. J.T. Groves, T.E. Nemo, J. Am. Chem. Soc. 105 (1983) 6243.
523. J.P. Collmman, T. Kodadek, S.A. Raybuck, J.I. Brauman, L.M. Papazian, J. Am. Chem. Soc. 107 (1985) 4343.
524. F.L. Benedito, S. Nakagaki, A.A. Sack, P.G. Peralta-Zamora, C.M.M. Costa, Appl. Catal. A Gen. 250 (2003) 1.
525. W. Nam, M.H. Lim, W. Shin, Angew. Chem. Int. Ed. 20 (2000) 39.
526. J.T. Groves, Z. Gross, M.K. Stern, Inorg. Chem. 33 (1994) 5065.
527. D.A. Skoog, D.M. West, F.J. Holler, Fundamentals of Analytical Chemistry, Saunders College Publishing, Orlando, 1996, p. 370.
528. H.I. Abdel-Shafy, J.M. El-Gamal, M.F. Abdel-Sabour, O. Abo-El-Wafa, Environ. Prot. Eng. 15 (1989) 63.
529. D.A. Blake, R.C. Blake, M. Khosraviani, A.R. Pavlov, Anal. Chim. Acta 376 (1998) 13.
530. R.E. Clement, G.A. Eiceman, C. Koester, Anal. Chem. 67 (1995) 21R.
531. V.K. Gupta, C.K. Jain, I. Ali, M. Sharma, V.K. Saini, Water Res. 37 (2003) 4038.
532. S. Bailey, T. Olin, M. Bricka, D. Adrian, Water Res. 33 (1999) 2469.
533. B.E. Reed, W. Lin, M.R. Matsumoto, J.N. Jensen, Water Environ. Res. 69 (1997) 444.
534. B.B. Saha, M. Iglesias, I.W. Cumming, M. Streat, Solvent Extr. Ion. Exch. 18 (2000) 133.
535. A.C.S. Costa, L. Lopes, M.A. Korn, J.G. Portela, J. Braz. Chem. Soc. 13 (2002) 674.
536. G.N. Manju, A. Anoop Krishnan, V.P. Vinod, T.S. Anirudhan, J. Hazard. Mater. B 91 (2002) 221.
537. K.N. Ghimire, K. Inoue, K. Makino, T. Miyajima, Sep. Sci. Technol. 37 (2002) 2785.
538. G. Bereket, A.Z. Aroğus, M.Z. Özel, J. Colloid Interface Sci. 187 (1997) 338.
539. L. Dupont, E. Guillon, Environ. Sci. Technol. 37 (2003) 4235.
540. B.M. Novak, Adv. Mater. 6 (1993) 422.
541. G. Schottner, Chem. Mater. 13 (2001) 3422.
542. L.L. Hench, J.K. West, Chem. Rev. 90 (1990) 33.
543. M.M. Collison, Trends Anal. Chem. 21 (2002) 1.
544. S. Bibmam, J. Medlar, A. Kabir, C. Shende, A. Ali, A. Malik, Anal. Chem. 74 (2002) 752.
545. M. Cichna, D. Knopp, R. Niessner, Anal. Chim. Acta 339 (1997) 241.
546. K.D. Behringer, J. Blumel, Inorg. Chem. 35 (1996) 1814.
547. E.M. Soliman, Anal. Lett. 30 (1997) 1739.
548. H.-J. Im, Y. Yang, L.R. Allain, C.E. Barnes, S. Dai, Z. Xue, Environ. Sci. Technol. 34 (2000) 2209.
549. I.M. El-Nahhal, B.A. El-Shetary, K.A.R. Salib, N.M. El-Ashgar, A.M. El-Hashash, Anal. Lett. 34 (2001) 2189.
550. A.R. Cestari, E.F.S. Vieira, C. Airoldi, J.A. Simoni, Thermochim. Acta 348 (2000) 25.
551. F.A. Pavan, W.F. de Magalhães, M.A. de Luca, C.C. Moro, T.M.H. Costa, E.V. Benvenutti, J. Non-Cryst. Solids 311 (2002) 54.
552. F.A. Pavan, Y. Gushikem, C.C. Moro, T.M.H.S. Costa, E.V. Benvenutti, Colloid Polym. Sci. 283 (2003) 173.
553. E.P. Barret, L.G. Joyner, P.P. Halenda, J. Am. Chem. Soc. 73 (1951) 373.
554. H.A. Flaschka, EDTA Tritation, An Introduction to Theory and Practice, second ed., Pergamon, Oxford, 1967.
555. M.E. Mahmound, Anal. Lett. 35 (2002) 1251.
556. M. Ajmal, R.A.K. Rao, R. Ahmad, J. Ahmad, J. Hazard. Mater. 79 (2000) 117.
557. R.J.P. Corriu, Eur. J. Inorg. Chem. (2001) 1109.
558. K.A. Shujaat, S. Arif raza Zaidi, F. Mahmood, M.Y. Khokhar, Sep. Sci. Technol. 37 (2002) 3099.
559. C. Airoldi, R.F. de Farias, Quim. Nova 23 (2000) 496.

560. Jr. Strelko, D.J. Malik, J. Colloid Interface Sci. 250 (2002) 213.
561. O.A.C. Monteiro Jr., C. Airoldi, J. Colloid Interface Sci. 212 (1999) 212.
562. L. Mercier, C. Detellier, Environ. Sci. Technol. 29 (1995) 1318.
563. L.M. Nunes, C. Airoldi, Thermochim. Acta 328 (1999) 295.
564. C. Airoldi, L.N.H. Arakaki, J. Colloid Interface Sci. 249 (2002) 290.
565. L.N.H. Arakaki, A.N. de Sousa, J.G.P. Espínola, S.F. de Oliveira, C. Airoldi, J. Colloid Interface Sci. 249 (2002) 290.
566. L.N.H. Arakaki, A.N. de Sousa, J.G.P. Espínola, S.F. de Oliveira, J.M.P. de Freitas, A.G. de Souza, C. Airoldi, Colloids Surf. A 203 (2002) 129.
567. G.S. Prado, L.N.H. Arakaki, C. Airoldi, Green Chem. 4 (2002) 42.
568. S.F. de Oliveira, J.G.P. Espínola, W.E.S. Lemus, A.G. de Souza, C. Airoldi, Colloids Surf. A 136 (1998) 151.
569. J.G.P. Espínola, J.M. de Freitas, S.F. de Oliveira, C. Airoldi, Colloids Surf. A 87 (1994) 33.
570. R.K. Iler, The Chemistry of Silica, Wiley, New York, 1979.
571. H.M. Iamail, S.A. Monsour, M.I. Azaki, Thermochim. Acta 202 (1992) 269.
572. T. Takei, E. Eriguchi, M. Fuji, T. Watanabe, M. Chikazawa, Thermochim. Acta 308 (1998) 139.
573. C. Airoldi, L.S. Santos Jr., Thermochim. Acta 104 (1986) 111.
574. M. Ito, K. Yamauchi, K. Matsuzawa, Colloids Surf. A 74 (1993) 107.
575. K. Hamdani, K.L. Cheng, Colloids Surf. A 63 (1992) 29.
576. A. Trujillo, H. Fryi, Chem. Anal. 56 (1997) 90.
577. L.T. Kubota, Y. Gushiken, S. Castro, J.C. Moreira, Colloids Surf. 57 (1991) 11.
578. M. Björklund, W.M.T. Hearn, J. Chromatogr. A 728 (1996) 149.
579. C.E. Lin, F.K. Li, C.H. Lin, J. Chromatogr. A 722 (1996) 211.
580. M. Mifune, Y. Shimomura, Y. Saito, Y. Mori, M. Onoda, A. Iwado, N. Motohashi, J. Haginaka, Bull. Chem. Soc. Jpn. 71 (1998) 1825.
581. A.J. Butterworth, J.H. Clark, P.H. Walton, S.J. Barlow, Chem. Commun. (1996) 1859.
582. J.P. Collman, J.A. Belmont, J.I. Brauman, J. Am. Chem. Soc. 105 (1983) 7288.
583. K. Wilson, J.H. Clark, Chem. Commun. (1998) 2135.
584. L.H. Wang, Z.S. Chen, Electroanalysis 16 (1997) 1294.
585. J.L. Fang, Y. Wu, K.P. Han, Plat. Surf. Finish. 84 (1997) 91.
586. J.M. Parmar, J.J. Modha, A.R. Parikh, Indian J. Chem., Sect. B: Org. Chem. Incl. Med. Chem. 3304 (1998) 440.
587. D. Panias, M. Taxiarchou, I. Paspaliaris, R.A. Kontopoulos, Hydrometallurgy 42 (1996) 257.
588. A.B. Zarcinas, Commun. Soil Sci. Plant Anal. 26 (1995) 713.
589. U. Bartels, Fresenius Z. Anal. Chem. 320 (1985) 56.
590. R.A. Hummel, E.B. Sandell, Anal. Chim. Acta 7 (1952) 308.
591. A. Krata, K. Pyrzynska, E. Bulska, Anal. Bioanal. Chem. 377(4) (2000) 735.
592. L.S. Liebeskind, J. Srogl, C. Savarin, C. Polanco, Pure Appl. Chem. 74 (2002) 115.
593. S.K. Dutta, D.B. MacConville, W.J. Yiangs, P. Antonovichi, A.B. Blank, Inorg. Chem. 36 (1997) 2517.
594. V.M. Bhuchar, Nature 191 (1961) 489.
595. L.N.H. Arakaki, J.G.P. Espínola, S.F. Oliveira, Curr. Top. Colloid Interface Sci. 4 (2001) 219.
596. E.M. Soliman, M.E. Mahmeud, S.A. Ahmed, Int. J. Environ. Anal. Chem. 82 (2002) 403.
597. S. Brunauer, P. Emett, E. Teller, J. Am. Chem. Soc. 60 (1938) 309.
598. R.G. Pearson, J. Chem. Educ. 45 (1968) 581.
599. R.G. Pearson, Inorg. Chem. 27 (1988) 734.
600. C. Airoldi, E.F.C. Alcântara, Colloids Surf. 39 (1989) 291.
601. M. Taramasso, G. Perego, B. Notari, US Patent 4,410,501 (1983).
602. X. Liu, J.K. Thomas, Chem. Commun. (1996) 1435.
603. S.M. Kuznicki, US Patent 4,853,202 (1989).
604. S.M. Kuznicki, K.A. Thrush, European Patent 0,405,978-A1 (1990).
605. M.W. Anderson, O. Terasaki, T. Oshuna, P.J. O'Malley, A. Philippou, S.P. Mackay, A. Ferreira, J. Rocha, S. Lidin, Philos. Mag. B 71 (1995) 813.
606. A. Philippou, M. Naderi, N. Pervaiz, J. Rocha, M.W. Anderson, J. Catal. 178 (1998) 174.

607. X. Yang, R.E. Truitt, J. Phys. Chem. 100 (1996) 3713.
608. J. Rocha, A. Ferreira, Z. Lin, M.W. Anderson, Microporous Mesoporous Mater. 23 (1998) 253.
609. X. Yang, P.W. Blosser, Zeolites 17 (1996) 237.
610. S. Ganapathy, T.Kr. Das, R. Vetrivel, S.S. Ray, T. Sen, S. Sivasanker, L. Delevoye, C. Fernandez, J.P. Amoreux, J. Am. Chem. Soc. 120 (1998) 4752.
611. T.Kr. Das, A.J. Chandwadkar, S. Sivasanker, Chem. Commun. (1996) 1105.
612. C.C. Lee, R.J. Gorte, W.E. Farneth, J. Phys. Chem. B 101 (1997) 3811.
613. S.G. Izmailova, I.V. Karetina, S.S. Khvoshchev, M.A. Shubaeva, J. Colloid Interface Sci. 165 (1994) 318.
614. L.A. Doronina, S.G. Izmailova, I.V. Karetina, N.N. Feoktistova, S.S. Khvoshchev, Russ. Chem. Bull. 44 (1995) 1857.
615. D.J. Parrillo, R.J. Gorte, W.E. Farneth, J. Am. Chem. Soc. 115 (1993) 12441.
616. A.T. Aguayo, A.G. Gayubo, J. Ereña, M. Olazar, J.M. Arandes, J. Bilbao, J. Chem. Technol. Biotechnol. 60 (1994) 141.
617. S. Savitz, A.L. Myers, R.J. Gorte, J. Phys. Chem. B 103 (1999) 3687.
618. Y. Israëli, C. Taviot-Guého, J. Besse, J. Morel, N. Desrosiers, J. Chem. Soc., Dalton Trans. (2000) 791.
619. B. Lee, Y. Kim, H. Lee, J. Yi, Microporous Mesoporous Mater. 50 (2001) 77.
620. S. Samal, S. Acharya, R.K. Dey, A.R. Ray, Talanta 57 (2002) 1075.
621. A.A. Mozeto, W.F. Jardim, Quim. Nova 25 (2002) 7.
622. J.S. Kim, J. Yi, Sep. Sci. Technol. 34 (1999) 2957.
623. J.S. Kim, J. Yi, J. Chem. Technol. Biotechnol. 74 (1999) 544.
624. M.A. Zolfigol, T. Madrakian, E. Ghaemi, A. Afkhami, S. Aziziam, S. Afshar, Green Chem. 4 (2002) 611.
625. O.A. Zaporozhets, T.E. Keda, L.E. Seletskaya, V.V. Sukhan, J. Anal. Chem. 55 (2000) 635.
626. O.A. Zaporozhets, L.S. Ivan_ko, I.V. Marchenko, V.V. Sukhan, J. Anal. Chem. 55 (2000) 640.
627. R.E. Morrid, E.L. Chang, Petrol. Sci. Technol. 18 (2000) 1147.
628. C. Borgo, Y. Gushikem, J. Colloid. Interface Sci. 246 (2002) 343.
629. L. Mercier, T.J. Pinnavaia, Microporous Mesoporous Mater. 20 (1998) 101.
630. J. Brown, L. Mercier, T.J. Pinnavaia, Chem. Commun. (1999) 69.
631. A.M. Lazarin, Y. Gushikem, J. Braz. Chem. Soc. 13 (2002) 88.
632. A.G.S. Prado, C. Airoldi, Anal. Chim. Acta 432 (2001) 401.
633. A. Seifert, S. Spange, H. Muller, S. Hesse, C. Jager, J. Sol-Gel Sci. Technol. 26 (2003) 77.
634. S. Spange, A. Seifert, H. Muller, S. Hesse, C. Jager, Angew. Chem., Int. Ed. 41 (2002) 1729.
635. O.A. Zaporozhets, L.Y. Tsyukalo, Talanta 58 (2002) 861.
636. S.T. Fujiwara, C.A. Pessoa, Y. Gushikem, Anal. Lett. 35 (2002) 1117.
637. L.N.H. Arakaki, A.N. Sousa, J.G.P. Espinola, S.F. Oliveira, C. Airoldi, J. Colloid Interface Sci. 249 (2002) 290.
638. J.A.A. Sales, A.G. Souza, J.G.P. Espinola, E.E.S. Teotonio, C.D. Pinheiro, An. Assoc. Bras. Quim. 47 (1998) 342.
639. A.G.S. Prado, J.A.A. Sales, C. Airoldi, J. Therm. Anal. Calorim. 70 (2002) 191.
640. H.J. Hoorn, P. Joode, W.L. Driessen, J. Reedjk, Recl. Trav. Chim. Pays-Bas 115 (1996) 191.
641. K. Kosuge, P.S. Singh, Chem. Mater. 13 (2001) 2476.
642. J.A.A. Sales et al., Polyhedron 23 (2004) 719.
643. M. Przybyciel, R.E. Majors, LC GC 20 (2002) 516.
644. R.E. Majors, LC GC 21 (2003) 240.
645. A. Nomura, J. Yamada, K. Tsunoda, Anal. Sci. 3 (1987) 209.
646. T.L. Ascah, B. Feibush, J. Chromatogr. 506 (1990) 357.
647. B. Feibush, US Patent 5,137,627-A (1990).
648. B. Buszewski, J. Schmid, K. Albert, E. Bayer, J. Chromatogr. 552 (1991) 415.
649. J. Schmid, K. Albert, E. Bayer, J. Chromatogr. A 694 (1995) 333.
650. B. Buszewski, M. Jaroniec, R.K. Gilpin, J. Chromatogr. A 668 (1994) 293.
651. T.L. Ascah, K.M.L. Kallury, C.A. Szafranski, S.D. Corman, F. Lui, J. Liq. Chromatogr. Relat. Technol. 19 (1996) 3049.

652. J.E. O'Gara, B.A. Alden, T.H. Walter, J.S. Petersen, C.L. Niederländer, U.D. Neue, Anal. Chem. 67 (1995) 3809.
653. U.D. Neue, C.L. Niederländer, J.S. Petersen, US Patent 5,374,755-A (1994).
654. J.E. O'Gara, D.P. Walsh, B.A. Alden, P. Casellini, T.H. Walter, Anal. Chem. 71 (1999) 2992.
655. C.R. Silva, C. Airoldi, Brazilian Patent PI 9, 903, 110-8 (1999).
656. C.R. Silva, I.C.S.F. Jardim, C. Airoldi, J. Chromatogr. A 913 (2001) 65.
657. E. Cruz, M.R. Euerby, C.M. Johnson, C.A. Hackett, Chromatographia 44 (1997) 151.
658. H. Engelhardt, R. Grüner, M. Scherer, Chromatographia 5 (2001) S154.
659. G. Sacchero, S. Apone, C. Sarzanini, E. Mentasti, J. Chromatogr. A 668 (1994) 365.
660. H.A. Claessens, M.A. van Straten, J.J. Kirkland, J. Chromatogr. A 797 (1998) 111.
661. D.A. Wardle, A. Ghani, Soil Biol. Biochem. 27 (1994) 821.
662. M. Alexander, Science 211 (1981) 132.
663. B.P. Degens, L.A.L. Schipper, G.P. Sparling, Soil Biol. Biochem. 32 (1999) 189.
664. S. Aikio, H. Vare, R. Strommer, Soil Biol. Biochem. 32 (2000) 1091.
665. S. Dumontet, A. Mazzatura, C. Casucci, P. Perucci, Biol. Fertil. Soils 34 (2001) 411.
666. C.P. Kushwaha, S.K. Tripathi, K.P. Singh, Appl. Soil Ecol. 16 (2001) 229.
667. B.L. Turner, A.W. Bristow, P.M. Haygarth, Soil Biol. Biochem. 33 (2001) 913.
668. E.G. Gregorich, B.C. Liang, C.F. Drury, A.F. Mackenzie, W.B. McGill, Soil Biol. Biochem. 32 (2000) 581.
669. D.A. Klein, M.W. Paschke, Appl. Soil Biol. 14 (2000) 257.
670. B.J. Bridge, A.J. Rixon, J. Soil Sci. 27 (1976) 279.
671. E.A. Kaiser, T. Mueller, R.G. Joergensen, H. Insam, O. Heinemeyer, Soil Biol. Biochem. 24 (1992) 675.
672. G.P. Sparling, C.W. Feltham, J. Reynolds, A.W. West, P. Singleton, Soil Biol. Biochem. 22 (1990) 301.
673. J.M. Wolf, A.H. Brown, D.R. Goddart, Plant Physiol. 27 (1952) 70.
674. A.A. Rodella, L.V. Saboya, Soil Biol. Biochem. 31 (1999) 2059.
675. G.P. Sparling, Soil Biol. Biochem. 13 (1981) 93.
676. K. Ljungholm, B. Noren, R. Skold, I. Wadso, Oikos 33 (1979) 15.
677. M.I. Barja, J. Proupin, L. Núnez, Thermochim. Acta 303 (1997) 155.
678. A.E. Beezer, Biological Microcalorimetry, Academic Press, London, 1980.
679. Soil Surfey Staff, Soil Taxonomy: A Basic System of Soil Classification for Making and Interpreting Surveys, Washington, DC, 1975.
680. W.J. Price, Spectrochemical Analysis by Atomic Absorption, Wiley, New York, 1979.
681. A.I. Vogel, A Textbook of Quantitative Inorganic Chemistry Including Instrumental Analysis, Longman, London, 1978.
682. K. Ljungholm, B. Noren, B.I. Wadso, Oikos 33 (1979) 24.
683. L. Gustafsson, Thermochim. Acta 193 (1991) 145.
684. M.A. Hammer, Q. Sun, Appl. Catal. A 221 (2001) 45.
685. K. Tanabe, W.F. Hoelderich, Appl. Catal. A 181 (1999) 434.
686. Y. Mori, T.J. Pinnavaia, Chem. Mater. 13 (2001) 2173.
687. W.M. Van Rhijin, D.E. De Vos, B.F. Sels, W.D. Bossaert, P.A. Jacob, Chem. Commun. (1998) 317.
688. M.A. Harmer, W.E. Farneth, Q. Sun, Adv. Mater. 15 (1998) 1255.
689. D. Margolese, J.A. Melero, S.C. Christiansen, B.F. Chmelka, G.D. Stucky, Chem. Mater. 12 (2000) 2448.
690. A.G.S. Prado, C. Airoldi, Green Chem. 4 (2002) 288.
691. S.T. Fujiwara, Y. Gushikem, J. Braz. Chem. Soc. 10 (1999) 389.
692. E.S. Ribeiro, Y. Gushikem, Electrochim. Acta 44 (1999) 3589.
693. H.G.M. Edwards, A.F. Johnson, E.E. Lawson, J. Molec. Struct. 351 (1995) 51.
694. E.E. Chuf_an, J.C. Pedregosa, J. Borras, Vib. Spectrosc. 15 (1997) 191.
695. P.D. Padilha, L.A.D. Gomes, C.C.F. Padilha, J.C. Moreira, N.L. Dias, Anal. Lett. 32 (1999) 1807.
696. L.A.D. Gomes, P.D. Padilha, J.C. Moreira, N.L. Dias, Y. Gushikem, J. Braz. Chem. Soc. 9 (1998) 494.

697. N.L. Dias, J. Colloid Interface Sci. 206 (1998) 131.
698. E.R. Kenawy, D.C. Sherrington, Eur. Polym. J. 28 (1992) 841.
699. A. Akelah, Mater. Sci. Eng. C 4 (1996) 83.
700. C. Airoldi, A.G.S. Prado, Thermochim. Acta 394 (2002) 163.
701. E.M. Soliman, M.E. Mohmoud, Talanta 54 (2001) 243.
702. J. Brown, R. Richer, T.J. Pinnavaia, Chem. Commun. (1999) 69.
703. A.G.S. Prado, C. Airoldi, J. Environ. Monit. 3 (2001) 394.
704. P.M. Padilha, L.A.M. Gomes, C.C.F. Padilha, J.C. Moreira, N.L. Dias Filho, Anal. Lett. 32 (1999) 1807.
705. E.M. Soliman, M.E. Mohmoud, S.A. Ahmed, Talanta 54 (1997) 2681.
706. A.G.S. Prado, C. Airoldi, Green Chem. 4 (2002) 488.
707. P.S. Roldan, I.L. Alcantara, C.R. Castro, J.C. Rocha, C.C.F. Padilha, P.M. Padilha, Anal. Bioanal. Chem. 375 (2003) 574.
708. P.M. Padilha, C.C.F. Padilha, J.C. Rocha, Quim. Anal. 18 (1999) 299.
709. N.L. Dias Filho, Mikrochim. Acta 130 (1999) 233.
710. Y. Okuno, T. Yokoyama, S. Yokoyama, S. Kamikado, S. Mashiko, J. Am. Chem. Soc. 124 (2002) 7218.
711. W. Rudzinski, R. Charmas, W. Piasecki, F. Thomas, F. Vllieras, B. Prelot, J.M. Cases, Langmuir 14 (1998) 5210.
712. F.S.C. Anjos, E.F.S. Vieira, A.R. Cestari, J. Colloid Interface Sci. 253 (2002) 243.
713. P. Trivedi, L. Axe, Environ. Sci. Technol. 34 (2000) 2215.
714. R. Lehrmann, J. Seelig, Biochim. Biophys. Acta 1189 (1994) 89.
715. M.N. Dimitrova, H. Matsumura, A. Dimitrova, V.Z. Neitchev, Int. J. Biol. Macromol. 27 (2000) 187.
716. W. Rudzinski, R. Charmas, W. Piasecki, J.M. Cases, M. Francois, F. Villieras, L.J. Michot, Colloids Surf. A 137 (1998) 57.
717. R.G. Pearson, J. Chem. Educ. 64 (1987) 561.
718. T. Schilling, P. Schramel, B. Michalke, G. Knapp, Mikrochim. Acta 116 (1994) 83.
719. W. Szczepaniak, A. Szymanski, Chem. Anal. 41 (1996) 193.
720. P. Tundo, P. Anastas, D.S. Black, J. Breen, T. Collins, S. Memoli, J. Miyamoto, M. Polyakoff, W. Tumas, Pure Appl. Chem. 72 (2000) 1207.
721. P.D. Verweij, M.J. Haanepen, J.J. de Ridder, W.L. Driessen, J. Reedijk, Recl. Trav. Chim. Pays-Bas 11 (1992) 371.
722. D.D. Perrin, W.L.F. Armarego, D.R. Perrin, Purification of Laboratory Chemicals, Pergamon, Oxford, 1988.
723. J.A.A. Sales, C. Airoldi, J. Non-Cryst. Solids 330 (2003) 142–149
724. M.R.M.C. Santos, C. Airoldi, J. Colloid Interface Sci. 183 (1996) 416.
725. C. Airoldi, Y. Gushikem, J.G.P. Espinola, Colloids Surf. A 17 (1986) 317.
726. D.E. Leyden, W.T. Collins, Silylated Surfaces, Gordon and Breach, New York, 1980.
727. M.G. Fonseca, J.G.P. Espínola, S.F. Oliveira, L.C. Ramos, A.G. Souza, C. Airoldi, Colloids Surf. A 133 (1998) 205.
728. M.G. Fonseca, C.R. Silva, C. Airoldi, Langmuir 15 (1999) 48.
729. C. Airoldi, E.F.C.J. Alcântara, J. Chem. Thermodyn. 27 (1995) 623.
730. C. Airoldi, E.F.C. Alcântara, Thermochim. Acta 259 (1995) 95.
731. A.R. Cestari, C. Airoldi, Langmuir 13 (1997) 2681.
732. M.G. Fonseca, C. Airoldi, J. Mater. Chem. 9 (1999) 1375.
733. U. Johansson, A. Holmgren, W. Forsling, R.L. Forst, Clays Miner. 34 (1999) 239.
734. H. Brunner, T. Vallant, U. Mayer, H. Hoffmann, Langmuir 15 (1999) 1899.
735. F. Tian, X. Xiao, M.M.T. Toy, Langmuir 15 (1999) 244.
736. J. Wang, J. Merino, P. Aranda, J.-C. Galván, E. Ruiz-Hitzky, J. Mater. Chem. 9 (1999) 161.
737. V. Porta, C. Sarzanini, O. Abolino, E. Mentasti, E.J. Cartini, Anal. Spectrom. 7 (1992) 19.
738. K.M.R. Kallury, W.E. Lee, M. Thompson, Anal. Chem. 65 (1993) 2459.
739. J.M. Fraire, J.I. Garcia, J.A. Mayoral, E. Vispe, J. Catal. 204 (2001) 1090.
740. T.M. Roshchina, N.K. Shonia, A.A. Kazmina, K.B. Gurevich, A.Y. Fadeev, J. Chromatogr. A 931 (2001) 119.

741. B. Buszewski, M. Jezierska, M. Welmak, D. Berek, J. High Resol. Chromatogr. 21 (1998) 267.
742. C.R.M. Peixoto, L.T. Kubota, Y. Gushikem, Anal. Proc. 32 (1995) 503.
743. K.K. Unger, Porous Silica, Elsevier, Amsterdam, 1979.
744. C.J. Brinker, G.W. Scherer, Sol–Gel Science, Academic Press, San Diego, 1990.
745. B.R. Guidotti, E. Herzog, F. Bangetter, W.R. Caseri, U.W. Suter, J. Colloid Interface Sci. 191 (1997) 209.
746. H. Suquet, Clays Clay Miner. 37 (1989) 439.
747. E. Gruska, J.L. Kikta, Anal. Chem. 49 (1977) 1004.
748. S. Yariv, L. Heller-Kallai, Clays Clay Miner. 23 (1975) 145.
749. C. Jolicoeur, P. Roberge, J.-L. Fortier, Can. J. Chem. 59 (1981) 1140.
750. D. Balduci, F. Valerio, Int. J. Environ. Anal. Chem. 27 (1986) 315.
751. D. Cozak, C. Barbeau, F. Gauvin, J.P. Barry, C. DeBlois, R. DeWolf, F. Kimmerle, Can. J. Chem. 61 (1983) 2753.
752. J.A. Benett, E.A. Schweikert, D. Poisson, C. Joliceur, Surf. Interface Anal. 15 (1990) 651.
753. K. Nakamoto, Infrared Spectra of Inorganic and Coordination Compounds, fourth ed., Wiley, New York, 1986.
754. D.L. Pavia, G.M. Lampman, G.S. Kriz, Introduction to Spectroscopy, second ed., Saunders, New York, 1996.
755. J. Esquena, C. Solans, Colloids Surf. 180 (2001) 85.
756. A.R. Cestari, E.F.S. Vieira, A.J.P. Nascimento, F.J.R. de Oliveira, R.E. Bruns, C. Airoldi, J. Colloid Interface Sci. 241 (2001) 45.
757. M. Mifune, M. Onada, T. Takatsuki, T. Kanai, A. Iwado, N. Motohashi, J. Haginaka, Y. Saito, Chem. Pharm. Bull. 47 (1999) 346.
758. M. Mifune, Y. Shimomura, Y. Saito, Y. Mori, M. Onada, A. Iwado, N. Motohashi, J. Haginaka, Chem. Pharm. Bull. 71 (1998) 1825.
759. E.F.M. Barradas, A.R. Cestari, C. Airoldi, R. Buffon, Braz. J. Chem. Eng. 15 (1998) 146.
760. C. Giacomini, A. Villarino, L. Franco-Fraguas, F. Batista-Vieira, J. Mol. Catal. B: Enzym. 4 (1998) 313.
761. L.K. Koopal, Y. Yang, A.J. Minaard, P.L.M. Theunissen, W.H. van Riemsdijk, Colloids Surf. 141 (1998) 385.
762. E.F.S. Vieira, A.R. Cestari, J. de A. Simoni, C. Airoldi, Thermochim. Acta 328 (1999) 247.
763. J. de A. Simoni, C. Airoldi, E.F.S. Vieira, A.R. Cestari, Colloids Surf. A 166 (2000) 109.
764. G.K. Dirk, B. Thomas, Langmuir 9 (1993) 2965.
765. G.D. Parfitt, C.H. Rochester, Adsorption from Solution at the Solid/Liquid Interfaces, Academic Press, London, 1983.
766. L.K. Koopal, Dev. Miner. Process 12 (1992) 37.
767. D.H. Everett, A.J.P. Fletcher, J. Chem. Soc., Faraday Trans. 1 (82) (1986) 2605.
768. H. Ahmad, Sci. Int. (Lahore) 2 (1990) 101.
769. W.J. Nord, J. Dispersion Sci. Technol. 13 (1992) 363.
770. T.K. Pal, A. Majumder, T.K. Raha, F. Fetting, Chem. Eng. Technol. 13 (1990) 298.
771. A.A. Lopatkin, Pure Appl. Chem. 61 (1989) 1989.
772. S. Brunauer, P. Emett, E. Teller, J. Am. Chem. Soc. 101 (1979) 159.
773. C. Airoldi, P.L.O. Volpe, J.M.M.M. Lira, Polyhedron 2 (1983) 1125.
774. W. Wasiak, Chromatographia 23 (1987) 423.
775. A.R. Cestari, C. Airoldi, J. Therm. Anal. 44 (1995) 79.
776. A.W. Adamson, Physical Chemistry of Surfaces, fifth ed., Wiley, New York, 1990.
777. R.T. Morrison, R.N. Boyd, Organic Chemistry, fifth ed., Allyn & Bacon, New York, 1976.
778. J.N. Kinkel, K.K. Unger, J. Chromatogr. A 316 (1984) 193.
779. U.D. Neue, E. Serowik, P. Iraneta, B.A. Alden, T.H. Walter, Materials 13 (2001) 1874.
780. L.R. Snyder, J.J. Kirkland, J.L. Glajch, in:Practical HPLC Method Development, second ed., John Wiley, New York, 1997.
781. D.L. Hjeresen, D.L. Schutt, J.M. Boese, J. Chem. Educ. 77 (2000) 1543.
782. N. Winterton, Green Chem. 3 (2001) 73.
783. J.H. Clark, D.J. Macquarrie, Chem. Soc. Rev. (1996) 303.

784. P.M. Padilha, L.A. de Melo Gomes, C.C.F. Padilha, J.C. Moreira, N.L. Dias Filho, Anal. Lett. 32 (1999) 1807.
785. T.W.G. Solomons, Fundamentals of Organic Chemistry, fourth ed., Wiley, New York, 1994.
786. A. Tuel, H. Hommel, A.P. Legrand, M.F. Gonnord, E. Mincsovics, A.M. Siouffi, J. Chim. Phys. 89 (1992) 477.
787. K.D. Behringer, J. Blumel, J. Liq. Chromatogr. Relat. Technol. 19 (1996) 2753.
788. C. Merckle, J. Blumel, Chem. Mater. 13 (2001) 3617.
789. E.R. Kenawy, D.C. Sherrington, Eur. Polym. J. 8 (1992) 841.
790. L. Cox, O. Celis, M.C. Hermosin, J. Cornejo, J. Agric. Food Chem. 48 (2000) 93.
791. S.R. Templeton, D. Zilerman, S.J. Yoo, Environ. Sci. Technol. 3 (1998) 1340.
792. W.F. Ritter, R.W. Scarborough, A.E. Chrirnside, J. Contam. Hydrol. 15 (1994) 73.
793. E.M. Vieira, A.G.S. Prado, M.O.O. Rezende, Quim. Nova 22 (1999) 305.
794. A.G.S. Prado, E.M. Vieira, M.O.O. Rezende, J. Braz. Chem. Soc. 12 (2001) 485.
795. R. Calvet, Environ. Health Perspect. 83 (1989) 145.
796. A.G.S. Prado, E.M. Vieira, M.O.O. Rezende, An. Assoc. Bras. Quim. 47 (1998) 239.
797. A.G.S. Prado, C. Airoldi, Pest Manage. Sci. 57 (2001) 640.
798. C. Tomlin, The Pesticide Manual, Crop Protection Publications, New York, 1995.
799. R.F. Vieira, Pesqui. Agropec. Bras. 34 (1999) 897.
800. F.N. Kok, M.Y. Arika, O. Genar, K. Abak, V. Hasira, Pestic. Sci. 55 (1999) 1194.
801. L. Szente, J. Therm. Anal. Calorim. 51 (1998) 957.
802. A. Ferraz, J.A. Souza, F.T. Silva, A.R. Goncalves, R.E. Bruns, A.R. Cotrim, R.M. Wilkins, J. Agric. Food Chem. 45 (1997) 1001.
803. E. Esposito, S.M. Paulilo, G.P. Manfio, Chemosphere 37 (1998) 571.
804. A.G.S. Prado, C. Airoldi, Thermochim. Acta 332 (1999) 71.
805. M. Raubuch, F. Beese, Soil Biol. Biochem. 31 (1999) 332.
806. P.L.O. Volpe, J. Braz. Chem. Soc. 8 (1997) 343.
807. G. Welp, G.W. Brummer, Ecotoxicol. Environ. Saf. 37 (1997) 37.
808. M. Yamaguchi, X.X. Peng, Plant Soil 173 (1995) 21.
809. K. Kubodera, Y. Miyazawa, J. Nakano, K. Otsuka, Opt. Commun. 27(1978) 345.
810. J.E. Marion, M.J. Weber, Eur. J. Solid State Inorg. Chem. 28 (1991) 271.
811. M. Buijs, G. Blasse, Chem. Phys. Lett. 113 (1985) 384.
812. H.S. Kiliaan, F.P.V. Herwijnen, G. Blasse, J. Solid State Chem. 74 (1988) 39.
813. M. Bagieu-Beucher, Thèse Doctorat d'Etat, Grenoble, France, 1980.
814. G.I. Dorokhova, O.G. Karpov, Kristallografia 29 (1984) 677.
815. M. Rzaigui, N. Kbir-Ariguib, Bull. Soc. Chim. Belg. 94 (1985) 619.
816. N.Yu. Anisimova, V.K. Trunov, N.B. Karmanovskaya, N.N. Chudinova, Izv. Akad. Nauk SSSR, Neorg. Mater. 28 (1992) 441.
817. H.Y.P. Hong, Acta Crystallogr. B 30 (1974) 468.
818. J. Matuszewski, J. Kropiwnicka, T. Znamierowska, J. Solid State Chem. 75 (1988) 285.
819. A. Jouini, M. Ferid, J.C. Gacon, L. Grosvalet, A. Thozet, M. Trabelsi-Ayadi, Mater. Res. Bull. 38 (2003) 1613.
820. H.Y.P. Hong, Acta Crystallogr. B 30 (1974) 1857.
821. P.P. Melnikov, L.N. Komissarova, T.A. Butuzova, Izv. Akad. Nauk SSSR, Neorg. Mater. 17 (1981) 2110.
822. E.W.J.L. Oomen, G. Blasse, J. Solid State Chem. 75 (1988) 201.
823. E.W.J.L. Oomen, R.C.M. Peeters, W.M.A. Smit, G. Blasse, J. Solid State Chem. 73 (1988) 151.
824. A. Jouini, J.C. Gacon, A. Brenier, M. Ferid, M. Trabelsi-Ayadi, J. Lumin. 99 (2002) 365.
825. R. Reisfeld, C.K. Jorgensen, in:Handbook of the Physics and Chemistry of Rare Earths, Elsevier, Amsterdam, 1987.
826. S.J.L. Ribeiro, R.S. Hiratsuka, A.M.G. Massabni, M.R. Davolos, C.V. Santilli, S.H. Pulcinelli, J. Non-Cryst. Solids 147–148 (1992) 162.
827. A.O. Wright, M.D. Seltzer, J.B. Gruber, B.H.T. Chai, J. Appl. Phys. 78 (4) (1995) 2456.
828. R. Ternane, M.T. Ayedi, N.K. Ariguib, B. Piriou, J. Lumin. 81 (1999) 165.
829. R. Ternane, G. Panczer, M.Th. Cohen-Adad, C. Goutaudier, G. Boulon, N.K. Ariguib, M.T. Ayedi, Opt. Mater. 16 (1–2) (2001) 291.

830. D.K. Agrawal, W.B. White, J. Electrochem. Soc. 133 (1986) 1261.
831. M. Buijs, G. Blasse, J. Lumin. 39 (1988) 323.
832. R.D. Shannon, Acta Crystallogr. A 32 (1976) 751.
833. S. Shionoya, W.M. Yen, Phosphor Handbook, Phosphor Research Society, CRC Press, Boca Raton, 1998.
834. H. Schimidt, J. Sol-Gel Sci. Technol. 8 (1997) 557.
835. A.B. Seddon, Crit. Rev. CR 68 (1997) 143.
836. P. Judenstein, C. Sanchez, J. Mater. Chem. 6 (4) (1996) 511.
837. S.J.L. Ribeiro, Y. Messaddeq, R.R. Gonçalves, M. Ferrari, M. Montagna, M.A. Aegerter, Appl. Phys. Lett. 77 (2000) 3502.
838. G.F. de Sá, O.L. Malta, C.M. Donega, A.M. Simas, R.L. Longo, P.A. Santa-Cruz, E.F. da Silva, Coord. Chem. Rev. 196 (2000) 165.
839. C. Molina, K. Dahmouche, Y. Messaddeq, S.J.L. Ribeiro, M.A.P. Silva, V. de Zéa Bermudez, L.D. Carlos, J. Lumin. 104 (2003) 93.
840. W.DeW. Horrocks Jr., D.R. Sudnick, J. Am. Chem. Soc. 101 (1979) 334.
841. E.J. Nassar, R.R. Gonçalves, M. Ferrari, Y. Messaddeq, S.J.L. Ribeiro, J. Alloys Compd. 344 (2002) 221.
842. F. Rousseau, C. Poinsignon, J. Garcia, M. Popall, Chem. Mater. 7 (1995) 828.
843. L.D. Carlos, Y. Messaddeq, H.F. Brito, R.A. Sá Ferreira, V. de Zea Bermudez, S.J.L. Ribeiro, Adv. Mater. 12 (8) (2000) 594.
844. O.L. Malta, H.F. Brito, J.F.S. Menezes, F.R. Gonçalves, S. Alves Jr., F.S. Farias, A.V.M. Andrade, Chem. Phys. Lett. 282 (1998) 233.
845. M.H.V. Werts, R.T.F. Jukes, J.W. Verhoeven, Phys. Chem. Chem. Phys. 4 (2002) 1542.
846. M.F. Hazenkamp, G. Blasse, Chem. Mater. 2 (1990) 105.
847. T. Ogasawara, H. Yonehara, H. Okabayashi, C.J. O'Connor, Colloid. Surf. A: Phys. Eng. Asp. 168 (2000) 147.
848. C.H. Chiang, N. Liu, J.L. Koenig, J. Colloid Interface Sci. 86 (1) (1982) 26.
849. T.A. Kerr, H. Wu, L.F. Nazar, Chem. Mater. 8 (1996) 2005.
850. M. Lira-Cantú, P. Gómez-Romero, Chem. Mater. 10 (1998) 698.
851. D.J. Maia, M.-A. De Paoli, O.L. Alves, A.J.G. Zarbin, S. Neves, Quim. Nova 23 (2000) 204.
852. E.A.R. Duek, M.-A. De Paoli, M. Mastragostino, Adv. Mater. 4 (1992) 287.
853. H.P. Oliveira, C.F.O. Graeff, C.A. Brunello, E.M. Guerra, J. Non-Cryst. Solids 273 (2000) 193.
854. A.A. Khan, M.M. Alam, Anal. Chim. Acta 504 (2004) 253.
855. A.A. Khan, M.M. Alam, F. Mohammad, Elctrochim. Acta 48 (2003) 2463.
856. Q.Y. Wu, X.F. Xie, Mater. Chem. Phys. 77 (2003) 621.
857. P. Gomez-Romero, M. Chojak, K. Cuentas-Gallegos, J.A. Asensio, P.J. Kulesza, N. Casan-Pastor, M. Lira-Cantu, Electrochem. Commun. 5 (2003) 149.
858. T.P. Chou, C. Chandrasekaran, S.J. Limmer, S. Seraji, Y. Wu, M.J. Forbess, C. Nguyen, G.Z. Cao, J. Non-Cryst. Solids 290 (2001) 153.
859. G. Torres-Gomez, E.M. Tejada-Rosales, P. Gomez-Romero, Chem. Mater. 13 (2001)3693.
860. T. Asefa, C. Yoshina-Ishii, M.J. MacLachlan, G.A. Ozin, J. Mater. Chem. 10 (2000) 1751.
861. P. Gomez-Romero, Adv. Mater. 13 (2001) 163.
862. Q.G. Wang, N.J. Liu, X.H. Wang, J. Li, X.J. Zhao, F.S. Wang, Macromolecules 36 (2003) 5760.
863. W.M. de Azevedo, D.J. Brondani, J. Non-Cryst. Solids 296 (2001) 224.
864. S. Roux, P. Audebert, J. Pagetti, M. Roche, J. Sol-Gel Sci. Technol. 26 (2003) 435.
865. Y.J. Wang, X.H. Wang, J. Li, Z.S. Mo, X.J. Zhao, X.B. Jing, F.S. Wang, Adv. Mater. 13 (2001) 1582.
866. F. Leroux, J.P. Besse, Chem. Mater. 13 (2001) 3507.
867. S. Roux, P. Audebert, J. Pagetti, M. Roche, New J. Chem. 24 (2000) 885.
868. R.O.R. Costa, W.L. Vasconcelos, J. Non-Cryst. Solids 304 (2002) 84.
869. V. Mehrotra, J.L. Keddie, J.M. Miller, E.P. Gianellis, J. Non-Cryst. Solids 136 (1991) 97.
870. A.J.G. Zarbin, M.-A. De Paoli, O.L. Alves, Synth. Met. 99 (1999) 227.
871. D.J. Maia, A.J.G. Zarbin, O.L. Alves, M.-A. De Paoli, Adv. Mater. 7 (1995) 792.

872. P.T. Sotomayor, I.M. Raimundo, A.J.G. Zarbin, J.J.R. Rohwedder, G.O. Neto, O.L. Alves, Sens. Actuators, B 74 (2001) 157.
873. D.C. Schnitzler, M. Meruvia, I.A. Hummelguen, A.J.G. Zarbin, Chem. Mater. 15 (2003) 4658.
874. D.C. Schnitzler, A.J.G. Zarbin, J. Braz. Chem. Soc. 15 (2004) 378.
875. F.A. Beleze, A.J.G. Zarbin, J. Braz. Chem. Soc. 12 (2001) 542.
876. A.J.G. Zarbin, D.J. Maia, M.-A. De Paoli, O.L. Alves, Synth. Met. 102 (1999) 1277.
877. A.B. Gonçalves, A.S. Mangrich, A.J.G. Zarbin, Synth. Met. 114 (2000) 119.
878. M.M. Oliveira, D. Zanchet, D. Ugarte, A.J.G. Zarbin, Prog. Colloid Polym. Sci. 128 (2004) 126.
879. M.S. Amorim, A. Galembeck, H. Bazin, G. Mathis, C.M. Donegá, G.F. de Sá, S. Alves Jr., Mol. Cryst. Liq. Cryst. 374 (2002) 267.
880. A. Galembeck, S.B.C. Silva, J.A.P. Silva, J. Del Nero, Opt. Mater. 24 (2004) 637.
881. M. Draoui, P. Vast, G. Palavit, Rev. Chem. Miner. 22 (1985) 256.
882. E.C.O. Lima, F. Galembeck, J. Colloid Interface Sci. 166 (1994) 309.
883. E.G. Castro, A.J.G. Zarbin, H.P. Oliveira, A. Galembeck, Synth. Met. 146 (2004) 57.
884. K. Yakushi, L.J. Lauchlan, T.C. Clarke, G.B. Street, J. Chem. Phys. 79 (1983) 4774.
885. D.Y. Kim, J.Y. Lee, D.K. Moon, C.Y. Kim, Synth. Met. 69 (1995) 471.
886. Y.-C. Liu, B.-J. Hwang, W.-J. Jian, R. Santhanam, Thin Solid Films, 374 (2000) 85.
887. S. Ghosh, A.B. Bowmaker, C.J.M. Seakins, Synth. Met. 95 (1998) 63.
888. A. Mogus-Milankovic, A. Gajovic, A. Santic, D.E. Day, J. Non-Cryst. Solids 289 (2001) 204.
889. L. Montagne, G. Palavit, M. Draoui, J. Non-Cryst. Solids 155 (1993) 115.
890. J. Wang, K.G. Neoh, E.T. Kang, Thin Solid Films 446 (2004) 205.
891. A. Bartl, L. Dunsch, H. Naarmann, D. Schmeißer, W. Göpel, Synth. Met. 61 (1993) 167.
892. G. Larsen, G.L. Haller, M.J. Marquez, Phys. Chem. 96 (1992) 4145.
893. A. Galembeck, J. Braz. Chem. Soc. 14 (2003) 930.
894. C. Li, W.T. Wong, J. Chem. Soc., Chem. Commun. (2002) 2034.
895. C. Galaup, J. Azema, P. Tisnes, C. Picard, P. Ramos, O. Juanes, E. Brunet, J.C. Rodriguez-Ubis, Helv. Chim. Acta 85 (2002) 1613.
896. V.M. Mukkala, P. Litti, I. Hemmilla, H. Takalo, C. Matachescu, J. Kankare, Helv. Chim. Acta 79 (1996) 295.
897. C. Piguet, J.C.-G. Bünzli, G. Bernardinelli, G. Hopfgartner, S. Petoud, O. Schaad, J. Am. Chem. Soc. 118 (1996) 6681.
898. L.D. Carlos, C.D. Donega, R.Q. Albuquerque, S. Alves, J.F.S. Menezes, O.L. Malta, Mol. Phys. 101 (2003) 1037.
899. X.Z. Jiang, A.K.Y. Jen, G.D. Phelan, D.Y. Huang, T.M. Londergan, L.R. Dalton, R.A. Register, Thin Solid Films 416 (2002) 212.
900. M. Kanesato, H. Houjou, Y. Nagawa, K. Hiratani, Inorg. Chem. Commun. 5 (2002) 984.
901. J.-M. Lehn, M. Pietraszkiewicz, J. Karpiuk, Helv. Chim. Acta 73 (1990) 106.
902. B. Alpha, J.-M. Lehn, G. Mathis, Angew. Chem. Int. Ed. Engl. 26 (1987) 266.
903. A. Dadabhoy, S. Faulkner, P.G. Sammes, J. Chem. Soc., Perkin Trans. 2 (2002) 348.
904. H. Takalo, I. Hemmila, T. Sutela, M. Latva, Helv. Chim. Acta 79 (1996) 789.
905. C. Piguet, J.C.-G. Bünzli, G. Bernardinelli, C.G. Bochet, P. Froidevaux, J. Chem. Soc., Dalton Trans. 83 (1995).
906. G.W. Lu, Y. Li, W.H. Lu, Y.Z.Z. Xu, W.M. Du, Spectro. Spectral Anal. 23 (2003) 307.
907. S. Lis, Z. Hnatejko, S. But, A. Szyczewski, M. Elbanowski, Mol. Phys. 101 (2003) 977.
908. B.L. An, P.Y. Liu, J.X. Shi, M.L. Gong, Y.S. Yang, N.S. Xu, J. Mater. Sci.-Mater. Electron. 14 (2003) 125.
909. L. Prodi, M. Montalti, N. Zaccheroni, G. Pickaert, L. Charbonniere, R. Ziessel, New J. Chem. 27 (2003) 134.
910. D. Parker, P.K. Senanayake, J.A.G. Williams, J. Chem. Soc., Perkin Trans. 2 (1998) 2129.
911. A. Arduini, V. Boehmer, L. Delmau, J.F. Desreux, J.F. Dozol, M.A. Garcia Carrera, B. Lambert, C. Musigmann, A. Pochini, A. Shivaniuk, F. Ugozzoli, Chem. Eur. J. 6 (2000) 2135.
912. L.H. Delmau, N. Simon, M.J. Schwing-Weill, Sep. Sci. Technol. 34 (1999) 863.
913. S. Barboso, A.G. Carrera, S.E. Matthews, J. Chem. Soc., Perkin Trans. 2 (1999) 719.

914. F. Arnaud-Neu, J.K. Browne, D. Byrne, Chem. Eur. J. 5 (1999) 175.
915. C. Wieser, C.B. Dieleman, D. Matt, Coord. Chem. Rev. 165 (1997) 93.
916. V. Christou, O.V. Salata, T.Q. Ly, S. Capecchi, N.J. Bailey, A. Cowley, A.M. Chippindale, Synth. Met. 111 (2000) 7.
917. D.J. Cram, S. Karbach, H.E. Kim, C.B. Knobler, E.F. Maverick, J.L. Ericson, R.C. Helgeson, J. Am. Chem. Soc. 110 (1988) 2229.
918. S.H. Jang, M.G. Han, S.S. Im, Synth. Met. 110 (2000) 17.
919. B. Lebeau, C. Sanchez, Curr. Opin. Solid State Mater. Sci. 4 (1999) 11.
920. G.-H. Hsiue, R.-H. Lee, R.-J. Jeng, Polymer 40 (1999) 6417.
921. B. Boury, R.J.P. Corriu, Adv. Mater. 12 (2000) 989.
922. N. Kimizuka, T. Kunitake, Adv. Mater. 8 (1996) 89.
923. E.P. Giannelis, Adv. Mater. 8 (1996) 29.
924. M. Kryszewski, Synth. Met. 109 (2000) 47.
925. T. Bein, P. Enzel, Angew. Chem. Int. Ed. Engl. 28 (1989) 1692.
926. Y. Wei, J.-M. Yeh, D. Jin, X. Jia, J. Wang, G.-W. Jang, C. Chen, R.W. Gumbs, Chem. Maters. 7 (1995) 969.
927. T. Hori, N. Kuramoto, H. Tagaya, M. Karasu, J. Kadokawa, K. Chiba, J. Mater. Res. (1999) 147.
928. E. Ruiz-Hitzky, P. Aranda, An. Quim. Int. Ed. 93 (1997) 197.
929. C. Sanchez, G.J.D.A. Soler-Illia, F. Ribot, T. Lalot, C.R. Mayer, V. Cabuil, Chem. Mater. 13 (2001) 3061.
930. G. Kickelbick, Prog. Polym. Sci. 28 (2003) 83.
931. Z. Ahmad, J.E. Mark, Chem. Mater. 13 (2001) 3320.
932. J. Del Nero, J.A.P. Silva, S.B.C. Silva, A. Galembeck, Synth. Met. 135 (2003) 157.
933. A. Galembeck, J. Braz. Chem. Soc. 6 (2003) 930.
934. J.E.P. Silva, M.L.A. Temperine, S.I.C. Torresi, Electrochim. Acta 44 (1999) 1887.
935. Y. Furukawa, F. Ueda, Y. Hyodo, I. Harada, T. Nakajima, T. Kawagoe, Macormolecules 21 (1988) 1297.
936. G. Louarn, M. Lapskowski, S. Quillard, A. Pron, J.P. Buisson, S. Lefrant, J. Phys. Chem. 100 (1996) 6998.
937. Y. Furukawa, T. Hara, Y. Hyoto, I. Harada, Synth. Met. 16 (1986) 189.
938. N.S. Sariciftici, M. Bartonek, H. Kuzmany, Synth. Met. 29 (1989) 193.
939. Y. Xia, M. Wiesinger, A.G. MacDiarmid, Chem. Mater. 7 (1995) 443.
940. C.-T. Kuo, C.-H. Chen, Synth. Met. 99 (1999) 163.
941. A.G. MacDiarmid, A.J. Epstein, Synth. Met. 69 (1995) 85.
942. K. Pielichowski, Solid State Ionics 104 (1007) 123.
943. E. Houzé, M. Nechtschein, Phys. Rev. B 53 (1996) 14309.
944. E.C.O. Lima, J.M. Moita Neto, F. Galembeck, F.Y. Fujiwara, J. Colloid Interface Sci. 176 (1995) 388.
945. E.C.O. Lima, M.M. Beppu, F. Galembeck, J.F. Valente Filho, D.M. Soares, J. Braz. Chem. Soc. 7 (1996) 209.
946. D.J. Zhou, C.H. Huang, K.Z. Wang, G.X. Xu, X.S. Zhao, X.M. Xie, L.G. Xu, T.K. Li, Langmuir 10 (1994) 1910.
947. O.S. Wolfbeis, R. Reisfield, I. Ochme, in:C.K. Jorgensen, R. Reisfield (Eds.), Structure and Bonding, vol. 85, Springer, New York, 1996.
948. E. Yariv, S. Schultheiss, T. Saraidarov, R. Reisfield, Opt. Mater. 16 (2001) 29.
949. S. Chakrabarty, J. Sahu, M. Chakraborty, H.N. Acharya, J. Non-Cryst. Solids 180 (1994) 96.
950. W. Strek, J. Sokolniki, J. Legendziewicz, K. Maruzewski, R. Reisfield, T. Pavich, Opt. Mater. 13 (1999) 41.
951. A. Clark, V. Terpugov, F. Medrano, M. Cervantes, D. Soto, Opt. Mater. 13 (1999) 355.
952. V. Bekiari, G. Pistolis, P. Lianos, Chem. Mater. 11 (1999) 3189.
953. V. de Zea Bermudez, L.D. Carlos, L. Alcacer, Chem. Mater. 11 (1999) 569.
954. T. Ishizka, Y. Kurokawa, T. Makino, Y. Segawa, Opt. Mater. 15 (2001) 293.
955. P.P. de Abreu Filho, F. Galembeck, F.C.G. Gandra, M.L. Baesso, E.C. Silva, H. Vargas, Langmuir 6 (1990) 1013.

956. E.C.O. Lima, F. Galembeck, Colloid Surf. A 75 (1993) 65.
957. V.A.R. Monteiro, E.F. Souza, M.M.M. Azevedo, F. Gelembeck, J. Colloid Interface Sci. 217 (1999) 237.
958. E.F. Souza, C.C. Bezerra, F. Galembeck, Polymer 38 (1997) 6285.
959. K. Zlatanova, P. Markovsky, I. Spassova, G. Danev, Opt. Mater. 5 (1996) 279.
960. M.J.S. Dewar, W. Thiel, J. Am. Chem. Soc. 99 (1977) 4899.
961. M.J. Frisch, Gaussian 98, Gaussian, Pittsburgh, PA, 1998.
962. J. Ridley, M.C. Zerner, Theor. Chim. Acta 32 (1973) 111.
963. J. Ridley, M.C. Zerner, Theor. Chim. Acta 42 (1976) 223.
964. A.D. Bacon, M.C. Zerner, Theor. Chim. Acta 53 (1979) 21.
965. J.D. Head, M.C. Zerner, Chem. Phys. Lett. 122 (1985) 264.
966. J.D. Head, M.C. Zerner, Chem. Phys. Lett. 131 (1986) 359.
967. W.P. Anderson, W.D. Edwards, M.C. Zerner, Inorg. Chem. 25 (1986) 2728.
968. W.D. Edwards, M.C. Zerner, Theor. Chim. Acta 72 (1987) 347.
969. J. Del Nero, B. Laks, Synth. Met. 101 (1999) 440.
970. J. Del Nero, B. Laks, Synth. Met. 101 (1999) 379.
971. R.L. Doretto, J. Del Nero, B. Laks, Synth. Met. 101 (1999) 178.
972. J. Del Nero, C.P. de Melo, Synth. Met. 121 (2001) 1741.
973. J. Del Nero, B. Laks, Synth. Met. 84 (1997) 423.
974. P.W. Harben, The Industrial Minerals Handbook, Industrial Minerals Division, Metal Bulletin PLC, London, 1995, p. 109.
975. M.V. Meli, R.B. Lennox, Langmuir 19 (2003) 9097.
976. Y.X. Zhou, A.T. Johnson, Nano Lett. 3 (2003) 1371.
977. F. Balzer, J. Beermann, S.I. Bozhevolnyi, A.C. Simonsen, H.G. Rubahn, Nano Lett. 3 (2003) 1311.
978. S. Ferrero, A. Piednoir, C.R. Henry, Nano Lett. 1 (2001) 227.
979. M. Brust, N. Stuhr-Hansen, K. Norgaard, J.B. Christensen, L.K. Nielsen, T. Bjornholm, Nano Lett. 1 (2001) 189.
980. Z.Q. Lin, S. Granick, Langmuir 19 (2003) 7061.
981. P.K. Mukherjee, D. Chakravorty, J. Mater. Res. 17 (2002) 3127.
982. G. Schmidt, M.M. Malwitz, Curr. Opin. Colloid Interface Sci. 8 (2003) 103.
983. J.H. Chang, Y.U. An, D.H. Cho, E.P. Giannelis, Polymer 44 (2003) 3715.
984. Y. Imai, Y. Inukai, H. Tateyama, Polym. J. 35 (2003) 230.
985. S.S. Ray, K. Yamada, A. Ogami, M. Okamoto, K. Ueda, Macromol. Rapid Commun. 23 (2002) 943.
986. A.M. Thayer, Chem. Eng. News 78 (2000) 36.
987. J.B. Hedrick, Am. Ceram. Soc. Bull. 78 (1999) 136.
988. R.E. Grim, Clay Mineralogy, McGraw-Hill, New York, 1968.
989. E. Papirer, A. Eckhardt, F. Muller, J. Yvon, J. Mater. Sci. 25 (1990) 5109.
990. J.L. Pérez-Rodríguez, in: J.L. Pérez-Rodríguez (Ed.), Applied Study of Cultural Heritage and Clays, Servicio Publicaciones, Consejo Superior de Investigaciones Cientificas, Madrid, 2003.
991. L.A. Pérez-Maqueda, M.C. Jiménez de Haro, J. Poyato, J.L. Pérez Rodríguez, J. Mater. Sci. 39 (2004) 5347.
992. L.A. Perez-Maqueda, O.B. Caneo, J. Poyato, J.L. Perez-Rodriguez, Phys. Chem. Miner. 28 (2001) 61.
993. J.L. Perez-Rodriguez, F. Carrera, J. Poyato, L.A. Perez- Maqueda, Nanotechnology 13 (2002) 382.
994. A. Wiewiora, J.L. Perez-Rodriguez, L.A. Perez-Maqueda, J. Drapala, Appl. Clay Sci. 24 (2003) 51.
995. A. Wiewióra, Clay Miner. 25 (1990) 73.
996. K.O. Backhaus, S. Durovic, Clays Clay Miner. 32 (1984) 453.
997. Z. Weiss, A. Wiewióra, Clays Clay Miner. 34 (1986) 53.
998. A. Wiewióra, Z. Weiss, Clay Miner. 20 (1985) 231.
999. A. Justo, PhD Thesis, University of Seville, Sevilla, 1984.

1000. L. Smrcok, Z. Weiss, J. Appl. Cryst. 26 (1993) 140.
1001. S.M. Richardson, J.W.j. Richardson, Am. Mineral. 67 (1982) 69.
1002. H. Takeda, M. Ross, Am. Mineral. 60 (1975) 1020.
1003. H. Shirozu, S.W. Bailey, Am. Mineral. 51 (1966) 1124.
1004. J.D. Clemens, S. Circone, A. Navrotsky, P.F. McMillan, B.K. Smith, V.J. Wall, Geochim. Cosmochim. Acta 51 (1987) 2569.
1005. B. Velde, Am. Mineral. 63 (1978) 343.
1006. E. Loh, J. Phys. C: Solid State Phys. 6 (1973) 1091.
1007. D.A. McKeown, M.I. Bell, E.S. Etz, Am. Mineral. 84 (1999) 1041.
1008. D.A. McKeown, M.I. Bell, E.S. Etz, Am. Mineral. 84 (1999) 970.
1009. L.A. Perez-Maqueda, V. Balek, J. Poyato, J.L. Perez-Rodriquez, J. Subrt, I.M. Bountsewa, I.N. Beckman, Z. Malek, J. Therm. Anal. Calorim. 71 (2003) 715.
1010. J. Poyato, L.A. Perez-Maqueda, A. Justo, V. Balek, Clays Clay Miner. 50 (2002) 791.
1011. R.F. Giese, C.J. van Oss, Surfactant Science Series, vol. 105, Marcel Dekker, New York, 2002.
1012. R.W.G. Wyckoff, Crystal Structures, Miscellaneous Inorganic Compounds, Silicates, and Basic Structural Information, vol. 4, John Wiley and Sons, New York, 1968.
1013. L.L. Schramm, J.C.T. Kwak, Clays Clay Miner. 30 (1982) 40.
1014. I. Bérend, J.M. Cases, M. Francois, J.P. Uriot, L.J. Michot, A. Masion, F. Thomas, Clays Clay Miner. 43 (1995) 324–336.
1015. J.M. Cases, I. Bérend, M. Francois, J.P. Uriot, F. Thomas, J.E. Poirier, Langmuir 8 (1992) 2730–2739.
1016. J.M. Cases, I. Bérend, M. Francois, J.P. Uriot, L.J. Michot, F. Thomas, Clays Clay Miner. 45 (1997) 8–22.
1017. A. Delville, Langmuir 8 (1992) 1796–1805.
1018. E.S. Boek, P.V. Coveney, N.T. Skipper, J. Am. Chem. Soc. 117 (1995) 12608–12617.
1019. F.C. Chang, N.T. Skipper, G. Sposito, Langmuir 13 (1997) 2074.
1020. V. Marry, P. Turq, T. Cartailler, D. Levesque, J. Chem. Phys. 117 (2002) 3454–3463.
1021. A.V.C. de Siqueira, N.T. Skipper, P.V. Coveney, E.S. Boek, Mol. Phys. 92 (1997) 1–6.
1022. N. Malikova, V. Marry, J.-F. Dufrêche, C. Simon, P. Turq, E. Giffaut, Mol. Phys. 102 (2004) 1965.
1023. N. Malikova, V. Marry, J.-F. Dufrêche, P. Turq, Curr. Opin. Colloid Interface Sci. 9 (2004) 124.
1024. J.O. Titiloye, N.T. Skipper, Mol. Phys. 99 (2001) 899.
1025. F.C. Chang, N.T. Skipper, G. Sposito, Langmuir 11 (1995) 2734.
1026. R. Sutton, G. Sposito, J. Colloid Interface Sci. 237 (2001) 174.
1027. J. Swenson, R. Bergman, S. Longeville, J. Chem. Phys. 115 (2001) 11299.
1028. M. Molera, T. Eriksen, Radiochim. Acta 90 (2004) 753.
1029. C. Daqing, T. Eriksen, Radiochim. Acta 82 (1998) 287.
1030. J.J. Tuck, P.L. Hall, M.H.B. Hayes, D.K. Ross, C. Poinsignon, J. Chem. Soc., Faraday Trans. 80 (1984) 309.
1031. J.J. Tuck, P.L. Hall, M.H.B. Hayes, D.K. Ross, J.B. Hayter, J. Chem. Soc., Faraday Trans. 1 81 (1985) 833.
1032. D.J. Cebula, R.K. Thomas, J.W. White, Clays Clay Miner. 29 (1981) 241.
1033. C. Poinsignon, J. Estrade-Szwarckopf, J. Conard, A.J. Dianoux, in: Proceedings of the International Clay Conference, Denver 1985, 1987.
1034. J. Swenson, R. Bergman, W.S. Howells, J. Chem. Phys. 113 (2000) 2873.
1035. E. Mamontov, J. Chem. Phys. 121 (2004) 9193–9194.
1036. J. Swenson, R. Bergman, W.S. Howells, S. Longeville, J. Chem. Phys. 121 (2004) 9195.
1037. D. Smith, Langmuir 14 (1998) 5959.
1038. H.J.C. Berendsen, J.R. Grigera, T.P. Straatsma, J. Phys. Chem. 91 (1987) 6269.
1039. E.S. Boek, P.V. Coveney, N.T. Skipper, Langmuir 11 (1995) 4629.
1040. D. Guillaume, A. Neaman, M. Cathelineau, R. Mosser-Ruck, C. Peiffert, M. Abdelmoula, J. Dubessy, F. Villieras, A. Baronnet, N. Michau, Clay Miner. 38 (3) (2003) 281.
1041. F. Mezei, Z. Phys. 255 (1972) 146.

1042. F. Mezei, Lecture Notes in Physics, vol. 128, Springer Verlag, Berlin, 1980.
1043. R. Golub, R. Gähler, Phys. Lett. A 123 (1987) 43.
1044. J.P. Hansen, I.R. McDonald, Theory of Simple Liquids, Academic Press, Dordrecht, 1986.
1045. M. Arab, D. Bourgeard, K.S. Smirnov, Phys. Chem. Chem. Phys. 5 (2003) 4699.
1046. M. Arab, D. Bourgeard, K.S. Smirnov, Phys. Chem. Chem. Phys. 6 (2004) 2446.
1047. O. Borodin, F. Trouw, D. Bedrov, G.D. Smith, J. Phys. Chem. B 106 (2002) 5184.
1048. J. Teixeira, M.-C. Bellissent-Funel, S.-H. Chen, A.J. Dianoux, J. Phys. C 7 (1984) 65.
1049. M.-C. Bellissent-Funel, J. Teixeira, J. Mol. Struct. 250 (1991) 213.
1050. M.-C. Bellissent-Funel, K.F. Bradley, S.H. Chen, J. Lal, J. Teixeira, Physica A 201 (1993) 277.
1051. J. Teixeira, J.-M. Zanotti, M.-C. Bellissent-Funel, S.H. Chen, Physica B 234–236 (1997) 370.
1052. J.J. Ullo, Phys. Rev. A 36 (2) (1987) 816–826.
1053. S.-H. Chen, P. Gallo, F. Sciortino, P. Tartaglia, Phys. Rev. E 56 (1997) 4231.
1054. E. Mamontov, J. Chem. Phys. 121 (18) (2004) 9087.
1055. A.J. Dianoux, F. Volino, H. Hervet, Mol. Phys. 30 (1975) 1181.
1056. R.E. Lechner, Solid State Ionics 77 (1995) 280.
1057. C.E. Weaver, Am. Miner. 41 (1956) 202.
1058. P.L. Hall, D.K. Ross, Mol. Phys. 36 (1978) 1549.
1059. P.L. Hall, D.K. Ross, Mol. Phys. 42 (1981) 673.
1060. J. Breu, W. Seidl, A. Stoll, Z. Anorg. Allg. Chem. 629 (2003) 503.
1061. B.C. Raymahashay, J. Geol. Soc. India 30 (1987) 408.
1062. C.P. Schulthess, C.P. Huang, Soil Sci. Soc. Am. J. 54 (1990) 679.
1063. N.C. Das, M. Bandyopadhyay, Appl. Clay Sci. 6 (1991) 221.
1064. N.C. Das, M. Bandyopadhyay, Water Environ. Res. 64 (1992) 852.
1065. L. Alberga, T. Holm, G. Tiravanti, D. Petruzzelli, Environ. Technol. 15 (1994) 245.
1066. Z. Weiss, Z. Klika, P. Capková, D. Janeba, S. Kozubova, Phys. Chem. Miner. 25 (1998) 534.
1067. F. Barbier, G. Duc, M. Petit-Ramel, Colloids Surf. A: Physicochem. Eng. Asp. 166 (2000) 153.
1068. T. Undabeytia, S. Nir, G. Rytmo, C. Serban, E. Morrillo, C. Maqueda, Environ. Sci. Technol. 36 (2002) 2677.
1069. E.M. Murphy, J.M. Zachara, Geoderma 67 (1995) 103.
1070. G. Sposito, The Chemistry of Soils, Oxford University Press, New York, 1989.
1071. R. Kretzschmar, D. Hesterberg, H. Sticher, Soil Sci. Soc. Am. J., 61 (1997) 101.
1072. Q. Du, Z. Sun, W. Forsling, H. Tang, Water Res. 33 (1999) 693.
1073. L. Jelinek, K. Inoue, T. Miyajima, Chem. Lett. 1 (1999) 65.
1074. D. Schmitt, H.E. Taylor, G.R. Aiken, D.A. Roth, F.H. Frimmel, Environ. Sci. Technol. 36 (2002) 2932.
1075. M. Arias, M.T. Barral, J.C. Mejuto, Chemosphere 48 (2002) 1081.
1076. J. Wang, Analytical Electrochemistry, second ed., Wiley, New York, 2000.
1077. G. Abate, J.C. Masini, Colloids Surf. A: Physicochem. Eng. Asp. 226 (2003) 25.
1078. G. Abate, J.C. Masini, J. Braz. Chem. Soc. 12 (1) (2001) 109.
1079. R.L. Wershaw, D.J. Pinckney, E.C. Llaguno, V.V. Beckett, Anal. Chim. Acta 232 (1990) 31.
1080. B.K. Schroth, G. Sposito, Environ. Sci. Technol. 32 (1998) 1404.
1081. K. Lackovic, M.J. Angove, J.D. Wells, B.B. Johnson, J. Colloid Interface Sci. 257 (2003) 31.
1082. J.C. Echeverria, E. Churio, J.J. Carrido, Clays Clay Miner. 50 (2002) 614.
1083. O. Altin, H. Onder, T. Ozbelge, Dogu, J.Colloid Interface Sci. 198 (1998) 130.
1084. C.H. Lai, C.Y. Chen, B.L. Wei, S.H. Yeh, Water Res. 36 (2002) 4943.
1085. N. Senesi, G. Sposito, K.M. Holtzclaw, G.R. Bradford, J. Environ. Qual. 18 (1989) 186.
1086. G. Fu, H.E. Allen, Water Res. 26 (1992) 225.
1087. D.C. Nayak, C. Varadachari, K. Ghosh, Soil Sci. 149 (1990) 268.
1088. A. Liu, R.D. Gonzalez, Langmuir 16 (2000) 3902.
1089. G. Abate, J.C. Masini, Org. Geochem. 33 (2002) 1171.
1090. R. Dahn, A.M. Scheidegger, A. Manceau, M.L. Schlegel, B. Baeyens, M.H. Bradbury, M. Morales, Geochim. Cosmochim. Acta 66 (2002) 2335.
1091. P. O'Day, G.E. Brown, G.A. Parks, J. Colloid Interface Sci. 165 (1994) 269.
1092. R.G. Ford, D.L. Sparks, Environ. Sci. Technol. 34 (2000) 2479.

1093. D.G. Strawn, N.E. Palmer, L.J. Furnare, C. Goodell, J.E. Amonette, R.K. Kukkadapu, Clays Clay Miner. 52 (2004) 321.
1094. S.F. Cheah, G.E. Brown, G.A. Parks, Am. Mineral. 55 (2000) 118.
1095. J.D. Morton, J.D. Semrau, K.M. Hayes, Geochim. Cosmochim. Acta 65 (2001) 2709.
1096. C. Mosser, L.J. Michot, F. Villeras, M. Romeo, Clays Clay Miner. 45 (1997) 789.
1097. M. McBride, M. Mortland, Soil Sci. Soc. Am. Proc. 38 (1974) 408.
1098. J.A. Kittrick, Clays Clay Miner. 21 (1973) 479.
1099. J.W. Stucki, D.C. Golden, C.B. Roth, Clays Clay Miner. 32 (1984) 350.
1100. M. McBride, Clays Clay Miner. 24 (1976) 211.
1101. D.M. Clementz, T.J. Pinnavaia, M.M. Mortland, J. Phys. Chem. 77 (1973) 196.
1102. M.B. McBride, T.J. Pinnavaia, M.M. Mortland, J. Phys. Chem. 79 (1975) 2430.
1103. M.L. Schlegel, A. Manceau, D. Chateigner, L. Charlet, J. Colloid Interface Sci. 215 (1999) 140.
1104. R. Baddreddine, R. Le Dred, R. Prost, Clay Miner. 37 (2002) 59.
1105. C.F. Baes, R.E. Mesmer, The Hydrolysis of Cations, Krieger Publishing Co., Malabar, 1986.
1106. F.W. Lytle, R.B. Greegor, D.R. Sandstrom, E.C. Marques, J. Wong, C.L. Spiro, F.E. Huggins, Nucl. Instrum. Methods 226 (1984) 542.
1107. T. Ressler, J. Synchrotron Radiat. 5 (1998) 118.
1108. A.L. Ankudinov, B. Ravel, J.J. Rehr, S.D. Conradson, Phys. Rev. B 58 (1998) 7565.
1109. B. Ravel, J. Synchrotron Radiat. 8 (2001) 314.
1110. E.A. Stern, in:D.C. Koningsberger, R. Prins (Eds.), X-Ray Absorption: Principles, Applications, and Techniques of EXAFS, SEXAFS, and XANES, Wiley, New York, 1988.
1111. A. Manceau, M.L. Schlegel, Phys. Chem. Miner. 28 (2001) 52.
1112. M.L. Farquhar, J.M. Charnock, K.E. England, D.J. Vaughan, J. Colloid Interface Sci. 177 (1996) 561.
1113. L.A. Douglas, in:J.B. Dixon, S.B. Weed (Eds.), Minerals in Soil Environments, Soil Science Society of America, Madison, 1989.
1114. R.M. Hazen, C.W. Burnham, Am. Mineral. 58 (1973) 889.
1115. J.H. Choy, J.B. Yoon, H. Jung, J. Phys. Chem. B 106 (2002) 11120.
1116. S.E. Shadle, J.E. Penner-Hahn, H.J. Schugar, B. Hedman, K.O. Hodgson, E.I. Solomon, J. Am. Chem. Soc. 115 (1982) 767.
1117. D.M. Moore, R.C. Reynolds Jr., X-Ray Diffraction and the Identification and Analysis of Clay Minerals, Oxford University Press, New York, 1997.
1118. S.F. Cheah, G.E. Brown, G.A. Parks, J. Colloid Interface Sci. 208 (1998) 110.
1119. R. Daehn, A.M. Scheidegger, A. Manceau, M.L. Schlegel, B. Baeyens, M.H. Bradbury, D. Chateigner, Geochim. Cosmochim. Acta 67 (2003) 1.
1120. E.M. Khaled, J. Stucki, Soil Sci. Soc. Am. J. 55 (1991) 550.
1121. J.W. Stucki, K. Lee, L. Zhang, R.A. Larson, Pure Appl. Chem. 74 (2002) 2145–2158.
1122. A. Gil, L.M. Gandía, M.A. Vicente, Catal. Rev. – Sci. Eng. 42 (2000) 145.
1123. M. Nakatsuji, R. Ishii, Z.-M. Wang, K. Ooi, J. Colloid Interface Sci. 272 (2004) 158.
1124. M.E. Landis, B.A. Aufdembrink, P. Chu, I.D. Johnson, G.W. Kirker, M.K. Rubin, J. Am. Chem. Soc. 113 (1991) 3189.
1125. J.S. Dailey, T.J. Pinnavaia, Chem. Mater. 4 (1992) 855.
1126. K. Kosuge, A. Tsunashima, J. Chem. Soc., Chem. Commun. (1995) 2427.
1127. A. Galarneau, A. Barodawalla, T.J. Pinnavaia, Nature 374 (1995) 529.
1128. M.H. Lim, C.F. blanford, A. Stein, J. Am. Chem. Soc. 119 (1997) 4090.
1129. T. Asefa, M.J. MacLachlan, N. Coombs, G.A. Ozin, Nature 402 (1999) 867.
1130. M.C. Chon, T. Tsuru, H. Takahashi, Clay Sci. 5 (1978) 155.
1131. M.A. Harkonen, R.L. Keiski, Colloids Surf. 11 (1984) 323.
1132. H. Suquet, S. Chevalier, C. Marcilly, D. Barthomeuf, Clay Miner. 26 (1991) 49.
1133. H. Kaviratna, T.J. Pinnavaia, Clays Clay Miner. 42 (1994) 717.
1134. R. Mokaya, W. Jones, J. Chem. Soc., Chem. Commun. (1994) 929
1135. R. Mokaya, W. Jones, J. Catal. 153 (1995) 76.
1136. J. Bovey, F. Kooli, W. Jones, Clay Miner. 31 (1996) 501.
1137. F. Kooli, W. Jones, J. Mater. Chem. 8 (1998) 2119.
1138. J. Temuujin, K. Okada, K.J.D. MacKenzie, Appl. Clay Sci. 805 (2002) 1.

1139. R.H. Glaser, G.L. Wilkes, C.E. Bronnimann, J. Non-Cryst. Solids 113 (1989) 73.
1140. W.G. Fahrenholtz, D.M. Smith, D-Wu Hua, J. Non-Cryst. Solids 144 (1992) 45.
1141. J. Sanchez, A. McCormick, J. Phys. Chem. 96 (1992) 8973.
1142. C.A. Fyfe, P.A. Aroca, J. Phys. Chem. B 101 (1997) 9504.
1143. Y. Sugahara, T. Inoue, K. Kuroda, J. Mater. Chem. 7 (1997) 53.
1144. A. Jitianu, A. Britchi, C. Deleanu, V. Badescu, M. Zaharescu, J. Non-Cryst. Solids 319 (2003) 263.
1145. G. Engelhardt, D. Michel, High Resolution Solid-State NMR Silicates and Zeolites, Wiley, Chichester, 1987, p. 119.
1146. B.M. De Witte, D. Commers, J.B. Uytterhoeven, J. Non-Cryst. Solids 202 (1996) 35.
1147. M.C. Matos, L.M. Ilharco, R.M. Almeida, J. Non-Cryst. Solids 147–148 (1992) 232.
1148. A. Fidalgo, L.M. Ilharco, J. Non-Cryst. Solids 283 (2001) 144.
1149. K.S.W. Sing, D.H. Everett, R.A.W. Haul, L. Moscou, R.A. Pierotti, J. Rouquerol, T. Siemieniewska, Pure Appl. Chem. 57 (1985) 603.
1150. M. Jaroniec, M. Kruk, J.P. Olivier, Langmuir 15 (1999) 5410.
1151. M.L. Occelli, Catal. Today 2 (1988) 339.
1152. T. Takei, A. Yamazaki, T. Watanabe, M. Chikasawa, J. Colloid Interface Sci. 188 (1997) 409.
1153. H. El Rassy, P. Buisson, B. Bouali, A. Perrard, A.C. Pierre, Langmuir 19 (2003) 358.
1154. J. Xu et al., Eur. Polym. J. 41 (2005) 881.
1155. T.L. Porter, D. Pace, R. Whitehorse, M.P. Eastman, E. Bain, Mater. Chem. Phys. 76 (2002) 92.
1156. Y. Hu, L. Song, J. Xu, L. Yang, Z. Chen, W. Fan, Colloid Polym. Sci. 279 (2001) 819.
1157. K.E. Strawhecker, E. Manias, Chem. Mater. 12 (2000) 2943.
1158. R.F. de Farias, C. Airoldi, Solid State Sci. 5 (2003) 611.
1159. Y.-J. Liu, M.G. Kanatzidis, Inorg. Chem. 32 (1993) 2989.
1160. R. Bissessur, D.C. Degroot, J.L. Schindler, C.R. Kannewurf, M.G. Kanatzidis, J. Chem. Soc., Chem. Commun. 8 (1993) 687.
1161. R. Bissessur, J.L. Schindler, C.R. Kannewurf, M. Kanatzidis, J. Chem. Soc., Chem. Commun. 20 (1993) 1582.
1162. J.P. Lemmon, M.M. Lerner, Solid State Commun. 94 (1995) 533.
1163. E. Benavente, M.A. Santa Ana, F. Mendizábal, G. González, Coord. Chem. Rev. 224 (2002) 87.
1164. C.-G. Wu, D.C. DeGroot, H.O. Marcy, J.L. Schindler, C.R. Kannewurf, T. Bakas, V. Papaefthymiou, W. Hirpo, J.P. Yesinowski, Y.-J. Liu, M.G. Kanatzidis, J. Am. Chem. Soc. 117 (1995) 9229.
1165. R. Setton, Synth. Met. 23 (1988) 467.
1166. M. Inagaki, J. Mater. Res. 4 (6) (1989) 1560.
1167. M.S. Dresselhaus, G. Dresselhaus, Adv. Phys. 30 (2) (1981) 139.
1168. B. Marcus, Ph. Touzain, Synth. Met. 23 (1988) 13.
1169. Ch. Simon, I. Roseman, F. Batallan, G. Pepy, H. Lauter, Synth. Met. 23 (1988) 147.
1170. W.S. Hummers, R.E. Offeman, J. Am. Chem. Soc. 80 (1958) 1339.
1171. A. Lerf, H. He, M. Forster, J. Klinowski, J. Phys. Chem. B 102 (1998) 4477.
1172. Y. Matsuo, K. Tahara, Y. Sugie, Carbon 35 (1997) 113.
1173. Y. Matsuo, K. Hatase, Y. Sugie, Chem. Mater. 10 (1998) 2266.
1174. T. Kyotani, H. Moriyama, A. Tomita, Carbon 35 (1997) 1185.
1175. P. Liu, K. Gong, P. Xiao, M. Xiao, J. Mater. Chem. 10 (2000) 933.
1176. C.V. Nicholas, D.J. Wilson, C. Booth, J.R.M. Giles, Br. Polym. J. 20 (1988) 289.
1177. W. Xu, J.-P. Belieres, C.A. Angell, Chem. Mater. 13 (2001) 575.
1178. T. Nakajima, A. Mabuchi, R. Hagwara, Carbon 26 (1988) 357.
1179. Q. Zhao, E.T. Samulski, Macromolecules 36 (2003) 6967.
1180. Z.-H. Liu, Z.-M. Wang, X. Yang, K. Ooi, Langmuir 18 (2002) 4926.
1181. R. Bissessur, R.I. Haines, R. Brüning, J. Mater. Chem. 13 (2003) 44.
1182. R. Bissessur, P.K.Y. Liu, S.F. Scully, Synth. Met. 156 (2006) 1023.
1183. R. Bissessur, P.K.Y. Liu, W. White, S.F. Scully, Langmuir 22 (2006) 1729.
1184. T. Cassagneau, F. Guérin, J.H. Fendler, Langmuir 16 (2000) 7318.
1185. T. Cassagneau, J.H. Fendler, Adv. Mater. 10 (1998) 877.
1186. R. Ding, Y. Hu, Z. Gui, R. Zong, Z. Chen, W. Fan, Polym. Degrad. Stab. 81 (2003) 473.

1187. N.I. Kovtyukhova, P.J. Olliver, B.R. Martin, T.E. Mallouk, S.A. Chizhik, E.V. Buzaneva, A.D. Gorchinskiy, Chem. Mater. 11 (1999) 771. W.S. Hummers, R.E. Offeman, J. Am. Chem. Soc. 80 (1958) 1339.
1188. M. Xiao, L. Sun, J. Liu, Y. Li, K. Gong, Polymer 43 (2002) 2245.
1189. P. Xiao, M. Xiao, P. Liu, K. Gong, Carbon 38 (2000) 623.
1190. Y. Matsuo, S. Higashika, K. Kimura, Y. Miyamoto, T. Fukutsuka, Y. Sugie, J. Mater. Chem. 12 (2002) 1592.
1191. C.-G. Wu, M.G. Kanatzidis, H.O. Marcy, D.C. DeGroot, C.R. Kannewurf, Polym. Mater. Sci. Eng. 61 (1989) 969.
1192. L. Wang, J. Schindler, J. Albritton Thomas, C.R. Kannewurf, M.G. Kanatzidis, Chem. Mater. 7 (1995) 1753.
1193. M. Kanatzidis, R. Bissessur, D.C. DeGroot, J.L. Schindler, C.R. Kannewurf, Chem. Mater. 5 (1993) 595.
1194. R. Bissessur, R.I. Haines, D.R. Hutching, R. Brüning, Chem. Commun. 17 (2001) 1598.
1195. R. Bissessur, B.D. Wagner, R. Brüning, J. Mater. Sci. 39 (2004) 119.
1196. C.T. Dalton, K.M. Ryan, V.M. Wall, C. Bousquet, D.G. Gilheany, Top. Catal. 5 (1998) 75.
1197. K. Srinivasan, P. Michard, J.K. Kochi, J. Am. Chem. Soc. 108 (1986) 2309.
1198. W. Zhang, E.N. Jacobsen, J. Org. Chem. 56 (1991) 2296.
1199. R. Irie, K. Noda, Y. Ito, N. Matsumoto, T. Katsuki, Tetrahedron Asymmetry 2 (1991) 481.
1200. T. Katsuki, K.B. Sharpless, J. Am. Chem. Soc. 102 (1980) 5974.
1201. W. Zhang, J.L. Loebach, S.R. Wilson, E.N. Jacobsen, J. Am. Chem. Soc. 112 (1990) 2801.
1202. B.B. De, B.B. Lohray, S. Sivaram, P.K. Dhal, Tetrahedron Asymmetry 6 (1995) 2105.
1203. F. Minutolo, D. Pini, A. Petri, P. Salvadori, Tetrahedron Asymmetry 7 (1996) 2293.
1204. S.B. Ogunwumi, T. Bein, Chem. Commun. (1997) 901.
1205. L. Frunza, H. Rossick, H. Landmesser, E. Hoft, R. Fricke, J. Mol. Catal. A: Chem. 123 (1997) 179.
1206. Z.H. Fu, D.L. Yin, W. Zhao, Y.D. Chen, D.H. Yin, J.W. Guo, L.X. Zhang, Catal. Lett. 90 (2003) 205.
1207. B.B. De, B.B. Lohray, S. Sivaram, P.K. Dhal, J. Polym. Sci. Polym. Chem. Ed. 35 (1997) 1809.
1208. F. Minutolo, D. Pini, A. Petri, P. Salvadori, Tetrahedron Lett. 37 (1996) 3375.
1209. L. Canali, J.K. Karjalainee, D.C. Sherrington, O.E.O. Hormi, Chem. Commun. (1997) 123.
1210. P. Piaggio, C. Langham, P. Mcmorm, D. Bethell, P.C. Bulman-Page, F.E. Hancock, C. Sly, G.J. Hutchings, J. Chem. Soc., Perkin Trans. 2 (2002) 143.
1211. G.J. Kim, J.H. Shin, Tetrahedron Lett. 40 (1999) 6827.
1212. X.G. Zhou, X.Q. Yu, J.S. Huang, C.M. Chen, S.G. Li, L.S. Li, Chem.Commun. (1999) 1789.
1213. M.J. Sabater, A. Corma, A. Domenech, V. Fornés, H. Garcýá, Chem. Commun. (1997) 1285.
1214. L. Canali, D.C. Sherrington, Chem. Soc. Rev. 28 (1999) 85.
1215. C. Li, Catal. Rev. 46 (2004) 419.
1216. P. Sutra, D. Brunel, Chem. Commun. (1996) 2485.
1217. P. Piaggio, P. McMorn, C. Langham, D. Bethell, P.C.B. Page, F.E. Hancock, G.J. Hutchings, New J. Chem. (1998) 1167.
1218. R. Gedye, F. Smith, K. Westaway, H. Ali, L. Baldisera, L. Laberge, J. Rousell, Tetrahedron Lett. 27 (1986) 279.
1219. R.S. Varma, Green Chem. 1 (1999) 43.
1220. A. Loupy, A. Petit, J. Hamelin, F. Texier-Boullet, P. Jacquault, D. Mathe, Synthesis (1998) 1213.
1221. A.R. Silva, C. Freire, B. de Castro, New J. Chem. 28 (2004) 253.
1222. E.N. Prabhakaran, J.P. Nandi, J. Iqbal, Tetrahedron Lett. 42 (2001) 333.
1223. B. Bosnich, J. Am. Chem. Soc. 90 (1968) 627.
1224. S.M. Crawford, Spectrochim. Acta 19 (1963) 255.
1225. B. Zhu, R. Xue, J. Wen, F. Gao, G. Hu, Z. Hang, A. Zhi, Acta Chim. Sinica 62 (2004) 58.
1226. F. Coppin, G. Berger, A. Bauer, S. Castet, M. Loubet, Chem. Geol. 182 (2002) 57.
1227. C. Pitteloud, D.H. Powell, M.A. Gonzalez, G.J. Cuello, Coll. Surf. A, 217 (2003) 129.
1228. A. Kowal-Fouchard, R. Drot, E. Simoni, J.J. Ehrhardt, Environ. Sci. Technol. 38 (2004) 1399.
1229. R.W. Mooney, A.G. Keenan, L.A. Wood, J. Am. Chem. Soc. 74 (1952) 1371.
1230. N.A. Chapman, I.G. McKinley, The Geological Disposal of Nuclear Waste, John Wiley & Sons, New York, 1989.

1231. JNC, H12: Project to Establish the Scienti.c and Technical Basis for HLW Disposal in Japan, Supporting Report 3: Safety Assessment of Geological Disposal System, Japan Nuclear Fuel Cycle Development Institute, Tokyo, 2000.
1232. A. Atkinson, R.M. Guppy, UKAEA Technical Report R12961, AEA, Harwell, 1988.
1233. U.R. Berner, Waste Manage. 12 (1992) 201.
1234. A. Inoue, M. Kawano, Y. Kuwahara, T. Kozaki, H. Komine, T. Sato, K. Tsukimura, J. Clay Sci. Soc. Jpn. 45 (2004) 31.
1235. J.A. Chermark, Clays Clay Miner. 40 (1992) 650.
1236. J.A. Chermark, Clays Clay Miner. 41 (1993) 365.
1237. C.I. Steelfel, P.C. Lichtner, Geochim. Cosmochim. Acta 58 (1994) 3595.
1238. J.M. Soler, Mineral. Mag. 62A (1998) 1427.
1239. S. Nakayama, Y. Sakamoto, T. Yamaguchi, M. Akai, T. Tanaka, T. Sato, Y. Iida, Appl. Clay Sci. 27 (2004) 53.
1240. M.A. Berubé, M. Choquette, J. Locat, Appl. Clay Sci. 5 (1990) 145.
1241. M. Choquette, M.A. Berubé, J. Locat, Can. Mineral. 29 (1991) 163.
1242. D.D. Eberl, B. Velde, T.C. McCormick, Clay Miner. 28 (1993) 49.
1243. G. Montes-H, B. Fritz, A. Clement, N. Michau, Appl. Clay Sci. 29 (2005) 155.
1244. E.R. Andrew, Prog. Nucl. Magn. Reson. Spectrosc. 8 (1971) 1.
1245. K. Gaare, D. Akpoiaye, J. Phys. Chem. B 101 (1997) 48.
1246. J. Rocha, J. Klinowski, Phys. Chem. Miner. 17 (1990) 179.
1247. S.R. Drachman, G.E. Roch, M.E. Smith, Solid State Nucl. Magn. Reson. 9 (1997) 257.
1248. S.C. Kohn, C.M.B. Henderson, R. Dupree, Am. Miner. 82 (1997) 1133.
1249. A. Labouriau, T.J. Higley, E.W. Earl, J. Phys. Chem. B 102 (1998) 2897.
1250. G.S. Crosson, S. Choi, J. Chorover, M.K. Amistadi, P.G. O'Day, K.T. Mueller, J. Phys. Chem. B 110 (2006) 723.
1251. S. Choi, G. Crosson, K.T. Mueller, S. Seraphin, J. Chorover, Geochim. Cosmochim. Acta 69 (2005) 4437.
1252. L. Frydman, J.S. Harwood, J. Am. Chem. Soc. 117 (1995) 5367.
1253. C. Fernandez, J.P. Amoureux, Chem. Phys. Lett. 242 (1995) 449.
1254. P. Faucon, T. Charpentier, A. Nonat, J.C. Petit, J. Am. Chem. Soc. 120 (1998) 12075.
1255. S.C. Kohn, M.E. Smith, P.J. Dirken, E.R.H. Van Eck, A.P.M. Kentgens, D. Dupree, Geochim. Cosmochim. Acta 62 (1998) 79.
1256. D. Iuga, S. Simon, E. de Boer, A.P.M. Kengens, J. Phys. Chem. B 103 (1999) 7591.
1257. K. Kanehashi, K. Saito, H. Sugisawa, Chem. Lett. 29 (2000) 588.
1258. S.M.C. Menezes, V.C. Camorim, Y.L. Lam, R.A.S. San Gil, A. Bailly, J.P. Amoureux, Appl. Catal. A 207 (2001) 367.
1259. J.A. van Bokhoven, A.L. Roest, D.C. Koningsberger, J.T. Miller, G.H. Nachtegaal, A.P.M. Kentgents, J. Phys. Chem. B 104 (2000) 6743.
1260. K. Kanehashi, K. Saito, J. Mol. Struct. 602 (2002) 105.
1261. K.U. Gore, A. Abraham, H. Hegde, R. Kumar, J.P. Amoureux, S. Ganapathy, J. Phys. Chem. B 106 (2002) 6115.
1262. R. Yang, R.I. Walton, S. Antonijevic, S. Wimperis, A.C. Hannon, J. Phys. Chem. B 108 (2004) 8208.
1263. D.R. Neuville, L. Cormier, D. Massiot, Geochim. Cosmochim. Acta 68 (2004) 5071.
1264. J.R. Allwardt, B.C. Schmidt, J.F. Stebbins, Chem. Geol. 213 (2004) 137.
1265. T. Ohkubo, K. Kanehashi, K. Saito, Y. Ikeda, Clays Clay Miner. 51 (2003) 513.
1266. A¢[a]. Zazzi, T.K. Hirsch, E. Lenova, A. Kaikkonen, J. Grins, H. Annersten, M. Eden, Clays Clay Miner. 54 (2006) 252.
1267. H. Sato, Appl. Clay Sci. 29 (2005) 267.
1268. D. Massiot, C. Bessada, J.P. Coutures, F. Taulelle, J. Magn. Reson. 90 (1990) 231.
1269. H. Hilbig, F.H. Köhler, P. Schiel, Cem. Concr. Res. 36 (2006) 326.
1270. J.P. Amoureux, C. Fernandez, S. Steuernagel, J. Magn. Reson. A 123 (1996) 116.
1271. A. Bauer, B. Velde, G. Berger, Appl. Geochem. 13 (1998) 619.
1272. J.H. de Boer, B.C. Lippens, B.G. Linsen, A. Brorkho., J.C.P. van den Hevvel, Th.J. Osinger, J. Colloid Interface Sci. 21 (1966) 405.

1273. A.V. Neimark, Adsorpt. Sci. Technol. 7 (1990) 210.
1274. G. Jozefaciuk, D. Matyka-Sarzynska, Clays Clay Miner. 54 (2006) 220.
1275. O. Altin, H.Ö. Özbelge, T. Dogu, J. Colloid Interface Sci. 217 (1999) 19.
1276. D. Müller, W. Gessner, H.J. Behrens, G. Scheler, Chem. Phys. Lett. 79 (1981) 59.
1277. K.J.D. MacKenzie, M.E. Smith, Multinuclear Solid-State NMR of Inorganic Materials, Pergamon Materials Series, vol. 5, Pergamon, Oxford, 2002.
1278. J.P. Amoureux, C. Fernandez, Solid State Nucl. Magn. Reson. 10 (1998) 211.
1279. A. Bauer, G. Berger, Appl. Geochem. 13 (1998) 905.
1280. E. Lippmaa, M. Mägi, A. Samson, G. Engelhardt, A.R. Grimmer, J. Am. Chem. Soc. 102 (1980) 4889.
1281. N. Janes, E. Old.eld, J. Am. Chem. Soc. 107 (1985) 6769.
1282. J.B. Higgins, D.E. Woessner, Eos 63 (1982) 1139.
1283. A.R. Grimmer, R. Radeglia, Chem. Phys. Lett. 6 (1984) 262.
1284. J.V. Smith, C.S. Blackwell, G.L. Hovis, Nature 390 (1984) 140.
1285. A.R. Grimmer, Chem. Phys. Lett. 119 (1985) 416.
1286. N. Weiden, H. Rager, Z. Naturforsch. A 40 (1985) 126.
1287. Y. Uno, H. Takeshi, J. Miner. Soc. Jpn. 14 (1979) 90.
1288. W. Mozgawa, Z. Fojud, M. Handke, S. Jurga, J. Mol. Struct. 614 (2002) 281.
1289. R.V. Villa, J. Cuevas, S. Ramýrez, S. Leguey, Eur. J. Mineral. 13 (2001) 635.
1290. E. Gottardi, E. Galli, Natural Zeolite, Springer-Verlag, Heidelberg, New York, Tokyo, 1985.
1291. H. Ghobarkar, O. Schäf, Mater. Sci. Eng. B 60 (1999) 163.
1292. A.T. Bell, Colloids Surf. A: Physicochem. Eng. Asp. 158 (2003) 221.
1293. M.T. Dove, Geoderma 80 (1997) 353.
1294. J. Cama, J. Ganor, C. Aora, A. Lasaga, Geochim. Cosmochim. Acta 64 (2000) 2701.
1295. D.D. Engelhardt, D. Zeigan, H. Jancke, D. Hoebbel, W. Wieker, Z. Anorg. Allg. Chem. 418 (1975) 17.
1296. S.D. Kinrade, C.T.G. Knight, D.L. Pole, R.T. Syvitski, Inorg. Chem. 37 (1998) 4272.
1297. C.H. Cheng, D.F. Shantz, J. Phys. Chem. B 110 (2006) 313.
1298. H. Cho, A.R. Felmy, R. Craciun, J.P. Keenum, N. Shah, D.A. Dixon, J. Am. Chem. Soc. 128 (2006) 2324.
1299. A.R. Felmy, H. Cho, J.M. Rustad, M.J. Mason, J. Sol. Chem. 30 (2001) 509.
1300. S. Yokoyama, M. Kuroda, T. Sato, Clays Clay Miner. 53 (2005) 147.
1301. S. Köhler, F. Dufaud, E.H. Oelkers, Geochim. Cosmochim. Acta 67 (2003) 3583.
1302. J.J. Fitzgerald, A.I. Hamza, C.E. Bronnimann, S.F. Dec, J. Am. Chem. Soc. 119 (1997) 7105.
1303. K.L. Nagy, in:A.F. White, S.L. Brantley (Eds.), Chemical Weathering Rates of Silicate Minerals, Mineralogical Society of America, Washington, DC, 1995.
1304. E. Wieland, W. Stumm, Geochim. Cosmochim. Acta 56 (1992) 3339.
1305. P.V. Brady, J.V. Walther, Geochim. Cosmochim. Acta 53 (1989) 2823.
1306. A. Dyer, An Introduction to Zeolite Molecular Sieves, J. Wiley, New York, 1988.
1307. D.W. Breck, Zeolite Molecular Sieves: Structure, Chemistry, and Use, John Wiley and Sons, 1974.
1308. J. Weitkamp, H.G. Karge, H. Pfeifer, W. Holderich (Eds.), Zeolites and Related Microporous Materials: State of the Art 1994, Elsevier Science Ltd., 1994.
1309. R.M. Barrer, Zeolites and Clay Minerals as Sorbents and Molecular Sieves, Academic Press, 1978.
1310. M.L. Occelli, H. Robson (Eds.), Synthesis of Microporous Materials – Expanded Clays and Other Microporous Solids, vol. 1, Springer, 1992.
1311. M.L. Occelli, H. Robson (Eds.), Synthesis of Microporous Materials – Expanded Clays and Other Microporous Solids, vol. 2, Springer, 1992.
1312. W. Löwenstein, Am. Mineral. 39 (1954) 92.
1313. F.R. Ribeiro (Ed.), Zeolites: Science and Technology, NATO Science Series E, Kluwer Academic Publishers, 1984.
1314. M.E. Davis, Acc. Chem. Res., 26 (1993) 111.
1315. M.T. Weller, S.E. Dann, Curr. Opin. Solid State Mater. Sci. 3 (1998) 137.
1316. Y. Ma, W. Tong, H. Zhou, S.L. Suib, Microporous Macroporous Mater. 37 (2000) 243.

1317. J. Weitkamp, Solid State Ionics 131 (2000) 175.
1318. S.I. Zones, M.E. Davis, Curr. Opin. Solid State Mater. Sci. 1 (1996) 107.
1319. D.E. Vaughan, Chem. Eng. Progr. February (1988) 25–31.
1320. E. Costa, A.D. Lucas, M.A. Uguina, J.C. Ruiz, Ind. Eng. Chem. Res. 27 (1988) 1291.
1321. P.A. Howel, G. Island, N.A. Acara, Process for Producing Molecular Sieve Bodies, US Patent 3,119,660 (1964).
1322. J. Kärger, D.M. Ruthven, Diffusion in Zeolites and other Microporous Solids, John Wiley and Sons, New York, 1992.
1323. N.Y. Chen, T.F. Degnan Jr., C.M. Smith, Molecular Transport and Reaction in Zeolites, VCH Publishers, New York, 1994.
1324. C. Saravanan, F. Jousse, S.M. Auerbach, Phys. Rev. Lett. 80 (1998) 5754.
1325. J.F. Yao, H.T. Wang, K.R. Ratinac, S.P. Ringer, Chem. Mater. 18 (2006) 1394.
1326. J.D. Perry, K. Nagai, W.J. Koros, MRS Bull. 31 (2006) 745.
1327. H.T. Wang, B.A. Holmberg, Y.S. Yan, J. Mater. Chem. 12 (2002) 3640.
1328. W.J. Koros, R. Mahajan, J. Membr. Sci. 175 (2000) 181.
1329. C.M. Zimmerman, A. Singh, W.J. Koros, J. Membr. Sci. 137 (1997) 145.
1330. Y. Li, H.M. Guan, T.S. Chung, S. Kulprathipanja, J. Membr. Sci. 275 (2006) 17.
1331. S. Li, Z.J. Li, D. Medina, C. Lew, Y.S. Yan, Chem. Mater. 17 (2005) 1851.
1332. W. Song, G. Li, V.H. Grassian, S.C. Larsen, Environ. Sci. Technol. 39 (2005) 1214.
1333. M. Smaihi, E. Gavilan, J.O. Durand, V.P. Valtchev, J. Mater. Chem. 14 (2004) 1347.
1334. J.C. Jansen, M. Macchione, E. Drioli, J. Membr. Sci. 255 (2005) 167.
1335. P.H. Li, L. Wang, Adv. Synth. Catal. 348 (2006) 681.
1336. B.A. Holmberg, S.J. Hwang, M.E. Davis, Y.S. Yan, Microporous Mesoporous Mater. 80 (2005) 347.
1337. L.D. Rollmann, J.L. Schlenker, S.L. Lawton, C.L. Kennedy, G.J. Kennedy, D.J. Doren, J. Phys. Chem. B 103 (1999) 7175.
1338. K. Yamamoto, Y. Sakata, Y. Nohara, Y. Takahashi, T. Tatsumi, Science 300 (2003) 470.
1339. M. Hunger, E. Brunner, Mol. Sieves 4 (2004) 201.
1340. O.B. Rudakov, I.P. Sedishev, Russ. Chem. Bull. 52 (2003) 55.
1341. R.J. Stokes, D.F. Evans, Fundamentals of Interfacial Engineering, Wiley-VCH, New York, 1997.
1342. J. Liu, H.T. Wang, L.X. Zhang, Chem. Mater. 16 (2004) 4205.
1343. N.H. Tran, G.R. Dennis, A.S. Milev, G.S.K. Kannangara, P. Williams, M.A. Wilson, R.N. Lamb, J. Colloid Interface Sci. 297 (2006) 541.
1344. J.J. Qin, T.S. Chung, Y.M. Cao, Desalination 193 (2006) 8.
1345. R. Gleiter, D. Kratz, Angew. Chem. Int. Ed. Engl. 29 (1990) 276.
1346. D.J. Brien, A. Naiman, K.P.C. Vollhardt, J. Chem. Soc., Chem. Commun. (1982) 133.
1347. Y. Wakatsuki, H. Yamazaki, J. Chem. Soc. Dalton Trans. (1978) 1278.
1348. E. Breitmaier, E. Bayer, Angew. Chem. Int. Ed. Engl. 8 (1969) 765.
1349. E. Breitmaier, E. Bayer, Tetrahedron Lett. (1970) 3291.
1350. E. Breitmaier, S. Gassenmann, E. Bayer, Tetrahedron 26 (1970) 5907.
1351. A. Couture, C. Bochu, P. Grandclaudon, Tetrahedron Lett. 30 (1989) 6865.
1352. H. Chafetz et al., US patent 3,336,313 (1967).
1353. S.J. Kulkarni, Stud. Surf. Sci. Catal. 113 (1998) 151.
1354. Y.V. Subba Rao, S.J. Kulkarni, M. Subrahmanyam, A.V. Rama Rao, J. Chem. Soc., Chem. Commun. (1993) 1456.
1355. Y.V. Subba Rao, S.J. Kulkarni, M. Subrahmanyam, A.V. Rama Rao, J. Org. Chem. 59 (1994) 3998.
1356. S.J. Kulkarni, K.V.V. Krishna Mohan, V. Radha Rani, N. Narender, K.V. Raghavan, Stud. Surf. Sci. Catal. 154C (2004) 2781.
1357. J. Weitkamp, L. Puppe (Eds.), Catalysis and Zeolites: Fundamentals and Applications, Spinger-Verlag, Berlin, Heidelberg, 1999.
1358. A. Ortalm, J. Rathousky, G. Schulz–Eklo., A. Zukal, Microporous Mater. 6 (1996) 171.
1359. M. Blazsó, J. Anal. Appl. Pyrolysis 39 (1997) 1.
1360. K. László, A. Bóta, L.G. Nagy, Carbon 38 (2000) 1965.

1361. K. László, A. Szücs, Carbon 39 (2001) 1945.
1362. K. Nakagawa, S.R. Mukai, T. Suzuki, H. Tamon, Carbon 41 (2003) 823.
1363. J.B. Parra, C.O. Ania, A. Arenillas, F. Rubiera, J.J. Pis, Appl. Surf. Sci. 238 (2004) 304.
1364. I. Fernández-Morales, M.C. Almazán-Almazán, M. Pérez-Mendoza, M. Domingo-Garcýá, F.J. López-Garzón, Microporous and Mesoporous Mater. 80 (2005) 107.
1365. Y. Sakata, M.A. Uddin, K. Koizumi, K. Murata, Polym. Degrad. Stab. 53 (1996) 111.
1366. D. Paszun, T. Spychaj, Ind. Eng. Chem. Res. 36 (1997) 1373.
1367. G.P. Karayannidis, A.P. Chatziavgoustis, D.S. Achilias, Adv. Polym. Technol. 21 (2002) 250.
1368. A. Linares-Solano, Energy Fuels 10 (1996) 1108.
1369. G.M. Davies, N.A. Seaton, V.S. Vassiliadis, Langmuir 15 (1999) 8235.
1370. M. Pérez-Mendoza, J. Gonzalez, P.A. Wright, N.A. Seaton, Langmuir 20 (2004) 7653.
1371. D. Frenkel, B. Smit, Understanding Molecular Simulations: From Algorithms to Applications, Academic Press, San Diego, 2001.
1372. M.P. Allen, D.J. Tildesley, Computer Simulation of Liquids, Clarendon Press, Oxford, 1989.
1373. G.M. Davies, N.A. Seaton, Langmuir 15 (1999) 6263.
1374. V.Y. Gusev, J.A. O'Brien, N.A. Seaton, Langmuir 13 (1997) 2815.
1375. W.W.M. Wendlandt, Thermal Methods of Analysis, Interscience, New York, 1964.
1376. N.N. Greenwood, A. Earnshaw, Chemistry of the Elements, Pergamon Press, Oxford, 1984.
1377. A. Addoun, J. Dentzer, P. Ehrburger, Carbon 40 (2002) 1140.
1378. T. Otowa, Y. Nojima, T. Miyazaki, Carbon 35 (1997) 1325.
1379. M.A. Lillo-Ródenas, D. Cazorla-Amorós, A. Linares-Solano, Carbon 41 (2003) 267.
1380. F. Rodrýguez-Reinoso, A. Linares-Solano, in:P.A. Thrower (Ed.), Microporous Structure of Activated Carbons as Revealed by Adsorption Methods, vol. 21, Marcel Dekker, New York, 1989.
1381. J. Garrido, A. Linares-Solano, J.M. Martýn-Martýnez, M. Molina-Sabio, F. Rodrýguez-Reinoso, R. Torregrosa, Langmuir 3 (1987) 76.
1382. A. Vichnyakov, P.I. Ravikovitch, A.V. Neimark, Langmuir 15 (1999) 8736.
1383. N. Setoyama, T. Suzuki, K. Kaneko, Carbon 36 (1998) 1459.
1384. M.J. Sellés-Pérez, J.M. Martýń-Martýńez, Fuel 70 (1991) 877.
1385. K. Kaneko, C. Ishii, M. Ruike, H. Kuwabara, Carbon 30 (1992) 1075.
1386. I.V. Mishakov, A.F. Betilo, R.M. Richards, V.V. Chesnokov, V. Vladimir, I. Zaikovskii, R.A. Buyanov, K.J. Klabunde, J. Catal. 206 (2002) 40.
1387. Y.-X. Li, K.J. Klabunde, Langmuir 7 (1991) 1388.
1388. H. Gleitter, Prog. Mater. Sci. 33 (1989) 223.
1389. R.W. Siegel, in: G.l. Trigg (Ed.), Encyclopedia of Applied Physics, vol. 11, VCH, Weinheim, 1994.
1390. R.W. Siegel, in:A.S. Edelstein, R.C. Cammarata (Eds.), Nanomaterials: Synthesis, Properties, and Applications, IOP Publishing, Philadelphia, 1998.
1391. K.J. Klabunde, in:C. Mohs, L.V. Interrante, M.J. Hampden-Smith (Eds.), Chemistry of Advance Materials: An Overview, Willey-VCH Inc., 1998.
1392. K. Tanabe, Solid Acids and Bases, Academic Press, New York, 1970.
1393. Yi Ding, G. Zhang, H. Wu, N. Hai, L. Wang, Y. Qian, Chem. Mater. 13 (2001) 435.
1394. B.R. Botter, A.W. Rearcy, J. Am. Ceram. Soc. 70 (1987) 155.
1395. J.A. Wang, O. Novaro, X. Bokhimi, T. López, R. Gómez, J. Navarrete, M.E. Llanos, E. López-Salinas, J. Phys. Chem. 101 (1997) 7448.
1396. R. Portillo, T. Lopez, R. Gomez, Bokhimi, A. Morales, O. Novaro, Langmuir 12 (1996) 40.
1397. G.T. Chandrappa, N. Stenuou, J. Livage, Nature 416 (2002) 702.
1398. R.P.S. Chakradhar, B.M. Nagabhushana, G.T. Chandrappa, K.P. Ramesh, J.L. Rao, J. Chem. Phys. 121 (2004) 10250.
1399. A. Civera, M. Pavese, G. Saracco, V. Specchia, Catal. Today 83 (2003) 199.
1400. K.C. Patil, in:S.T. Aruna, P. Bera, M.S. Hedge (Eds.), Nanomaterials in Environment Protection and Remediation, Research Signpost, Trivandrum, India, 2003.
1401. WHO Guidelines for Drinking Water Quality, vol. 2, World Health Organization, Geneva, 1984.
1402. IS10500: Indian Standard Code for Drinking Water, BIS INDIA, 1991.

1403. GNDWM. Prevention and Control of Fluorosis in India: Water Quality and Defluoridation Techniques, vol. 2, Rajiv Gandhi National Drinking Water Mission, Ministry of Rural Development, New Delhi, 1993.
1404. S. Suma Latha, S.R. Ambika, J. Prasad, Cur. Sci. 7 (1999)730.
1405. D. Mohapatra, D. Mishra, S.P. Mishra, G. Roy Chaudhury, R.P. Das, J. Colloid Interface Sci. 275 (2004)355.
1406. W.G. Nawlakhe, D.N. Kulkarni, B.N. Pathak, K.R. Bulusu, Indian J. Environ. Health 17 (1975)26.
1407. M. Agarwal, K. Rai, R. Shrivastav, S. Dass, J. Clean. Prod. 11 (2003)439.
1408. S. Chidambaram, A.L. Ramanathan, S. Vasudevan, Water SA 29 (2003)339.
1409. Y. Wang, E.J. Reardon, Appl. Geochem. 16 (2001) 531.
1410. C.R. Nagendra Rao, Fluoride and environment – a review, in: M.J. Bunch, V.M. Suresh, T.V. Kumaran (Eds.), Proceedings of the Third International Conference on Environment and Health, York University, December 15–17, 2003.
1411. M. Muthuraman, N.A. Dhas, K.C. Patil, Bull. Mater. Sci. 17 (1994) 977.
1412. R.P. Sreekanth Chakradhar, B.M. Nagabhushana, G.T. Chandrappa, K.P. Ramesh, J.L. Rao, Mater. Chem. Phys. 95 (2006) 169.
1413. I.B. Singh, M. Prasad, Indian J. Chem. Technol. 11 (2004) 185.
1414. M.D. Jia, B. Chen, R.D. Nobel, J.L. Falconer, J. Membr. Sci. 90 (1994) 1.
1415. Y. Yan, M.E. Davis, G.R. Gavalas, Ind. Eng. Chem. Res. 34 (1995) 1652.
1416. Z.A.E.P. Vroon, K. Keizer, M.J. Gilde, H. Verweij, A.J. Burggraaf, J. Membr. Sci. 113 (1996) 293.
1417. E. Piera, A. Giroir-Fendler, J.A. Dalmon, H. Moueddeb, J. Coronas, M. Menendez, J. Santamaria, J. Membr. Sci. 142 (1998) 97.
1418. M. Matsukata, N. Nishiyama, K. Ueyama, J. Chem. Soc., Chem. Commun. (1994) 339.
1419. T. Matsufuji, N. Nishiyama, M. Matsukata, K. Uyama, J. Membr. Sci. 178 (2000) 25.
1420. S. Alfaro, M. Arruebo, J. Coronas, M. Menendez, J. Santamaria, Microporous Mesoporous Mater. 50 (2001) 195.
1421. M.C. Lovallo, M. Tsapatsis, AIChE J. 42 (1996) 3020.
1422. K. Kusakabe, T. Kuroda, A. Murata, S. Morooka, Ind. Eng. Chem. Res. 36 (1997) 649.
1423. J. Hedlund, S. Mintova, J. Sterte, Microporous Mesoporous Mater. 28 (1999) 185.
1424. X. Zhongqiang, C. Qingling, L. Guanzhong, M. Yi-Hua, Chin. J. Catal. 21 (2000) 151.
1425. G. Xomeritakis, S. Nair, M. Tsapatsis, Microporous Mesoporous Mater. 38 (2000) 61.
1426. S. Li, V.A. Tuan, J.L. Falconer, R.D. Noble, Ind. Eng. Chem. Res. 40 (2001) 1952.
1427. K. Okamoto, H. Kita, K. Horii, K. Tanaka, M. Kondo, Ind. Eng. Chem. Res. 40 (2001) 163.
1428. Y. Takata, T. Tsuru, Y. Toshinori, A. Tomohisa, M. Asaeda, Microporous Mesoporous Mater. 54 (2002) 257.
1429. O.O. Van der Biest, L.J. Vandeperre, Annu. Rev. Mater. Sci. 29 (1999) 327.
1430. T. Seike, M. Matsuda, M. Miyake, J. Mater. Chem. 12 (2002) 366.
1431. A. Berenguer-Murcia, E. Morallon, D. Cazorla-Amoros, A. Linares-Solano, Microporous Mesoporous Mater. 66 (2003) 331.
1432. V. Nikolakis, Curr. Opin. Colloid Interface Sci. 10 (2005) 203.
1433. B.J. Schoeman, J. Sterte, Kona 15 (1997) 150.
1434. V. Valtchev, J. Hedlund, B.J. Schoeman, J. Sterte, S. Mintova, Microporous Mater. 8 (1997) 93.
1435. Z. Lai, M. Tsapatsis, J.P. Nicolich, Adv. Funct. Mater. 14 (2004) 716.
1436. T.C. Bowen, R.D. Noble, J.L. Falconer, J. Membr. Sci. 245 (2004) 1.
1437. M.E. Davis, Nature 417 (2002) 813.
1438. X.-D. Zou, T. Conradsson, M. Klingstedt, M.S. Dadachov, M. O'Keee, Nature 437 (2005) 716.
1439. Y. Liu, W. Zhang, T.J. Pinnavaia, Angew. Chem., Int. Ed. 40 (2001) 1255.
1440. F.-S. Xiao, L. Wang, C. Yin, K. Lin, Y. Di, J. Li, R. Xu, D. Su, R. Schlogl, T. Yokoi, T. Tatsumi, Angew. Chem. Int. Ed. 45 (2006) 3090.
1441. M. Choi, H.S. Cho, R. Srivastava, C. Venkatesan, D.-H. Choi, R. Ryoo, Nat. Mater. 5 (2006) 718.

1442. B.T. Holland, L. Abrams, A. Stein, J. Am. Chem. Soc. 121 (1999) 4308.
1443. C.J.H. Jacobsen, C. Madsen, J. Houzvicka, I. Schmidt, A. Carlsson, J. Am. Chem. Soc. 122 (2000) 7116.
1444. I. Schmidt, A. Boisen, E. Gustavsson, K. Sta¢ªhl, S. Pehrson, S. Dahl, A. Carlsson, C.J.H. Jacobsen, Chem. Mater. 13 (2001) 4416.
1445. A. Sakthivel, S.-J. Huang, W.-H. Chen, Z.-H. Lan, K.-H. Chen, T.-W. Kim, R. Ryoo, A.S.T. Chiang, S.-B. Liu, Chem. Mater. 16 (2004) 3168.
1446. W. Song, R. Kanthasamy, V.H. Grassian, S.C. Larsen, Hexagonal, Chem. Commun. 17 (2004) 1920.
1447. A. Dong, Y. Wang, Y. Tang, N. Ren, Y. Zhang, Y. Yue, Z. Gao, Adv. Mater. 14 (2002) 926.
1448. H. Kato, T. Minami, T. Kanazawa, Y. Sasaki, Angew. Chem., Int. Ed. 43 (2004) 1251.
1449. R. Ryoo, S.H. Joo, S. Jun, J. Phys. Chem. B 103 (1999) 7743.
1450. S. Jun, S.H. Joo, R. Ryoo, M. Kruk, M. Jaroniec, Z. Liu, T. Ohsuna, O. Terasaki, J. Am. Chem. Soc. 122 (2000) 10712.
1451. U. Mueller, K.K. Unger, in: K.K. Unger, J. Rouquerol, K.S.W. Sing, H. Kral (Eds.), Characterization of Porous Solids III, Studies in Surface Science and Catalysis, vol. 39, Elsevier, Amsterdam, 1994.
1452. P.L. Llewellyn, Y. Grillet, F. Schüth, H. Reichert, K.K. Unger, Microporous Mater. 3 (1994) 345.
1453. M. Thommes, B. Smarsly, P.I. Ravokovitch, A.V. Neimark, Langmuir 22 (2006) 765.
1454. P.I. Ravikovitch, A.V. Neimark, Langmuir 18 (2002) 1550.
1455. P.I. Ravikovitch, A.V. Neimark, Langmuir 18 (2002) 9830.
1456. T. Ohsuna, O. Terasaki, D. Watanabe, M.W. Anderson, S.W. Carr, Chem. Mater. 6 (1994) 2201.
1457. S. van Donk, A.H. Janssen, J.H. Bitter, K.P. De Jong, Catal. Rev. 45 (2003) 297.
1458. D.W. Breck, Zeolite Molecular Sieves, John Wiley and Sons, New York, 1974.
1459. O. Talu, J. Li, R. Kumar, P.M. Mathias, J.D. Moyer, J.M. Schork, Gas Sep. Purif. 10 (1996) 149.
1460. L.Q. Zue, J.L. Tu, Y.J. Shi, Gas Sep. Purif. 5 (1991) 173.
1461. N.N. Dutta, G.S. Patil, Gas Sep. Purif. 9 (1995) 277.
1462. R.W. Triebe, F.H. Tezel, Gas Sep. Purif. 9 (1995) 223.
1463. G.C. Tustin, M. Rule, J. Catal. 147 (1994) 186.
1464. R.J. Davis, J. Catal. 216 (2003) 396.
1465. D. Barthomeuf, Catal. Rev. Sci. Eng. 38 (1996) 521.
1466. I.I. Ivanova, O.A. Ponomoreva, E.B. Pomakhina, E.E. Knyazeva, V.V. Yuschenko, M. Hunger, J. Weitkamp, Stud. Surf. Sci. Catal. 142 (2002) 659.
1467. A. Zecchina, C. Otero Areán, Chem. Soc. Rev. 25 (1996) 187.
1468. H. Knözinger, S. Huber, J. Chem. Soc., Faraday Trans. 87 (1998) 2047.
1469. C. Otero Areán, O.V. Manoilova, M. Rodrýguez Delgado, A.A. Tsyganenko, E. Garrone, Phys. Chem. Chem. Phys. 3 (2001) 4187.
1470. K.I. Hadjiivanov, G.N. Vayssilov, Adv. Catal. 47 (2002) 307.
1471. C. Otero Areán, A.A. Tsyganenko, E. Escalona Platero, E. Garrone, A. Zecchina, Angew. Chem. Int. Ed. 37 (1998) 3161.
1472. O.V. Manoilova, M. Penãrroya Mentruit, G. Turnes Palomino, A.A. Tsyganenko, C. Otero Areán, Vib. Spectrosc. 26 (2001) 107.
1473. A.A. Tsyganenko, E.V. Kondratieva, V.S. Yanko, P. Yu. Storozhev, J. Mater. Chem. 16 (2006) 2358.
1474. D. Nachtigallová, O. Bludský, C. Otero Areán, R. Bulánek, P. Nachtigall, Phys. Chem. Chem. Phys. 8 (2006) 4849.
1475. I. Salla, T. Montanari, P. Salagre, Y. Cesteros, G. Busca, Phys. Chem. Chem. Phys. 7 (2005) 2526.
1476. T. Montanari, P. Kozyra, I. Salla, J. Datka, P. Salagre, G. Busca, J. Mater. Chem. 16 (2006) 995.
1477. E. Garrone, R. Bulánek, K. Frolich, C. Otero Areán, M. Rodrýguez Delgado, G. Turnes Palomino, D. Nachtigallová, P. Nachtigall, J. Phys. Chem. B 110 (2006) 22542.
1478. S. Bordiga, G. Turnes Palomino, C. Pazé, A. Zecchina, Microporous Mesoporous Mater. 34 (2000) 67.

1479. E. Garrone, C. Otero Areán, Chem. Soc. Rev. 34 (2005) 846.
1480. C. Otero Areán, O.V. Manoilova, G. Turnes Palomino, M. Rodrýguez Delgado, A.A. Tsyganenko, B. Bonelli, E. Garrone, Phys. Chem. Chem. Phys. 4 (2002) 5713.
1481. B. Civalleri, A.M. Ferrari, M. Llunell, R. Orlando, M. Merawa, P. Ugliengo, Chem. Mater. 15 (2003) 3996.
1482. D. Nachtigallová, P. Nachtigall, O. Bludský, Phys. Chem. Chem. Phys. 6 (2004) 5580.
1483. P. Ugliengo, C. Busco, B. Civalleri, C.M. Zicovich-Wilson, Mol. Phys. 103 (2005) 2559.
1484. P. Nachtigall, R. Bulánek, Appl. Catal. A 307 (2006) 118.
1485. M.M.J. Treacy, J.B. Higgins, Collection of Simulated XRD Powder Patterns for Zeolites, Elsevier, Amsterdam, 2001.
1486. A.A. Tsyganenko, P. Yu. Storozhev, C. Otero Areán, Kinet. Catal. 45 (2004) 530.
1487. C. Otero Areán, O.V. Manoilova, A.A. Tsyganenko, G. Turnes Palomino, M. Penãrroya Mentruit, F. Geobaldo, E. Garrone, Eur. J. Inorg. Chem. (2001) 1739.
1488. O. Bludský, M. Silhan, P. Nachtigall, T. Bucko, L. Benco, J. Hafner, J. Phys. Chem. B 109 (2005) 9631.
1489. P.A. Vaughan, Acta Crystallogr. 21 (1966) 983.
1490. D. Nachtigallová, P. Nachtigall, M. Sierka, J. Sauer, Phys. Chem. Chem. Phys. 1 (1999) 2019.
1491. C. Otero Areán, G. Turnes Palomino, E. Garrone, D. Nachtigallová, P. Nachtigall, J. Phys. Chem. B 110 (2006) 395.
1492. G. Kresse, J. Hafner, Phys. Rev. B 48 (1993) 13115.
1493. G. Kresse, J. Hafner, Phys. Rev. B 49 (1994) 14251.
1494. G. Kresse, J. Furthmuller, Phys. Rev. B 54 (1996) 11169.
1495. G. Kresse, J. Furthmuller, Comput. Mater. Sci. 6 (1996) 15.
1496. J.P. Perdew, K. Burke, M. Ernzerhof, Phys. Rev. Lett. 77 (1996) 3865.
1497. J.P. Perdew, K. Burke, M. Ernzerhof, Phys. Rev. Lett. 78 (1997) 1396.
1498. P.E. Blöchl, Phys. Rev. B 50 (1994) 17953.
1499. G. Kresse, D. Joubert, Phys. Rev. B 59 (1999) 1758.
1500. O. Bludský, M. Silhan, D. Nachtigallová, P. Nachtigall, J. Phys. Chem. A 107 (2003) 10381.
1501. O. Bludský, M. Silhan, P. Nachtigall, J. Chem. Phys. 117 (2002) 9298.
1502. T.H. Dunning, J. Chem. Phys. 90 (1989) 1007.
1503. D.E. Woon, T.H. Dunning, J. Chem. Phys. 98 (1993) 1358.
1504. M.A. Iron, M. Oren, J.M.L. Martin, Mol. Phys. 101 (2003) 1345.
1505. A. Schafer, H. Horn, R. Ahlrichs, J. Chem. Phys. 97 (1992) 2571.
1506. S.F. Boys, F. Bernardi, Mol. Phys. 19 (1970) 553.
1507. M.J. Frisch et al., Gaussian 03, Gaussian Inc., Pittsburgh, PA, 2003.
1508. A.A. Tsyganenko, E. Escalona Platero, C. Otero Areán, E. Garrone, A. Zecchina, Catal. Lett. 61 (1999) 187.
1509. M. Davidová, D. Nachtigallová, P. Nachtigall, J. Sauer, J. Phys. Chem. B 108 (2004) 13674.
1510. M. Davidová, D. Nachtigallová, R. Bulánek, P. Nachtigall, J. Phys. Chem. B 107 (2003) 2327.
1511. M. Katoh, T. Yamazaki, S. Ozawa, Bull. Chem. Soc. Jpn. 67 (1994) 1246.
1512. M. Katoh, T. Yamazaki, S. Ozawa, J. Colloid Interface Sci. 203 (1998) 447.
1513. E.F. Vansant, Pore Size Engineering in Zeolites, Wiley, New York, 1990.
1514. M.A. Djieugoue, A.M. Prakash, L. Kevan, J. Phys. Chem. B 104 (2000) 6452.
1515. K. Kusakabe, T. Kuroda, S. Morooka, J. Membr. Sci. 148 (1998) 13.
1516. Y. Hasegawa, K. Watanabe, K. Kusakabe, S. Morooka, Sep. Purif. Technol. 22–23 (2001) 319.
1517. K. Kusakabe, T. Kuroda, K. Uchino, Y. Hasegawa, S. Morooka, AIChe J. 45 (1999) 1220.
1518. Y. Hasegawa, K. Watanabe, K. Kusakabe, S. Morooka, J. Membr. Sci. 208 (2002) 415.
1519. K. Aoki, V.A. Tuan, J.L. Falconer, R.D. Noble, Micropor. Mesopor. Mater. 39 (2000) 485.
1520. G. Guan, K. Kusakabe, S. Morooka, Sep. Sci. Technol. 36 (2001) 2233.
1521. G.Q. Guan, K. Kusakabe, S. Morooka, Micropor. Mesopor. Mater. 50 (2001) 109.
1522. J.C. Poshusta, V.A. Tuan, J.L. Falconer, R.D. Noble, Ind. Eng. Chem. Res. 37 (1998) 3924.
1523. J.C. Poshusta, V.A. Tuan, E.A. Pape, R.D. Noble, J.L. Falconer, AIChe J. 46 (2000) 779.
1524. S.G. Li, J.L. Falconer, R.D. Noble, J. Membr. Sci. 241 (2004) 121.
1525. S.G. Li, J.L. Falconer, R.D. Noble, Adv. Mater. 18 (2006) 2601.

1526. B.M. Lok, C.A. Messina, R.L. Patton, R.T. Gajek, T.R. Cannan, E.M. Flanigen, US Patent 4,440,871 (1984).
1527. T. Ishihara, M. Kagawa, F. Hadama, Y. Takita, J. Catal. 169 (1997) 93.
1528. A. Frache, B.I. Palella, M. Cadoni, R. Pirone, H.O. Pastore, L. Marchese, Top. Catal. 22 (2003) 53.
1529. M. Briend, R. Vomscheid, M.J. Peltre, P.P. Man, D. Barthomeuf, J. Phys. Chem. 99 (1995) 8270.
1530. M. Zamadics, X.H. Chen, L. Kevan, J. Phys. Chem. 96 (1992) 2652.
1531. K.S. Walton, M.B. Abney, M.D. LeVan, Microporous Mesoporous Mater. 91 (2006) 78.
1532. D. Barthomeuf, Microporous Mesoporous Mater. 66 (2003) 1.
1533. F.D.P. Mees, L.R.M. Martens, M.J.G. Janssen, A.A. Verberckmoes, E.F. Vansant, Chem. Commun. (2003) 44.
1534. F. Mees, E. Vansant, M.J. Janssen, L.R.M. Martens, US Patent 20,030,149,321 (2003).
1535. J.C. Poshusta, R.D. Noble, J.L. Falconer, J. Membr. Sci. 186 (2001) 25.
1536. P.A. Jacobs, R. von Ballmoos, J. Phys. Chem. 86 (1982) 3050.
1537. I. Kiriesi, C. Flego, G. Pazzuconi, W.O. Parker Jr., R. Millini, C. Perego, G. Bellussi, J. Phys. Chem. 98 (1994) 4627.
1538. E. Loeer, U. Lohse, Ch. Peuker, G. Oehlmann, L.M. Kustov, V.L. Zholobenko, V.B. Kazansky, Zeolites 10 (1990) 266.
1539. P.M. Chunjuan, P. Massiani, Dense Barthomeuf, J. Chem. Soc., Faraday Trans. 89 (1993) 3559.
1540. M.W. Anderson, J. Klinowsky, Zeolites 6 (1986) 455.
1541. J.A. Lobo, Caracterización de cuatro zeolitas MFI protonicas mediante espectroscopia infrarroja y piridina como molécula sonda, Thesis, Universidad de Los Andes, Mérida, 2002.
1542. J.A. Lobo, P. Hoffmann, A. Cien. Ven. 51 (2000) 234.
1543. P. Hoffmann, E. Knözinger, Surf. Sci. 188 (1987) 181.
1544. V.L. Zholobenko, L.M. Kustov, V. Yu. Borovkov, V.B. Kazansky, Zeolites 8 (1988) 175.
1545. E. Bourgeat-Lami, P. Massiani, F. Di Renzo, P. Espiau, F. Fajula, Appl. Catal. 72 (1991) 139.
1546. A. Janin, M. Maache, J.C. Lavalley, J.F. Joly, F. Raatz, N. Szydlowski, Zeolites 11 (1991) 391.
1547. M.A. Makarova, A.F. Ojo, K. Karim, M. Hunger, J. Dwyer, J. Phys. Chem. 98 (1994) 3619.
1548. R.S. McDonald, J. Phys. Chem. 62 (1958) 1168.
1549. J.W. Ward, J. Catal. 9 (1967) 225.
1550. A.J. McFarlan, B.A. Morrow, J. Phys. Chem. 95 (1991) 5388.
1551. H. Knözinger, in: P. Shuster, G. Zundel, C. Sandorfy (Eds.), The Hydrogen Bond, vol. 3, North Holland, 1976.
1552. O.M. Yaghi, M. O'Keefe, N.W. Ockwig, H.K. Chae, M. Eddaoudi, J. Kim, Nature 423 (2003) 705.
1553. R.Q. Snurr, J.T. Hupp, S.T. Nguyen, AIChE J. 50 (2004) 1090.
1554. D.N. Dybtsev, H. Chun, S.H. Yoon, D. Kim, K. Kim, J. Am. Chem. Soc. 126 (2004) 32.
1555. S.S.-Y. Chui, S.M.-F. Lo, J.P.H. Charmant, A.G. Orpen, I.D. Williams, Science 283 (1999) 1148.
1556. A. Vishnyakov, P.I. Ravikovitch, A.V. Neimark, M. Bulow, Q.M. Wang, Nano Lett. 3 (2003) 713.
1557. L. Pan, M.B. Sander, X.Y. Huang, J. Li, M. Smith, E. Bittner, B. Bockrath, J.K. Johnson, J. Am. Chem. Soc. 126 (2004) 1308.
1558. J.L.C. Rowsell, A.R. Millward, K.S. Park, O.M. Yaghi, J. Am. Chem. Soc. 126 (2004) 5666.
1559. J.-Y. Lee, L. Pan, S.K. Kelly, J. Jagiello, T.J. Emge, J. Li, Adv. Mater. 17 (2005) 2703.
1560. L. Pan, H. Liu, X. Lei, X. Huang, D. Olson, N.J. Turro, J. Li, Angew. Chem. Int. Ed. 42 (2003) 542.
1561. K. Schlichte, T. Kratze, S. Kaskel, Microporous Mesoporous Mater. 73 (2004) 81.
1562. L. Pan, D.H. Olson, L.R. Ciemnolonski, R. Heddy, J. Li, Angew. Chem. Int. Ed. 45 (2006) 616.
1563. L. Pan, B. Parker, X.Y. Huang, D.H. Olson, J.-Y. Lee, J. Li, J. Am. Chem. Soc. 128 (2006) 4180.

1564. A.J. Fletcher, E.J. Cussen, T.J. Prior, M.J. Rosseinsky, C.J. Kepert, K.M. Thomas, J. Am. Chem. Soc. 123 (2001) 10001.
1565. A.J. Fletcher, E.J. Cussen, D. Bradshaw, M.J. Rosseinsky, K.M. Thomas, J. Am. Chem. Soc. 126 (2004) 9750.
1566. J.T. Culp, C. Matranga, M. Smith, E.W. Bittner, B. Bockrath, J. Phys. Chem. B 110 (2006) 8325.
1567. S. Natesakhawat, J.T. Culp, C. Matranga, B. Bockrath, J. Phys. Chem. C 111 (2007) 1055.
1568. L. Sarkisov, T. Duren, R.Q. Snurr, Mol. Phys. 102 (2004) 211.
1569. J. Jiang, S.I. Sandler, Langmuir 22 (2006) 5702.
1570. E.W. Bittner, M.R. Smith, B.C. Bockrath, Carbon 41 (2003) 1231.
1571. D.M. Ruthven, Principles of Adsorption and Adsorption Processes, John Wiley and Sons, New York, 1984.
1572. M. Eic, D.M. Ruthven, Zeolites 8 (1988) 40.
1573. F. Brandani, D. Ruthven, C.G. Coe, Ind. Eng. Chem. Res. 42 (2003) 1451.
1574. K. Banas, F. Brandani, D.M. Ruthven, F. Stallmach, J. Kärger, Magn. Reson. Imaging 23 (2005) 227.
1575. S.F. Zaman, K.F. Loughlin, S.S. Al-Khattaf, Ind. Eng. Chem. Res. 44 (2005) 2027.
1576. S. Brandani, Chem. Eng. Sci. 53 (1998) 2719.
1577. N.S. Osborne, D.C. Ginnings, J. Res. NBS 39 (1947) 453.
1578. A. Corma, V. Fornés, S.B.C. Pergher, Th.L. Maesen, J.G. Buglass, Nature 396 (1998) 353.
1579. A. Corma, V. Fornés, F. Rey, Adv. Mater. 14 (2002) 71.
1580. S. Ching, Catal. Today 49 (1999) 303.
1581. A. Corma, V. Fornés, J.M. Guil, S. Pergher, Th.L.M. Maesen, J.G. Buglass, Microporous Mesoporous Mater. 38 (2000) 301.
1582. A. Corma, U. Dyáz, M.E. Domine, V. Fornés, Angew. Chem. 8 (2000) 112.
1583. A. Corma, V. Fornés, U. Dýáz, Chem. Commun. 6 (2001) 2642.
1584. M.E. Leonowicz, J.A. Lawton, S.L. Lawton, M.K. Rubin, Science 264 (1994) 1910.
1585. I. Rodriguez, M.J. Climent, S. Iborra, V. Fornés, A. Corma, J. Catal. 192 (2000) 441.
1586. A. Corma, A. Garcyá, J. Miralles, Microporous Mesoporous Mater. 43 (2001) 161.
1587. M.V. Goméz, A. Cantýn, A. Corma, A. de la Hoz, J. Mol. Catal. A: Chem. 240 (2005) 16.
1588. A. Corma, P. Botella, C. Mitchell, Chem. Commun. 17 (2004) 2008.
1589. A. Corma, A. Martýnez, V. Martýnez-Soriay, J. Catal. 200 (2001) 259.
1590. M.J. Climent, A. Corma, S. Iborra, J. Catal. 233 (2005) 308.
1591. M.J. Climent, A. Corma, A. Velty, Appl. Catal. A: Gen. 263 (2004) 155.
1592. V. Ayala, A. Corma, M. Iglesias, J.A. Rincón, F. Sánchez, J. Catal. 224 (2004) 170.
1593. R. Schenkel, J.-O. Barth, J. Kornatowski, J.A. Lercher, in: R. Aiello, F. Testa, G. Giordano (Eds.), Impact of Zeolites and Other Porous Materials on the New Technologies at the Beginning of the Newn Millennium, Studies in Surface Science and Catalysis, vol. 142, Elsevier, Amsterdam, 2002.
1594. P. Frontera, F. Testa, R.B. Aiello, J. Nagy, in: J. Ceika, N. Zilkova, P. Natchigall (Eds.), Molecular Sieves: From Basic Research to Industrial Application, Studies in Surface Science and Catalysis, Elsevier, Amsterdam, 2005.
1595. M.K. Rubin, P. Chu, US Patent 4,954,325 (1990).
1596. A. Corma, C. Correl, J. Perez-Pariente, Zeolites 15 (1995) 2.
1597. L. Puppe, J. Weisser, US Patent 4,439,409 (1984).
1598. R.M. Dessau, E.W. Valyocsik, N.H. Goeke, Zeolites 12 (1992) 776.
1599. S.L. Lawton, A.S. Fung, G.J. Kennedy, L.B. Alemany, C.D. Chang, G.H. Hatzibos, D. Lissy, M.K. Rubin, H.K.C. Trimben, S. Steuernagel, D.E. Woessner, J. Phys. Chem. 100 (1996) 3788.
1600. R. Aiello, F. Crea, F. Testa, G. Demortier, P. Lentz, M. Wiame, J.B. Nagy, Microporous Mesoporous Mater. 35–36 (2000) 585.
1601. G.J. Kennedy, S.L. Lawton, A.S. Feng, M.K. Rubin, S. Steurnage, Catal. Today 49 (1999) 385.
1602. I. Mochida, S. Eguchi, M. Hironaka, S. Nagao, Zeolites 18 (1997) 142.
1603. S.L. Lawton, M.E. Leonowicz, R.D. Patdrige, P. Chu, M.K. Rubin, Microporous Mesoporous Mater. 23 (1998) 109.

1604. A.C. Voegtlin, A. Matijasic, J. Patarin, C. Sauerland, Y. Grillet, L. Huve, Microporous Mater. 10 (1997) 137.
1605. A. Corma, V. Diaz, V. Fornés, J.M. Guil, J. Martinez-Triguero, E.J. Creyghton, J. Catal. 191 (2000) 218.
1606. C.T. Kresge, M.E. Leonowicz, W.J. Roth, J.C. Vartuli, J.S. Beck, Nature 359 (1992) 710.
1607. D.Y. Zhao, J. Feng, Q. Huo, N. Melosh, G.H. Fredrickson, B.F. Chmelka, G.D. Stucky, Science 279 (1998) 548.
1608. D. Zhao, Q. Huo, J. Feng, B.F. Chmelka, G.D. Stucky, J. Am. Chem. Soc. 120 (1998) 6024.
1609. A. Sayari, Y. Yang, M. Kruk, M. Jaroniec, J. Phys. Chem. B 103 (1999) 3651.
1610. W.H. Zhang, L. Zhang, J. Xiu, Z. Shen, Y. Li, P. Ying, C. Li, Microporous Mesoporous Mater. 89 (2006) 179.
1611. J.L. Blin, C. Otjacques, G. Herrier, B. Su, Langmuir 16 (2000) 4229.
1612. J.L. Blin, B. Su, Langmuir 18 (2002) 5303.
1613. J.R. Matos, M. Kruk, L.P. Mercuri, M. Jaroniec, L. Zhao, T. Kamiyama, O. Terasaki, T.J. Pinnavaia, Y. Liu, J. Am. Chem. Soc. 125 (2003) 821.
1614. J. Fan, C. Yu, J. Lei, Q. Zhang, T. Li, B. Tu, W. Zhou, D. Zhao, J.Am. Chem. Soc. 127 (2005) 10794.
1615. J. Sun, H. Zhang, D. Ma, Y. Chen, X. Bao, A. Klein-Hoffmann, N. Pfänder, D. Su, Chem. Commun. (2005) 5343.
1616. J. Sun, H. Zhang, R. Tian, D. Ma, X. Bao, D. Su, H. Zou, Chem. Commun. (2006)1322.
1617. A.S. Maria Chong, X.S. Zhao, J. Phys. Chem. B 107 (2003) 12650.
1618. P. Schmidt-Winkel, W.W. Lukens, D. Zhao, P. Yang, B.F. Chmelka, G.D. Stucky, J. Am. Chem. Soc. 121 (1999) 254.
1619. J.S. Lettow, Y.J. Han, P. Schmidt-Winkel, P. Yang, D. Zhao, G.D. Stucky, J.Y. Ying, Langmuir 16 (2000) 8291.
1620. Q. Huo, D.I. Margolese, G.D. Stucky, Chem. Mater. 8 (1996) 1147.
1621. J. Shah, T.J. Pinnavaia, Chem. Commun. (2005)1598.
1622. C. Boissiére, A. Larbot, A. van der Lee, P.J. Kooyman, E. Prouzet, Chem. Mater. 12 (2000) 2902.
1623. M. Kruk, M. Jaroniec, S.H. Joo, R. Ryoo, J. Phys. Chem. B 107 (2003) 2205.
1624. J. Liu, Q. Yang, L. Zhang, D. Jiang, X. Shi, J. Yang, H. Zhong, C. Li, Adv. Funct. Mater. 17 (2007) 569.
1625. Y.J. Han, W.T. Jordan, G.D. Stucky, A. Butler, J. Mol. Catal. Enzym. 17 (2002) 1.
1626. S.J. Hyun, B.S. Kang, J. Am. Ceram. Soc. 79 (1997) 279.
1627. X. Ju, P. Huang, N. Xu, J. Shi, J. Membr. Sci. 202 (2002) 63.
1628. X. Ju, P. Huang, N. Xu, J. Shi, J. Membr. Sci. 166 (2000) 41.
1629. J.S. Beck, J.C. Vartuli, W.J. Roth, M.E. Leonowicz, C.T. Kresge, K.D. Schmidtt, C.T-W Chu, D.H. Olson, E.W. Sheppard, S.B. MeCullen, J.B. Higgins, J.L. schlenker, J. Am. Chem. Soc. 114 (1992) 10834.
1630. H. Yang, N. Coombs, I. Sokolov, G.A. Ozin, Nature 381 (1996) 589.
1631. H. Yang, A. Kuperman, N. Coombs, S. Mamich-Afara, G.A. Ozin, Nature 379 (1996) 703.
1632. M. Ogawa, T. Kikuchi, Adv. Mater. 10 (1998) 1077.
1633. R. Ryoo, C.H. Ko, S.J. Cho, J.M. Kim, J. Phys. Chem. B 101 (1997) 10610.
1634. H. Miyata, K. Kuroda, Chem. Mater. 12 (2000) 49.
1635. Z. Hua, J. Shi, L. Wang, W. Zhang, J. Non-Cryst. Solids 292 (2001) 177.
1636. G. Xomeritakis, C.M. Braunbarth, B. Smarsly, N. Liu, R. Köhn, Z. Klipowicz, C.J. Brinker, Microporous Mesoporous Mater. 66 (2003) 91.
1637. O. de la Iglesia, M. Pedernera, R. Mallada, Z. Lin, J. Rocha, J. Coronas, J. Santamaria, J. Membr. Sci. 280 (2006) 867.
1638. B.A. McCool, N. Hill, J. DiCarlo, W.J. DeSisto, J. Membr. Sci. 218 (2003) 55.
1639. P. Kumar, J. Ida, S. Kim, V.V. Guliants, J.Y.S. Lin, J. Membr. Sci. 279 (2006) 539.
1640. N. Nishiyama, A. Koide, Y. Egashira, K. Ueyama, Chem. Commun. (1998)2147.
1641. N. Nishiyama, D.H. Park, A. Koide, Y. Egashira, K. Ueyama, J. Membr. Sci. 182 (2001) 235.
1642. D.H. Park, N. Nishiyama, Y. Egashira, K. Ueyama, Ind. Eng. Chem. Res. 40 (2001) 6105.
1643. N. Nishiyama, H. Saputra, D.H. Park, Y. Egashira, K. Ueyama, J. Membr. Sci. 218 (2003) 165.

1644. N. Nishiyama, D.H. Park, Y. Egashira, K. Ueyama, Sep. Purif. Technol. 32 (2003) 127.
1645. D.H. Park, N. Nishiyama, Y. Egashira, K. Ueyama, Microporous Mesoporous Mater. 66 (2003) 69.
1646. Y. Lu, R. Ganguli, C.A. Drewien et al., Nature 389 (1997) 364.
1647. C.Y. Tsai, S.Y. Tam, Y. Lu, B.C. Jerey, J. Membr. Sci. 169 (2000) 255.
1648. C.Y. Liu, J.Q. Wang, Z.H. Rong, J. Membr. Sci. 287 (2007) 6.
1649. F. Kleitz, W. Schmidt, F. Schüth, Microporous Mesoporous Mater. 65 (2003) 1.
1650. J. Weitkamp, M. Hunger, U. Rymsa, Microporous Mesoporous Mater. 48 (2001) 255.
1651. G. Marciniak, A. Delgado, G. Leclerc, J. Velly, N. Decker, J. Schwartz, J. Med. Chem. 32 (1989) 1402.
1652. E. Angelescu, O.D. Pavel, R. Birjega, R. Zavoianu, G. Costentin, M. Che, Appl. Catal. A: Gen. 308 (2006) 13.
1653. H. Hattori, Chem. Rev. 95 (1995) 537.
1654. L.F. Tietze, U. Beifuss, in: B.M. Trost, I. Fleming (Eds.), Comprehensive Organic Synthesis: Selectivity, Strategy and E.ciency in Modern Organic Chemistry, vol. 2, Pergamon Press, Oxford, 1991.
1655. S.E. Sen, S.M. Smith, K.A. Sullivan, Tetrahedron 55 (1999) 12657.
1656. K. Tanabe, W.F. Hölderich, Appl. Catal. A: Gen. 181 (1999) 399.
1657. L. Martins, D. Cardoso, Quim. Nova 29 (2006) 358.
1658. C.-M. Yang, K.-J. Chao, J. Chin. Chem. Soc. 49 (2002) 883.
1659. F. Fajula, D. Brunel, Microporous Mesoporous Mater. 48 (2001) 119.
1660. C.N. Pérez, E. Moreno, C.A. Henriques, S. Valange, Z. Gabelica, J.L.F. Monteiro, Microporous Mesoporous Mater. 41 (2000) 137.
1661. S. Ernst, T. Bongers, C. Casel, S. Munsch, in: I. Kiricsi, G. Pál-Borbély, J.B. Nagy, H.G. Karge (Eds.), Porous Materials in Environmentally Friendly Processes, Studies in Surface Science and Catalysis, vol. 125, Elsevier, Amsterdam, 1999.
1662. Y. Kubota, Y. Nishizaki, H. Ikeya, M. Saeki, T. Hida, S. Kawazu, M. Yoshida, H. Fujii, Y. Sugi, Microporous Mesoporous. Mater. 70 (2004) 135.
1663. Y. Kubota, H. Ikeya, Y. Sugi, T. Yamada, T. Tatsumi, J. Mol. Catal. A: Chem. 249 (2006) 181.
1664. L. Martins, T.J. Bonagamba, E.R. Azevedo, P. Bargiela, D. Cardoso, Appl. Catal. A: Gen. 312 (2006) 77.
1665. R. Srivastava, D. Srinivas, P. Ratnasamy, Appl. Catal. A: Gen. 289 (2005) 128.
1666. J. Guo, A.-J. Han, H. Yu, J.-P. Dong, H. He, Y.-C. Long, Microporous Mesoporous Mater. 94 (2006) 166.
1667. L. Martins, D. Cardoso, BR Patent PI 0,505,706 (2005).
1668. L. Martins, R.T. Boldo, D. Cardoso, Microporous Mesoporous Mater. 98 (2007) 166.
1669. M. Grün, K.K. Unger, A. Matsumoto, K. Tsutsumi, Microporous Mesoporous Mater. 27 (1999) 207.
1670. M.T. Anderson, J.E. Martin, J.G. Odinek, P.P. Newcomer, Chem. Mater. 10 (1998) 1490.
1671. P. Selvam, S.K. Bhatia, C.G. Sonwane, Ind. Eng. Chem. Res. 40 (2001) 3237.
1672. K. Fontell, A. Khan, M.D. Lindström, S. Puang-Ngern, Colloid Polym. Sci. 269 (1991) 727.
1673. M.T. Anderson, J.E. Martin, J.G. Odinek, P.P. Newcomer, Chem. Mater. 10 (1998) 311.
1674. D. Myers, Surfactant Science and Technology, second ed., VCH Publishers, 1992.
1675. G.D. Stucky, Q. Huo, A. Firouzi, B.F. Chmelka, in: H. Chon, S.-K. Ihm, Y.S. Uh (Eds.), Progress in Zeolite and Microporous Materials, Studies in Surface Science and Catalysis, vol. 105, Elsevier, Amsterdam, 1997.
1676. N. Igarashi, K.A. Koyano, Y. Tanaka, S. Nakata, K. Hashimoto, T. Tatsumi, Microporous Mesoporous Mater. 59 (2003) 43.
1677. X.S. Zhao, G.Q. Lu, A.K. Whittaker, G.J. Millar, H.Y. Zhu, J. Phys. Chem. B 101 (1997) 6525.
1678. M. Kruk, M. Jaroniec, A. Sayari, J. Phys. Chem. B 101 (1997) 583.
1679. H.P. Lin, C.P. Kao, C.Y. Mou, Microporous Mesoporous Mater. 48 (2001) 135.
1680. Y.R. Cheng, H.P. Lin, C.Y. Mou, Phys. Chem. Chem. Phys. 1 (1999) 5051.

1681. Y. Kubota, Y. Nishizaki, H. Ikeya, J. Nagaya, Y. Sugi, in: A. Sayari, M. Jaroniec (Eds.), Nanoporous Materials III, Studies in Surface Science and Catalysis, vol. 141, Elsevier, Amsterdam, 2002.
1682. Z.H. Li, S.J. Roy, Y.Q. Zou, R.S. Bowman, Environ. Sci. Technol. 32 (1998) 2628.
1683. R. Pool, P.G. Bolhuis, J. Phys. Chem. B 109 (2005) 6650.
1684. R. Srivastava, D. Srinivas, P. Ratnasamy, Tetrahedron Lett. 47 (2006) 4213.
1685. A. Stein, Adv. Mater. 15 (2003) 763.
1686. J.Y. Ying, C.P. Mehnert, M.S. Wong, Angew. Chem., Int. Ed. 38 (1999) 56.
1687. M. Vallet-Regi, A. Ramila, R.P. del Real, J. Perez-Pariente, Chem. Mater. 13 (2001) 308.
1688. B. Munoz, A. Ramila, J.Pérez-Pariente, I. Diaz, M.V. Regi, Chem. Mater. 15 (2003) 500.
1689. A.L. Doadrio, E.M. Sousa, J.C. Doadrio, J.P. Pariente, I.I. Barba, M. Vallet-Regi, J. Controlled Release 97 (2004) 125.
1690. W. Xia, J. Chang, J. Controlled Release 110 (2006) 522.
1691. Y. Zhu, J. Shi, W. Shen, X. Dong, J. Feng, M. Ruan, Y. Li, Angew. Chem. Int. Ed. 44 (2005) 5083.
1692. X.Y. Yan, C. Zhong, X. Tang, D. Zhao, Angew. Chem. Int. Ed. 43 (2004) 5980.
1693. M. Vallet-Regý, I.I. Barba, A. Rámila, J. Pérez-Pariente, F. Babonneau, J.M. González-Calbet, Solid State Sci. 7 (2005) 233.
1694. A. Lopez-Noriega, D. Arcos, I. Izquierdo-Barba, Y. Sakamoto, O. Terasaki, M. Vallet-Regi, Chem. Mater. 18 (2006) 3173.
1695. M. Vallet-Regi, C.V. Ragel, A.J. Salinas, Eur. J. Inorg. Chem. (2003) 1029.
1696. M.I. Alemany, P. Velasquez, M.A. de la Casa-Lillo, P.N. De Aza, J. Non-Cryst. Solids 351 (2005) 1716.
1697. Z. Gou, J. Chang, J. Eur. Ceram. Soc. 24 (2004) 93.
1698. P. Karandikar, K.R. Patil, A. Mitra, B. Kakade, A.J. Chandwadkar, Microporous Mesoporous Mater. 98 (1–3) (2007) 189.
1699. S.D. Bhagat, Y.-H. Kim, Y.-S. Ahn, J.-G. Yeo, Microporous Mesoporous Mater. 96 (2006) 237.
1700. A.S. Dias, M. Pillinger, A.A. Valente, Microporous Mesoporous Mater. 94 (2006) 214.
1701. M.R. Jamali, Y. Aassadi, F. Shemirani, M.R.M. Hosseini, R.R. Kozani, M.M. Farahani, M.S. Niasani, Anal. Chem. Acta 579 (1) (2006) 68.
1702. A.V. Rao, N.D. Hegde, H. Hirashima, J. Colloid Interface Sci. 305 (2007) 124.
1703. Q. Tang, Y. Xu, D. Wu, Y. Sun, J. Solid State Chem. 179 (2006) 1513.
1704. S.A. Bagshaw, E. Prouzet, T.J. Pinnavaia, Science 269 (1995) 1242.
1705. F. Zang, Y. Yan, Y. Meng, Y. Xia, B. Tu, D. Zhao, Microporous Mesoporous Mater. 98 (2007) 6.
1706. M. Park, S. Komarneni, Microporous Mesoporous Mater. 25 (1998) 75.
1707. Y. Zhang, J.M. Kim, D. Wu, Y. Sun, D. Zhao, S. Peng, J. Non-Cryst. Solids 351 (2005) 777.
1708. K. Moller, T. Bein, Chem. Mater. 10 (1998) 2950.
1709. B.J. Melde, B.T. Holland, C.F. Blanford, A. Stein, Chem. Mater. 11 (1999) 3302.
1710. S.D. Bhagat, C.-S. Oh, Y.-H. Kim, Y.-S. Ahn, J.-G. Yeo, Microporous Mesoporous Mater. 100 (2007) 350.
1711. S.S. Kistler, Nature 127 (1931) 741.
1712. A.V. Rao, S.D. Bhagat, Solid State Sci. 6 (2004) 945.
1713. J. Fricke, Sci. Am. 256 (1998) 68.
1714. A.P. Rao, A.V. Rao, G.M. Pajonk, J. Sol-Gel Sci. Technol. 36 (2005) 285.
1715. A.V. Rao, S.D. Bhagat, H. Hirashima, G.M. Pajonk, J. Colloid Interface Sci. 300 (1) (2006) 279.
1716. M. Casu, M.F. Casula, A. Corrias, G. Paschina, J. Non-Cryst. Solids 315 (2003) 97.
1717. A.C. Pierre, E. Elaloui, G.M. Pajonk, Langmuir 14 (1998) 66.
1718. R. Deshpande, D.M. Smith, C.J. Brinker, J. Non-Cryst. Solids 144 (1992) 32.
1719. D.M. Smith, G.W. Scherer, J.M. Anderson, J. Non-Cryst. Solids 188 (1995) 191.
1720. D.M. Smith, D. Stein, J.M. Anderson, W. Ackerman, J. Non-Cryst. Solids 186 (1995) 104.
1721. G. Carlson, D.L. McKinley, J. Richardson, T. Tilloston, J. Non- Cryst. Solids 186 (1995) 372.

1722. F. Schwertfeger, D. Frank, M. Schimidt, J. Non-Cryst. Solids 225 (1998) 24.
1723. S.S. Prakash, C.J. Brinker, A.J. Hurd, S.M. Rao, Nature 374 (1995) 439.
1724. H. Yokogawa, H. Yokoyama, J. Non-Cryst. Solids 186 (1995) 23.
1725. S.D. Bhagat, A.V. Rao, Appl. Surf. Sci. 252 (12) (2006) 4289.
1726. C.J. Lee, G.S. Kim, S. Hyun, J. Mater. Sci. 37 (2002) 2237.
1727. A.P. Rao, G.M. Pajonk, A.V. Rao, J. Mater. Sci. 40 (2005) 3481.
1728. M.M. Dubinin, E.G. Zhukovskaia, M.L. Lukianovich, K.O. Murdmaa, E.F. Polstianov, E.E. Senderov, Izv. Akad. Nauk SSSR, Ser. Khim. (1965) 1500.
1729. D.E.W. Vaughan, in: L.B. Sand, F.A. Mumpton (Eds.), Natural Zeolites; Occurrence, Properties, Use, Pergamon Press, Oxford, 1978.
1730. M.W. Ackley, S.U. Rege, Himanshu Saxena, Microporous Mesoporous Mater. 61 (2003) 25.
1731. T.G. Andronikashvili, G.V. Tsitsishvili, T.A. Chumburidze, N.J. Skhirtladze, L.G. Eprikashvili, in: Proceedings of the Soviet-Bulgarian Symposium on Studies of Physico-Chemical Properties of Natural Zeolites, Tbilisi, Georgia, October 29–31, 1976, Ed. Mietsniereba, 1979.
1732. X. Liu, H. Qing Yao, Zh. Ma, Q. Zhen Kong, in: J.C. Hansen, L. Moscou, M.F.M. Post (Eds.), Eighth International Zeolite Conference, Amsterdan, July 10–14, 1989.
1733. M. Turco, G. Bagnasco, L. Lisi, G. Russo, D. Sanni, P. Ciambelli, in: D.W. Ming, F.A. Mumpton (Eds.), International Committee on Natural Zeolite, Brockport, New York, 1993.
1734. M.A. Hernández, L. Corona, F. Rojas, Adsorption 6 (2000) 33.
1735. R.M. Barrer, D.L. Peterson, Proc. R. Soc. A 280 (1974) 466.
1736. P.E. Eberly Jr., J. Phys. Chem. 67 (1963) 2404.
1737. A. Cointot, P. Cartraud, M.C. Veyssiere, Bull. Soc. Chim. 1 (1973) 21.
1738. T. Takaishi, A. Yusa, F. Amakasu, Trans. Faraday Soc. 67 (1971) 3565.
1739. M. Autie, A. Diaz, Inf. Técnicas 5 (1978) 3.
1740. M. Autie, R. Roque, Rev. CNIC Cienc. Quýmicas 16 (1985) 129.
1741. M. Autie, Rev. Tecnológica, XVI 2 (1986) 43.
1742. M. Autie, R. Roque, Revista Tecnológica, XIV 0 (1984) 12.
1743. R. Moreno-Tost, J. Santamarýa-González, E. Rodrýguez-Castellón, A. Jýménez-López, M.A. Autie, E. González, M. Carreras, C. de las Pozas, Appl. Catal. B: Environ. 50 (2004) 279.
1744. R. Moreno-Tost, J. Santamarýa-González, E. Rodrýguez-Castellón, A. Jýménez-López, M.A. Autie, M. Carreras-Glacial, G. Autié Castro, M. Guerra, Z. Anorg. Allg. Chem. 611 (2005) 2253.
1745. J. Balmaseda, E. Reguera, A. Gómez, B. Dýaz, M. Autie, Microporous Mesoporous Mater. 54 (2002) 285.
1746. J. Balmaseda, E. Reguera, A. Gómez, J. Roque, C. Vazquez, M. Autie, J. Phys. Chem. B 107 (2003) 11360.
1747. E. Reguera, J. Balmaseda, J. Rodrýguez-Hernández, M. Autie, A. Gordillo, H. Yee-Madeira, J. Porous Mater. 11 (2004) 219.
1748. M.M. Dubinin, E.D. Zaverina, L. Raduhkievich, V. Zh. Fiz. Khimii 21 (1947) 1351.
1749. S. Kennet, W. Sing, Carbon 27 (1) (1989) 5.
1750. F. Rodrýguez-Reinoso, J. Garrido, J.M. Martýn-Martýnez, M. Molina-Sabio, R. Torregrosa, Carbon 27 (1) (1989) 23.
1751. M.M. Dubinin, V.A. Astakhov, Izv. Akad. Nauk SSSR, Ser. Khim. (1971) 5.
1752. B.P. Bering, M.M. Dubinin, V.V. Serpinski, J. Colloid Interface Sci. 21 (1966) 378.
1753. S. Brunauer, S. Deming, L.S. Deming, E. Teller, J. Am. Chem. Soc. 62 (1940) 1723.
1754. Y.Y. Huang, J. Catal. 30 (1973) 187.
1755. Y. Xie, X. Tang, Adv. Catal. 37 (1990) 1.
1756. Y. Xie, J. Zhang, J. Qiu, J.F. Tong, H. Yang, X. Tang, Adsorption 3 (1996) 27.
1757. C. Torre-Abreu, M.F. Ribeiro, C. Henriques, G. Delahay, Appl. Catal. B: Environ. 14 (1997) 261.
1758. T. Kyotani, Carbon 38 (2000) 269.
1759. R. Ryoo, S.H. Joo, M. Kruk, M. Jaroniec, Adv. Mater. 13 (2001) 677.
1760. A.H. Lu, W. Schmidt, B. Spliethо, F. Schüth, Adv. Mater. 15 (2003) 1602.
1761. W. Lu, D.D.L. Chung, Carbon 35 (1997) 427.

1762. A.B. Fuertes, F. Pico, J.M. Rojo, J. Power Sources 133 (2004) 329.
1763. G.Y. Liu, F.Y. Kang, B.H. Li, Z.H. Huang, X.Y. Chuan, J. Phys. Chem. Solids 67 (2006) 1186.
1764. Z.H. Hu, M. Srinvasan, Y.M. Ni, Adv. Mater. 12 (2000) 62.
1765. A.P. Wang, F.Y. Kang, Z.H. Huang, Z.C. Guo, Clays Clay Miner. 54 (2006) 485.
1766. G.Y. Liu, Z.H. Huang, F.Y. Kang, New Carbon Mater. 20 (2005) 13.
1767. M. Kruk, M. Jaronic, J. Phys. Chem. B 104 (2000) 7960.
1768. S.H. Joo, S. Jun, R. Ryoo, Microporous Mesoporous Mater. 44–45 (2001) 153.
1769. J. Lee, J. Kim, T. Hyeon, Chem. Commun. 10 (2003) 1138.
1770. S. Han, K.T. Lee, S.M. Oh, T. Hyeon, Carbon 41 (2003) 1049.
1771. Z.G. Shi, Y.Q. Feng, L. Xu, S.L. Da, M. Zhang, Carbon 42 (2004) 1677.
1772. A.B. Fuertes, Microporous Mesoporous Mater. 67 (2004) 273.
1773. K. Böhme, W.D. Einicke, O. Klepel, Carbon 43 (2005) 1918.
1774. X.C. Lu, X.Y. Chuan, A.P. Wang, F.Y. Kang, Acta Geol. Sin. 80 (2006) 278.
1775. J.P. Oliver, J. Porous Mater. 2 (1995) 9.
1776. S.R. Levis, P.B. Deasy, Int. J. Pharm. 243 (2002) 125.
1777. P.M. Barata-Rodringues, T.J. Mays, G.D. Moggridge, Carbon 41 (2003) 2231.
1778. B. Notari, Catal. Today 18 (1993) 163.
1779. A.J.H.P. van der Pol, A.J. Verduyn, J.H.C. van Hoo, Appl. Catal. A. 92 (1992) 113.
1780. K. Johannsen, A. Boisen, M. Brorson, I. Schmidt, C.J.H. Jacobsen, Stud. Surf. Sci. Catal. 142 (2002) 109.
1781. R. Schodel, P. Birke, R. Geyer, P. Kraak, W. Muller, H.D. Neubauer, R. Pester, F. Vogt, K.P. Wendlandt, US Patent 5,736,479 (1998).
1782. R. Schodel, R. Pester, W. Muller, R. Geyer, P. Birke, DE Patent 4,240,698 (1994).
1783. P. Birke, P. Kraak, R. Pester, R. Schodel, F. Vogt, Stud. Surf. Sci. Catal. 83 (1994) 425.
1784. G. Li, C.Z. Jin, X.S.Wang, CN Patent 1,554,483 (2004).
1785. P. Serp, M. Corrias, P. Kalck, Appl. Catal. A. 253 (2003) 337.
1786. M.J. Ledoux, R. Vieira, C.P. Huu, N. Keller, J. Catal. 216 (2003) 333.
1787. P. Li, T. Li, J.H. Zhou, Z.J. Sui, Y.C. Dai, W.K. Yuan, D. Chen, Microporous Mesoporous Mater. 95 (2006) 1.
1788. K.P. de Jong, J.W. Geus, Catal. Rev. Sci. Eng. 42 (2000) 481.
1789. N.M. Rodriguez, J. Mater. Res. 8 (1993) 3233.
1790. P. Li, T.J. Zhao, J.H. Zhou, Z.J. Sui, Y.C. Dai, W.K. Yuan, Carbon 43 (2005) 2701.
1791. A. Thangaraj, R. Kumar, S.P. Mirajkar, P. Ratnasamy, J. Catal. 130 (1991) 1.
1792. T.J. Zhao, PhD Thesis, East China University of Science and Technology, 2004.
1793. Q. Zhao, X.H. Bao, X.W. Han, X.M. Liu, D.L. Tan, L.W. Lin, X.W. Guo, G. Li, X.S. Wang, Mater. Chem. Phys. 66 (2000) 41.
1794. R.S. Drago, S.C. Dias, J.M. McGilvray, A.L.M.L. Mateus, J. Phys. Chem. B 102 (1998) 1508.
1795. Q. Zhao, X.W. Han, X.M. Liu, R.S. Zhai, L.W. Lin, X.H. Bao, X.W. Guo, G. Li, X.S. Wang, Chin. J. Catal. 1 (1999) 55.
1796. M.A. Uguina, D.P. Serrano, G. Ovejero, R. Van Grieken, M. Camacho, Appl. Catal. A. 124 (1995) 391.
1797. T.G. Ros, A.J. van Dillen, J.W. Geus, D.C. Koningsberger, Chem-Phys Chem 2 (2002) 209.
1798. D.M. Ruthven, Stud. Surf. Sci. Catal. 97 (1995) 223.
1799. S. Vasenkov, W. Boehlmann, P. Galvosas, O. Geier, H. Liu, J.Kaerger, J. Phys. Chem. B 105 (2001) 5922.
1800. P. Grenier, F. Meunier, P.G. Gray, J. Kaerger, Z. Xu, D.M. Ruthven, Zeolites 14 (1994) 242.
1801. M. Bulow, P. Struve, W. Mietk, M. Kocirik, J. Chem. Soc. Faraday Trans. I 80 (1984) 813.
1802. J. Wloch, Microporous Mesoporous Mater. 62 (2003) 81.
1803. J. Wloch, J. Kornatowski, Langmuir 20 (2004) 1180.
1804. A. Micke, M. Bulow, M. Kocirik, P. Struve, J. Phys. Chem. 98 (1994) 12337.
1805. M. Kocirik, A. Micke, Langmuir 11 (1995) 3042.
1806. M. Eic, A. Micke, M. Kocirik, M. Jama, A. Zikanova, Adsorption 8 (2002) 15.
1807. J. Kornatowski, M. Sychev, V.V. Goncharuk, W.H. Baur, Stud. Surf. Sci. Catal. 65 (1991) 581.

1808. J. Kornatowski, B. Wichterlova, M. Rozwadowski, W.H. Baur, Stud. Surf. Sci. Catal. 84 (1994) 117.
1809. J. Kornatowski, Zeolites 8 (1988) 77.
1810. V. Masarik, P. Novak, A. Zikanova, J. Kornatowski, J. Maixner, M. Kocirik, Collect. Czech. Chem. Commun. 63 (1998) 321.
1811. R.L. June, A.T. Bell, D.N. Theodorou, J. Phys. Chem. 96 (1992) 1051.
1812. E.R. Geus, J.C. Jansen, H. van Bekkum, Zeolites 14 (1994) 32.
1813. C. Weidenthaler, R.X. Fisher, R.D. Shannon, J. Phys. Chem. 98 (1994) 12687.
1814. O. Geier, S. Vasekov, E. Lehmann, J. Karger, U. Schemmert, R.A. Rakoczy, J. Weitkamp, J. Phys. Chem. B 105 (2001) 10217.
1815. J.R. Agger, N. Hanif, C.S. Cundy, A.P. Wade, S. Dennison, P.A. Rawlinson, M.W. Anderson, J. Am. Chem. Soc. 125 (2003) 830.
1816. L. Brabec, M. Kocirik, Mater. Chem. Phys. 102 (2007) 67.
1817. M. Kocirik, J. Kornatowski, V. Masarik, P. Novak, A. Zikanova, J. Maixner, Microporous Mesoporous Mater. 23 (1998) 295.
1818. E. Lehmann, S. Vasenkov, J. Kaerger, G. Zadrozna, J. Kornatowski, Oe. Weiss, F. Schueth, J. Phys. Chem. B 107 (2003) 4685.
1819. J. Kornatowski, Stud. Surf. Sci. Catal. 158 (2005) 303.
1820. J. Crank, The Mathematics of Diffusion, Oxford University Press, Oxford, 1956.
1821. J. Caro, M. Noack, J. Richter-Mendau, F. Marlow, D. Petersohn, M. Griepentrog, J. Kornatowski, J. Phys. Chem. 97 (1993) 13685.
1822. J.B. Donnet, T.K. Wang, J.C.M. Peng, S. Rebouilla, Carbon Fibers, third ed., Dekker, New York, 1998.
1823. T.J. Mays, Activated carbonfibers, in: T.D. Burchell (Ed.), Carbon Materials for Advanced Technologies, Pergamon, Oxford, 1999.
1824. M. Suzuki, Carbon 32 (1994) 577.
1825. R.N. Mcnair, G.N. Arons, Sorptive textile systems containing activated carbon fibers, in: P.N. Cheremisino, F. Ellerbusch (Eds.), Carbon Adsorption Handbook, Ann Arbor Science, Ann Arbor, MI, 1980.
1826. S.N. Ahmed, R. Baldwin, F. Derbyshire, B. McEnaney, J. Stencel, Fuel 72 (1993) 287.
1827. H. Teng, E.M. Suuberg, J.M. Calo, Energy Fuels 6 (1992) 398.
1828. M.J. Illán-Gómez, A. Linares-Solano, L.R. Radovic, C. Salinas-Martinéz de Lecea, Energy Fuels 10 (1996) 158.
1829. K. Kaneko, N. Fukuzaki, K. Kakei, T. Suzuki, S. Ozeki, Langmuir 5 (1989) 960.
1830. K. Kaneko, Y. Nakahigahi, K. Nagata, Carbon 26 (1988) 327.
1831. C.M. Yang, K. Kaneko, J. Colloid Interface Sci. 255 (2002) 236.
1832. K. Otto, M. Shelef, J. Catal. 18 (1970) 184.
1833. K. Kaneko, Langmuir 3 (1987) 357.
1834. K. Kaneko, N. Shindo, Carbon 27 (1989) 815.
1835. Z.M. Wang, T. Suzuki, N. Uekawa, K. Asakura, K. Kaneko, J. Phys. Chem. 96 (1992) 10917.
1836. J. Alcañiz-Monge, M.A. Lillo-Ródenas, A. Bueno-López, M. Illán-Gómez, Microporous Mesoporous Mater. 100 (2007) 202.
1837. J. Alcañiz-Monge, D. Cazorla-Amorós, A. Linares-Solano, S. Yoshida, A. Oya, Carbon 32 (1994) 1277.
1838. F. Rodrýguez-Reinoso, A. Linares-Solano, Microporous structure of activated carbons as revealed by adsorption methods, in: P.A. Thrower (Ed.), Chemistry and Physics of Carbon, vol. 21, Dekker, New York, 1988.
1839. H. Teng, E.M. Suuberg, J. Phys. Chem. 97 (1993) 478.
1840. M.J.D. Low, Chem. Rev. 60 (1960) 267.
1841. G.F. Cerofolini, Adsorption kinetics on real surfaces, in: A. Darbrowski, V.A. Tertykh (Eds.), Adsorption on New and Modified Inorganic Adsorbents, Elsevier, Amsterdam, 1995.
1842. A.S. Porter, F.C. Tompkins, Proc. R. Soc. Lond. A 217 (1953) 529.
1843. C.E. Brown, P.G. Hall, Trans. Faraday Soc. 67 (1971) 3558.
1844. K. Kaneko, C. Ishii, Colloids Surf. 67 (1992) 203.
1845. M. Sato, T. Sukegawa, T. Suzuki, K. Kaneko, J. Phys. Chem. B. 101 (1997) 1845.

1846. H. Birnholtz, A. Nir, N. Lotan, C. Aharoni, Can. J. Chem. Eng. 62 (1984) 233.
1847. I. Mochida, M. Ogaki, H. Fujitsu, Y. Komatsubara, Fuel 64 (1985) 1054.
1848. H. Teng, E.M. Suuberg, Ind. Eng. Chem. Res. 32 (1993) 416.
1849. J. Imai, T. Suzuki, K. Kaneko, Catal. Lett. 20 (1993) 133.
1850. K. Kaneko, A. Kobayashi, T. Suzuki, S. Ozeki, K. Kakei, N. Kosugi, H. Kuroda, J. Chem. Soc. Faraday Trans. I. 84 (1988) 1795.
1851. Y. Nishi, T. Suzuki, K. Kaneko, J. Phys. Chem. B. 101 (1997) 1938.
1852. K. Kaneko, S. Ozeki, K. Inouye, Colloid Polym. Sci. 265 (1987) 1018.
1853. I. Aarna, E.M. Suuberg, Fuel 76 (1997) 475.
1854. A. Newkirk, Anal. Chem. 32 (1960) 1558.
1855. M.J. Illán-Gómez, A. Linares-Solano, C. Salinas-Martýnez de Lecea, Energy Fuels 9 (1995) 976.
1856. J.D. Fast, Interaction of Metals and Gases, Academic Press, New York, 1965, pp. 61–70.
1857. K. Kaneko, K. Inouye, Adsorpt. Sci. Technol. 3 (1986) 11.
1858. J. Imai, M. Souma, S. Ozeki, T. Suzuki, K. Kaneko, J. Phys. Chem. 95 (1991) 9955.
1859. Z.M. Wang, N. Shindo, Y. Otake, K. Kaneko, Carbon 32 (1994) 515.
1860. H.W. Roesky, K.K. Pandey, Adv. Inorg. Chem. Radiochem. 26 (1983) 337.
1861. F.A. Cotton, G. Wilkinson, C.A. Murillo, M. Bochmann, Advanced Inorganic Chemistry, sixth ed., Wiley, New York, 1999.
1862. A.M. Rich, R.S. Armstrong, P.J. Ellis, P.A. Lay, J. Am. Chem. Soc. 120 (1998) 10827.
1863. D.H. Young, A.O. Crowell, Physical Adsorption of Gases, Butterworth, London, 1962.
1864. B.M. Hoffman, N.J. Nelson, J. Chem. Phys. 50 (1969) 2598.
1865. A. Corma, Chem. Rev. 97 (1997) 2373.
1866. M. Lindén, S. Schacht, F. Schüth, A. Steel, K. Unger, J. Porous Mater. 5 (1998) 177.
1867. G. Øye, J. Sjöblom, M. Stöcker, Adv. Colloid Interface Sci. 89–90 (2001) 439.
1868. F. Di Renzo, A. Galarneau, P. Trens, F. Fajula, in: F. Schüth, K.S.W. Sing, J. Weitkamp (Eds.), Handbook of Porous Solids, Wiley-VCH, Weinheim, 2002.
1869. T. Lissen, K. Cassiers, P. Cool, E.F. Vansant, Adv. Colloid Interface Sci. 103 (2003) 121.
1870. A. Tagushi, F. Schüth, Microporous Mesoporous Mater. 77 (2005) 1.
1871. G. Øye, W.R. Glomm, T. Vra¢ªlstad, S. Volden, H. Magnusson, M. Stöcker, J. Sjöblom, Adv. Colloid Interface Sci. 123–126 (2006) 17.
1872. C.-Y. Chen, H.-X. Li, M.E. Davis, Microporous Mater. 2 (1993) 17.
1873. J.M. Kim, J.H. Kwak, S. Jun, R. Ryoo, J. Phys. Chem. 99 (1995) 16742.
1874. M.T.J. Keene, R.D.M. Gougeon, R. Denoyel, R.K. Harris, J. Rouquerol, P.L. Llewellyn, J. Mater. Chem. 9 (1999) 2843.
1875. R. Mokaya, J. Phys. Chem. B 103 (1999) 10204.
1876. Y. Inaki, H. Yoshida, K. Kimura, S. Inagaki, Y. Fukushima, T. Hattori, Phys. Chem. Chem. Phys. 2 (2000) 5293.
1877. A. Galarneau, D. Desplantier, F. DiRenzo, F. Fajula, Catal. Today 68 (2001) 191.
1878. M. Broyer, S. Valange, J.P. Bellat, O. Bertrand, G. Weber, Z. Gabelica, Langmuir 18 (2002) 5083.
1879. K. Cassiers, T. Linssen, M. Mathieu, M. Benjelloun, K. Schrijnemakers, P. Van Der Voort, P. Cool, E.F. Vansant, Chem. Mater. 14 (2002) 2317.
1880. H. Chen, Y. Wang, Ceram. Int. 28 (2002) 541.
1881. T.R. Gaydhankar, U.S. Taralkar, R.K. Jha, P.N. Joshi, R. Kumar, Catal. Commun. 6 (2005) 361.
1882. C.-Y. Chen, S.-Q. Xiao, M.E. Davis, Microporous Mater. 4 (1995) 1.
1883. A. Matsumoto, T. Sasaki, N. Nishimiya, K. Tsutsumi, Colloids Surf. A: Physicochem. Eng. Asp. 203 (2002) 185.
1884. E.B. Celer, M. Kruk, M. Jaroniec, Adsorption 11 (2004) 698.
1885. M. Kruk, E.B. Celer, M. Jaroniec, Chem. Mater. 16 (2004) 698.
1886. C.H. Rhee, J.S. Lee, Catal. Lett. 40 (1996) 261.
1887. L.Y. Chen, S. Jaenicke, G.K. Chuah, Microporous Mater. 12 (1997) 323.
1888. Z. Shan, J.C. Jansen, L. Marchese, T. Maschmeyer, Microporous Mesoporous Mater. 41 (2001) 181.

1889. V. Zelenják, V. Hornebecq, S. Mornet, O. Schäf, P. Llewellyn, Chem. Mater. 18 (2006) 3184.
1890. V.Y. Gusev, X. Feng, Z. Bu, G.L. Haller, J.A. O'Brien, J. Phys. Chem. 100 (1996) 1985.
1891. T. Tatsumi, K.A. Koyano, Y. Tanaka, S. Nakata, Chem. Lett. 26 (1997) 469.
1892. T. Tatsumi, K.A. Koyano, Y. Tanaka, S. Nakata, J. Porous Mater. 6 (1999) 13.
1893. J. Wu, X. Liu, S.H. Tolbert, J. Phys. Chem. B 104 (2000) 11837.
1894. M.A. Springuel-Huet, J.L. Bonardet, A. Gédéon, Y. Yue, V.N. Romannikov, J. Fraissard, Microporous Mesoporous Mater. 44–45 (2001) 745.
1895. N. Bai, Y. Chi, Y. Zou, W. Pang, Mater. Lett. 54 (2002) 37.
1896. K.A. Koyano, T. Tatsumi, Y. Tanaka, S. Nakata, J. Phys. Chem. B 101 (1997) 9436.
1897. M. Hartmann, C. Bischof, J. Phys. Chem. B 103 (1999) 6230.
1898. T. Ishikawa, M. Matsuda, A. Yasukawa, K. Kandori, S. Inagaki, T. Fukushima, S. Kondo, J. Chem. Soc. Faraday Trans. 92 (1996) 1985.
1899. A. Vinu, V. Murugesan, M. Hartmann, Chem. Mater. 15 (2003) 1385.
1900. N. Igarashi, K. Hashimoto, T. Tatsumi, J. Mater. Chem. 12 (2002) 3631.
1901. T.N. Silva, J.M. Lopes, F. Ramôa Ribeiro, M.R. Carrott, P.C. Galacho, M.J. Sousa, P. Carrott, React. Kinet. Catal. Lett. 77 (2002) 83.
1902. C. Galacho, M.M.L. Ribeiro Carrott, P.J.M. Carrott, Microporous Mesoporous Mater. 1–3 (2007) 312.
1903. A. Matsumoto, H. Chen, K. Tutsumi, M. Grün, K. Unger, Microporous Mesoporous Mater. 32 (1999) 55.
1904. F. Rouquérol, J. Rouquérol, K.S.W. Sing, Adsorption by Powders and Porous Solids, Academic Press, London, 1999.
1905. M.M.L. Ribeiro Carrott, A.J. Candeias, P.J.M. Carrott, P.I. Ravikovitch, A.V. Neimark, A.D. Sequeira, Microporous Mesoporous Mater. 47 (2001) 323.
1906. M.M.L. Ribeiro Carrott, F.L. Conceição, J.M. Lopes, P.J.M. Carrott, C. Bernardes, J. Rocha, F. Ramôa Ribeiro, Microporous Mesoporous Mater. 92 (2006) 270.
1907. M.M.L. Ribeiro Carrott, C. Galacho, F.L. Conceição, P.J.M. Carrott, in: P.L. Llewellyn, J. Rouquerol, F. Rodriguez-Reinoso, N. Seaton (Eds.), Characterization of Porous Solids VII, Studies in Surface Science and Catalysis, vol. 160, Elsevier, Amsterdam, 2006.
1908. M.M.L. Ribeiro Carrott, A.J.E. Candeias, P.J.M. Carrott, K.K. Unger, Langmuir 15 (1999) 8895.
1909. Z. Luan, L. Kevan, J. Phys. Chem. B 101 (1997) 2021.
1910. L. Marchese, T. Maschmeyer, E. Gianotti, S. Coluccia, J.M. Thomas, J. Phys. Chem. B 101 (1997) 8836.
1911. L. Marchese, E. Gianotti, V. Dellarocca, T. Maschmeyer, F. Rey, S. Coluccia, J.M. Thomas, Phys. Chem. Chem. Phys. 1 (1999) 585.
1912. P. Ratnasamy, D. Srinivas, H. Knözinger, Adv. Catal. 48 (2004) 1.
1913. W. Zhang, M. Fröba, J. Wang, P.T. Tanev, J. Wong, T.J. Pinnavaia, J. Am. Chem. Soc. 118 (1996) 9164.
1914. Z. Zhang, T.J. Pinnavaia, J. Am. Chem. Soc. 124 (2002) 12294.
1915. F. Vaudry, S. Khodabandeh, M.E. Davis, Chem. Mater. 8 (1996) 1451.
1916. S. Cabrera, J.E. Haskouri, J. Alamo, A. Beltrán, S. Mendioroz, M.D. Macros, P. Amorós, Adv. Mater. 11 (1999) 379.
1917. Z. Zhang, R.W. Hicks, T.R. Pauly, T.J. Pinnavaia, J. Am. Chem. Soc. 124 (2002) 1592.
1918. R.W. Hick, T.J. Pinnavaia, Chem. Mater. 15 (2003) 78.
1919. H.C. Lee, H.J. Kim, S.H. Chung, K.H. Lee, H.C. Lee, J.S. Lee, J. Am. Chem. Soc. 125 (2003) 2882.
1920. D. Kuang, Y. Fang, H. Liu, C. Frommen, D. Fenske, J. Mater. Chem. 13 (2003) 660.
1921. T.-Z. Ren, Z.-Y. Yuan, B.-L. Su, Langmuir 20 (2004) 1531.
1922. L. Qu, C. He, Y. Yang, Y. He, Z. Liu, Mater. Lett. 59 (2005) 4034.
1923. J. Aguado, J.M. Escola, M.C. Castro, B. Paredes, Microporous Mesoporous Mater. 83 (2005) 181.
1924. K. Niesz, P. Yang, G.A. Somorjai, Chem. Commun. (2005)1986.
1925. J.A. Toledo, X. Bokhimi, C. Lopez, C. Angeles, F. Hernandez, J.J. Fripiat, J. Mater. Res. 20 (2005) 2947.

1926. T.F. Baumann, A.E. Gash, S.C. Chinn, A.M. Sawvel, R.S. Maxwell, J.H. Satcher Jr., Chem. Mater. 17 (2005) 395.
1927. Z. Ruihong, G. Fen, H. Yongqi, Z. Huanqi, Microporous Mesoporous Mater. 93 (2006) 212.
1928. J. Cejka, P.J. Kooyman, L. Veselá, J. Rathouský, A. Zukal, Phys. Chem. Chem. Phys. 4 (2002) 4823.
1929. J. Cejka, Appl. Catal. A Gen. 254 (2003) 327.
1930. T.J Pinnavaia, Z. Zhang, R.W. Hicks, Nanoporous materials-IV, in: A. Sayari, M. Jaroniec (Eds.), Elsevier, Amsterdam, Stud. Surf. Sci. Catal. 156 (2005) 1.
1931. N. Zilková, A. Zukal, J. Cejka, Microporous Mesoporous Mater. 95 (2006) 176.
1932. X. Krokidis, P. Raybaud, A.-E. Gobichon, B. Rebours, P. Euzen, H. Toulhoat, J. Phys. Chem. B 105 (2001) 5121.
1933. H. Hou, Y. Xie, Q. Yang, Q. Guo, C. Tan, Nanotechnology 16 (2005) 741.
1934. J. Lee, S. Han, T. Hyeon, J. Mater. Chem. 14 (2004) 269.
1935. F. Schüth, Chem. Mater. 13 (2001) 3184.
1936. A.-H. Lu, F. Schüth, C. R. Chim. 14 (2005) 269.
1937. S. Tanaka, N. Nishiyama, Y. Egashira, K. Ueyama, Chem. Commun. (2005)2125.
1938. Y. Meng, D. Gu, F. Zhang, Y. Shi, H. Yang, Z. Li, C. Yu, B. Tu, D. Zhao, Angew. Chem. Int. Ed. 44 (2005) 7053.
1939. F. Zhang, Y. Meng, D. Gu, Y. Yan, C. Yu, B. Tu, D. Zhao, J. Am. Chem. Soc. 127 (2005) 13508.
1940. F. Schüth, W. Schmidt, Adv. Mater. 14 (2003) 629.
1941. A. Stein, Adv. Mater. 16 (2003) 763.
1942. H. Yang, D. Zhao, J. Mater. Chem. 15 (2005) 1217.
1943. A.-H. Lu, F. Schüth, Adv. Mater. 18 (2006) 1793.
1944. M. Choi, R. Ryoo, Nature Mater. 2 (2003) 473.
1945. A.-H. Lu, W. Li, Z. Hou, F. Schüth, Chem. Commun. (2007)1038.
1946. P. Vinke, M. Van der Eijk, M. Verbree, A.F. Voskamp, H. Van Bekkum, Carbon 32 (1994) 675.
1947. V. Gomez-Serrano, F. Piriz-Almeida, C.J. Duran-Valle, J. Pastor-Villegas, Carbon 37 (1999) 1517.
1948. X. Chen, M. Faber, Y. Gao, I. Kulaots, E.M. Suuberg, R.H. Hurt, Carbon 41 (2003) 1489.
1949. P.L. Walker Jr. (Ed.), Chemistry and Physics of Carbon, Marcel Dekker, New York, 1970.
1950. Z. Li, W. Yan, S. Dai, Langmuir 21 (2005) 11999.
1951. K. Lászlá, K. Josepovits, E. Tombácz, Anal. Sci. 17 (2001) 1741.
1952. F. Schüth, Annu. Rev. Mater. Res. 35 (2005) 209.
1953. M. Burghard, K. Balasubramanian, Small 1 (2005) 180.
1954. L.R. Radovic, F. Rodriquez-Reinoso, Chem. Phys. Carbon 25 (1997) 243.
1955. P. Chingombe, B. Saha, R.J. Wakeman, Carbon 43 (2005) 3132.
1956. B.K. Pradhan, N.K. Sandle, Carbon 37 (1999) 1323.
1957. S.S. Barton, M.J.B. Evans, E. Halliop, J.A.F. MacDonald, Carbon 35 (1997) 1361.
1958. A.-H. Lu, W.-C. Li, N. Muratova, B. Splietho., F. Schüth, Chem. Commun. (2005) 5184.
1959. W. Zhao, Ch. Song, P. Pehrsson, J. Am. Chem. Soc. 124 (2002) 12418.
1960. S.C. Tsang, Y.K. Chen, P.J.F. Harris, M.L.H. Green, Nature 372 (1994) 159.
1961. S. Jun, M. Choi, S. Ryu, H.-Y. Lee, R. Ryoo, Studies in Surface Science and Catalysis, vol. 37, Elsevier Science B.V., 2003.
1962. A.-H. Lu, W.-C. Li, W. Schmidt, W. Kiefer, F. Schüth, Carbon 42 (2004) 4303.
1963. A.C. Dillon, T. Gennet, K.M. Jones, J.L. Alleman, P.A. Parilla, M.J. Heben, Adv. Mater. 11 (1999) 1354.
1964. J. Zhang, H. Zou, Q. Quing, Y. Yang, Q. Li, Z. Liu, X. Gu, Z. Du, J. Phys. Chem. B 107 (2003) 3712.
1965. Y.H. Li, S. Wang, Z. Luan, J. Ding, C. Xu, D. Wu, Carbon 41 (2003) 1057.
1966. D. Chattopadhyay, I. Galeska, F. Papadimitrakopoulos, Carbon 40 (2002) 985.
1967. R. Yu, L. Chen, Q. Liu, J. Lin, K.-L. Tan, S.C. Ng, H.S.O. Chan, G.-Q. Xu, T.S.A. Hor, Chem. Mater. 10 (1998) 718.
1968. A.-H. Lu, W. Li, W. Schmidt, F. Schüth, Microporous Mesoporous Mater. 80 (2005) 117.

1969. J.L. Figureido, M.F.R. Pereira, M.M.A. Freitag, J.J.M. Orfão, Carbon 37 (1998) 1379.
1970. S. Haydar, C. Moreno-Castilla, M.A. Ferro-Garcia, F. Carrasco-Marin, J. Rivera-Utrilla, Carbon 38 (2000) 1297.
1971. J. Garcia-Martin, R. Lopez-Garzón, M. Luz Godino-Salido, M. Dolores Gutiérrez Valero, P. Arranz-Mascarós, R. Cuesta, F. Carrasco-Marin, Langmuir 21 (2005) 6908.
1972. Y. Otake, R.G. Jenkins, Carbon 31 (1993) 109.
1973. G. Tremblay, F.J. Vastola, P.L. Walker Jr., Carbon 16 (1978) 35.
1974. P.-Z. Cheng, H. Teng, Carbon 41 (2003) 2057.
1975. C. Nozaki, C.G. Lugmair, A.T. Bell, T.D. Tilley, J. Am. Chem. Soc. 124 (2002) 13194.
1976. F. Lefebvre, J.M. Basset, J. Mol. Catal. A: Chem. 146 (1999) 3.
1977. J.M. Thomas, Top. Catal. 15 (2001) 85.
1978. R.K. Grasselli, Top. Catal. 15 (2001) 93.
1979. J.C. Volta, Top. Catal. 15 (2001) 121.
1980. A. Ma, W. Grünert, Chem. Commun. (1999)71.
1981. Q. Sun, Z.X. Gao, H.Y. Chen, W.M.H. Sachtler, J. Catal. 201 (2001) 88.
1982. Q.L. Richard, T.Y. Ralph, J. Am. Chem. Soc. 121 (1999) 5595.
1983. M. Schwidder, M.S. Kumar, K. Klementiev, M.M. Pohl, A. Brückner, W. Grünert, J. Catal. 231 (2005) 314.
1984. G.I. Panov, M.A. Uriarte, M.A. Rodkin, V.I. Sobolev, Catal. Today 41 (1998) 365.
1985. V.I. Sobolev, K.A. Dubkov, O.V. Panna, G.I. Panov, Catal. Today 24 (1995) 251.
1986. A. Szegedi, G. Pal-Borbely, K. Lazar, React. Kinet. Catal. Lett. 74 (2001) 277.
1987. L. Brabec, M. Jeschke, R. Klik, J. Nováková, L. Kubelková, J. Meusinger, Appl. Catal. A Gen. 170 (1998) 105.
1988. J.O. Petunchi, W.K. Hall, J. Catal. 78 (1982) 327.
1989. Y.M. Kang, B.Z. Wan, Catal. Today 35 (1997) 379.
1990. A. Szegedi, M. Hegedús, J.L. Margitfalvi, I. Kiricsi, Chem. Commun. (2005)1441.
1991. N. Kritchayanon, N. Thanabodeekij, S. Jitkarnka, A.M. Jamieson, S. Wongkasemjit, Appl. Organomet. Chem. 20 (2006) 155.
1992. L.J. Lobree, I.C. Hwang, J.A. Rimer, A.T. Bell, J. Catal. 186 (2000) 242.
1993. P. Marturano, L. Drozdova, A. Kogelbauer, J. Catal. 192 (2000) 236.
1994. R. Joyner, M. Stockenhuber, J. Phys. Chem. B 103 (1999) 5963.
1995. T. Maschmeyer, F. Rey, G. Sankar, J.M. Thomas, Nature 378 (1995) 159.
1996. X. Wang, H. Xu, X. Fu, P. Liu, F. Lefebvre, J.M. Basset, J. Mol. Catal. A: Chem. 238 (2005) 185.
1997. X. Wang, W. Lian, X. Fu, J. Catal. 238 (2006) 13.
1998. M. Iwamoto, H. Kusano, S. Kagawa, Inorg. Chem. 22 (1983) 3365.
1999. M. Iwamoto, S. Nakamura, H. Kusano, S. Kagawa, J. Phys. Chem. 90 (1986) 5244.
2000. J.B. Nagy, M.V. Eenoo, E.G. Derouane, J. Catal. 58 (1979) 230.
2001. G.A. Ozin, J. Godber, J. Phys. Chem. 93 (1989) 878.
2002. A. Borvornwattananont, K. Moller, T. Bein, J. Phys. Chem. 93 (1989) 4205.
2003. K. Moller, A. Borvornwattananont, T. Bein, J. Phys. Chem. 93 (1989) 4562.
2004. F.R. Sarria, O. Marie, J. Saussey, J. Phys. Chem. B 109 (2005) 1660.
2005. J.S. Bodenheimer, W. Low, Spectrochim. Acta A 29 (1973) 1733.
2006. M.S. Kumar, M. Schwidder, W. Grünter, A. Brückner, J. Catal. 227 (2004) 384.
2007. S. Bordiga, R. Buzzoni, F. Geobaldo, C. Lamberti, E. Giamello, A. Zecchina, G. Ledanti, G. Petrini, G. Tozzola, G. Vlaic, J. Catal. 158 (1996) 486.
2008. J.F. Moulder, W.F. Stickle, P.E. Sobol, K.D. Bomben, Handbook of X-ray Photoelectron Spectroscopy, Physical Electronic, Inc., United States of America, 1995.
2009. F. Heinrich, C. Schmidt, E. Löer, M. Menzel, W. Grünert, J. Catal. 212 (2002) 157.
2010. M. Wilke, F. Farges, P.E. Peltit, G.E.J. Brown, F. Martin, Am. Mineral. 86 (2001) 714.
2011. Y. Shiro, F. Sato, T. Suzuki, T. Iizuka, T. Matsushima, H. Oyanagi, J. Am. Chem. Soc. 112 (1990) 2921.
2012. C.Y. Yang, S.M. Heald, J.M. Tranquada, Y. Xu, Y.L. Wang, A.R. Moodenbaugh, D.O. Welch, M. Suenga, Phys. Rev. 39 (1989) 6681.
2013. A.L. Roe, D.J. Schneider, R.J. Mayer, J.W. Pyrz, J. Widom, L.J. Que, J. Am. Chem. Soc. 106 (1984) 1676.

2014. S.H. Choi, B.R. Wood, J.A. Ryder, A.T. Bell, J. Phys. Chem. B 107 (2003) 11843.
2015. S.H. Choi, B.R. Wood, A.T. Bell, M.T. Janicke, K.C. Ott, J. Phys. Chem. B 107 (2004) 8970.
2016. O.A. Carlos, T.P. Gemma, E.P. Estrella, P.M. Margarita, J. Chem. Soc., Dalton Trans. (1997) 873.
2017. M.A. Uddin, T. Komatsu, T. Yashima, J. Catal. 146 (1994) 468.
2018. A. Ribera, I.W.C.E. Arends, S. de Vries, J.P. Ramýrez, R.A. Sheldon, J. Catal. 195 (2000) 287.
2019. J.P. Ramyréz, G. Mul, F. Kapteijn, J. Moulijn, A.A.R. Overweg, A. Doménech, A. Ribera, I.W.C.E. Arends, J. Catal. 207 (2002) 113.
2020. R. Szostak, V. Nair, T.L. Thomas, J. Chem. Soc., Faraday Trans. 183 (1987) 487.
2021. G. Blasse, B.C. Grabmaier, Luminescent Materials, Springer, Heidelberg, 1994.
2022. R. Reisfeld, C.K. Jorgensen, Lasers and Excited States of Rare Earth, Springer, New York, 1977.
2023. T. Aitasalo, J. Hölsä, H. Jungner, M. Lastusaari, J. Niittykoski, J. Phys. Chem. B 110 (2006) 4589.
2024. C. Feldmann, T. Jüstel, C.R. Ronda, P.J. Schmidt, Adv. Funct. Mater. 13 (2003) 511.
2025. N. Rakov, F.E. Ramos, G. Hirata, M. Xiao, Appl. Phys. Lett. 83 (2003) 272.
2026. J. Hölsä, E. Antic-Fidancev, M. Lastusaari, A. Lupei, J. Solid State Chem. 171 (2003) 282.
2027. C.A. Kodaira, H.F. Brito, O.L. Malta, O.A. Serra, J. Lumin. 101 (2003) 11.
2028. W.M. Fautino, O.L. Malta, E.E.S. Teotônio, H.F. Brito, A.M. Simas, G.F. Sá, J. Chem. Phys. A 110 (2006) 2510.
2029. G.K. Chuah, S. Jaenicke, T.H. Xu, Microporous Mesoporous Mater. 37 (2000) 345.
2030. J. Sanches-Valente, X. Bokhimi, J.A. Toledo, Appl. Catal. A Gen. 264 (2004) 175.
2031. A.A. Kaplyanskii, A.B. Kulinkin, A.B. Kutsenko, S.P. Feolov, R.I. Zakharchenya, T.N. Vasilevskaya, Phys. Solid State 40 (1998) 1310.
2032. O. Ozuna, G.A. Hirata, J. McKittrick, Appl. Phys. Lett. 94 (2004) 1296.
2033. T. Yamamoto, T. Tanaka, T. Matsuyama, T. Funabiki, S. Yoshida, J. Phys. Chem. B 105 (2001) 1908.
2034. G. Hirata, N. Perea, M. Tejeda, J.A. Gonzalez-Ortega, J. McKittrick, Opt. Mater. 27 (2005) 1311.
2035. T. Ishizaka, R. Nozaki, Y. Kurokawa, J. Phys. Chem. Solids 63 (2002) 613.
2036. T. Ishizaka, Y. Kurokawa, J. Appl. Phys. 90 (2001) 243.
2037. J. Wrzyszcz, W. Mista, D. Hreniak, W. Strek, M. Zawadzki, H. Grabowska, J. Alloys Compd. 341 (2002) 358.
2038. A. Patra, E. Sominska, S. Ramesh, Y. Koltypin, Z. Zhong, H. Minti, R. Reisfeld, A. Gedanken, J. Phys. Chem. B 103 (1999) 3361.
2039. R. Kudrawiec, J. Misiewicz, L. Bryja, I.S. Molchan, N.V. Gaponenko, J. Alloy. Compd. 341 (2002) 211.
2040. M.U. Pechini, US Patent 3,330,697 (1967).
2041. S. Shikao, W. Jiye, J. Alloy. Compd. 327 (2001) 82.
2042. G.Y. Adachi, K.I. Machida, J. Shiokawa, J. Less-Common Met. 93 (1983) 389.
2043. B.D. Cullity, Elements of X-Ray Diffraction, Addison-Wesley Publishing Company Inc., Reading, Massachusetts, 1967.
2044. K. Nakamoto, Infrared and Raman Spectra of Inorganic and Coordination Compounds, Willey-Interscience Publication, 1986.
2045. M.T. Hernández, M. González, J. Eur. Ceram. Soc. 22 (2002) 2861.
2046. S. Desset, O. Spalla, P. Lixon, B. Cabane, Colloids Surface A: Phys. Eng. Asp. 196 (2002) 1.
2047. Q. Dai, G.N. Robinson, A. Freedman, J. Phys. Chem. B 101 (1997) 4940.
2048. C. Laberty-Robert, F. Ansart, C. Deloget, M. Gaudon, A. Rousset, Mater. Res. Bull. 36 (2001) 2083.
2049. S.T. Aruna, K.S. Rajam, Mater. Res. Bull. 39 (2004) 157.
2050. Y. Sarikaya, K. Ada, T. Alemdaroglu, I. Bozdogan, J. Eur. Ceram. Soc. 22 (2002) 1905.
2051. S. Cava, S.M. Tebcherani, S.A. Pianaro, C.A. Paskocimas, E. Longo, J.A. Varela, Mater. Chem. Phys. 97 (2006) 102.
2052. R.G. Burns, Mineralogical Applications of Crystal Field Theory, Cambridge University Press, 1993.

2053. O.L. Malta, H.F. Brito, J.F.S. Menezes, F.R.G. Silva, S. Alves Jr., F.S. Farias Jr., A.V.M. de Andrade, J. Lumin. 75 (1997) 255.
2054. G.L. Baugis, H.F. Brito, W. Oliveira, F.R. Castro, E.F. Souza-Aguiar, Microporous Mesoporous Mater. 49 (2001) 179.
2055. W.T. Carnall, H. Crosswhite, H.M. Crosswhite, Energy Structure and Transition Probabilities of the Trivalent Lanthanides in LaF3, Argonne National Laboratory Report, unnumbered, 1977.
2056. O.L. Malta, S.J.L. Ribeiro, M. Faucher, P. Porcher, J. Phys. Chem. Sol. 52 (1991) 587.
2057. R.W. Pekala, J. Mater. Sci. 24 (1989) 3221.
2058. R.W. Pekala, C.T. Alviso, Mater. Res. Soc. Symp. Proc. 270 (1992) 3.
2059. H. Tamon, H. Ishizaka, Carbon 36 (9) (1998) 1397.
2060. Yao-Jun Zhang, A. Maroto-Valiente, I. Rodriguez-Ramos, Qin Xin, A. Guerrero-Ruiz, Catal. Today 93–95 (2004) 619.
2061. C. Liang, G. Sha, S. Guo, J. Non-Cryst. Solids 271 (2000) 167.
2062. T. Yamamoto, T. Nishimura, T. Suzuki, H. Tamon, J. Non-Cryst. Solids 288 (2001) 46.
2063. S.Y. Kim, D.H. Yeo, J.W. Lim, K.P. Yoo, K.H. Lee, H. Kim, J. Chem. Eng. Jpn. 34 (2001) 216.
2064. R.W. Pekala, US Patent 4,997,804 (1991).
2065. K. Barral, J. Non-Cryst. Solids 225 (1998) 46.
2066. M.F. Grenier-Loustalot, S. Larroque, D. Grande, P. Grenier, D. Bedel, Polymer 37 (1996) 1363.
2067. T. Horikawa, J. Hayashi, K. Muroyama, Carbon 42 (2004) 1625.
2068. R.W. Pekala, C.T. Alviso, F.M. Kong, S.S. Hulsey, J. Non-Cryst. Solids 145 (1992) 90.
2069. R.W. Pekala, D.W. Schaefer, Macromolecules 26 (1993) 5487.
2070. G. Reichenauer, A. Emerling, J. Fricke, R.W. Pekala, J. Non-Cryst. Solids 225 (1998) 210.
2071. C. Lin, J.A. Ritter, Carbon 35 (9) (1997) 1271.
2072. R.W. Pekala, US Patent 4,873,218, October 10 (1989).
2073. R. Saliger, V. Bock, R. Petricevic, T. Tillotson, S. Geis, J. Fricke, J. Non-Cryst. Solids 221 (1997) 144.
2074. T. Yamamoto, T. Nishimura, T. Suzuki, H. Tamon, Carbon 39 (2001) 2369.
2075. T.F. Baumann, J.H. Satcher Jr., J. Non-Cryst. Solids 350 (2004) 120.
2076. H. Tamon, H. Ishizaka, T. Araki, M. Okazaki, Carbon 36 (9) (1998) 1257.
2077. H. Tamon, H. Ishizaka, M. Mikami, M. Okazaki, Carbon 35 (6) (1997) 791.
2078. M. Wiener, G. Reichenauer, T. Scherb, J. Fricke, J. Non-Cryst. Solids 350 (2004) 126.
2079. W.-C. Li, A.-H. Li, S.-C. Guo, J. Colloid Interface Sci. 254 (2002) 153–157.
2080. C. Moreno-Castilla, F.J. Maldonado-Hódar, Carbon 43 (2005) 455.
2081. D.J. Suh, J. Non-Cryst. Solids 350 (2004) 314.
2082. R. Brandt, J. Fricke, J. Non-Cryst. Solids 350 (2004) 131.
2083. Y. Hanzawa, H. Hatori, N. Yoshizawa, Y. Yamada, Carbon 40 (2002) 575.
2084. H. Marsh, W.F.K. Wynne-Jones, Carbon 1 (3) (1964) 269.
2085. F.J. Maldonado-Hódar, C. Moreno-Castilla, A.F. Pérez-Cárdenas, Microporous Mesoporous Mater. 69 (2004) 119.
2086. S.R. Mukai, T. Sugiyama, H. Tamon, Appl. Catal. A Gen. 256 (2003) 99.
2087. S.T. Wilson, B.M. Lok, C.A. Messina, T.R. Cannan, E.M. Flanigen, J. Am. Chem. Soc. 104 (1982) 1146.
2088. E.M. Flanigen, B.M. Lok, R.L. Patton, S.T. Wilson, in: Y. Murakami, A. Iijima, J.W. Ward (Eds.), New Developments in Zeolite Science and Technology, Elsevier, New York, 1986.
2089. L. Yang, Y. Aizen, X. Quinhua, Appl. Catal. 67 (1991) 169.
2090. S. Subranium, A. Mitra, C.V.V. Satyanarayana, D.K. Chakrabarty, Appl. Catal. Gen. 159 (1997) 229.
2091. J.M. Thomas, R. Raja, Chem. Commun. 675 (2001).
2092. S.T. Wilson and E.M. Flanigen, US Patent 4,567,029 (1986).
2093. M. Hartmann, L. Kevan, Chem. Rev. 99 (1999) 635.
2094. A.M. Prakash, V. Kurshev, L. Kevan, J. Phys. Chem. B 101 (1997) 9794.

2095. S. Ashtekar, A.M. Prakash, D.K. Chakrabarty, C.V.V. Satyanarayana, J. Chem. Soc., Faraday Trans. 92 (1996) 2481.
2096. A.M. Prakash, C.V.V. Satyanarayana, S. Ashtekar, D.K. Chakrabarty, J. Chem. Soc., Faraday Trans. 92 (1996) 1257.
2097. L. Fruza, J. Pelgrims, H. Leeman, P. Van Der Voort, E.F. Vansant, R.A. Schoonheydt, B.M. Weckhuysen, J. Phys. Chem. B 105 (2001) 2677.
2098. R.A. Sheldon, Chem. Tech. (1991) 566.
2099. T. Yamaguchi, Catal. Today 20 (1994) 199.
2100. K. Tanabe, T. Yamaguchi, Catal. Today 20 (1994) 185.
2101. M.K. Dongare, D.P. Sabde, R.A. Shaikh, K.R. Kamle, S.G. Hedge, Catal. Today 49 (1999) 267.
2102. P. Meriaudeau, V.A. Tuan, V.T. Naghiem, J. Fraissard, C. Naccache, in: Proceedings of 12th International Zeolite Conference, 1999.
2103. P. Mariaudeau, V.A. Tuan, L.N. Hung, F. Lefebvre, H.P. Nguyen, J. Chem. Soc., Faraday Trans. 93 (1997) 4201.
2104. K. Chaudhari, R. Bal, T.K. Das, A. Chandwadkar, D. Srinivas, S. Sivasanker, J. Phys. Chem. B 104 (2000) 11066.
2105. J. Ming, K.T. Ranjit, P. Rungrochaipon, L. Kevan, J. Phys. Chem. B 109 (2005) 29284.
2106. M.A. Zanjanchi, M.K. Rashidi, Spectrochem. Acta A 55 (1999) 947.
2107. C.S. Blackwell, R.L. Patton, J. Phys. Chem. 88 (25) (1984) 6135.
2108. C.S. Blackwell, R.L. Patton, J. Phys. Chem. 92 (1988) 3965.
2109. R. Dabestani, I.N. Ivanov, Photochem. Photobiol. 70 (1999) 10.
2110. A. Bjoerseth, T. Ramdhal, Handbook of Polycyclic Aromatic Hydrocarbons, vol. 2, Marcel Dekker, New York, 1985.
2111. A. Sayari, Stud. Surf. Sci. Catal. 102 (1996) 1.
2112. J.C. Vartuli, K.D. Schmitt, C.T. Kresge, W.J. Roth, M.E. Leonowicz, S.B. McCullen, S.D. Hellring, J.S. Beck, J.L. Schlenker, Chem. Mater. 6 (1994) 2317.
2113. M.J. Climent, A. Corma, S. Iborra, S. Miquel, J. Primo, F. Rey, J. Catal. 183 (1999) 76.
2114. K.S.W. Sing, Adv. Colloid Interface Sci. 76–77 (1998) 3.
2115. J. Choma, M. Jaroniek, W. Burakiewicz-Mortka, M. Kloske, Appl. Surf. Sci. 196 (2002) 216.
2116. M. McNall, R.L. Laurence, W. Curtis Conner, Microporous Mesoporous Mater. 44–45 (2001) 697.
2117. A.V. Neimark, P.I. Ravikovitch, Microporous Mesoporous Mater. 44–45 (2001) 709.
2118. P.J. Branton, P.G. Hall, K.S.W. Sing, H. Reichert, F. Schüth, K.K. Unger, J. Chem. Soc., Faraday. Trans. 90 (1994) 2956.
2119. R. Schmidt, M. Stöcker, E. Hansen, D. Akporiaye, O.H. Ellestad, Microporous Mater. 3 (1995) 443.
2120. N. Tanchoux, P. Trens, D. Maldonado, F. Di Renzo, F. Fajula, Colloids Surf. A: Physicochem. Eng. Asp. 246 (2004) 1.
2121. F. Courivaud, E.W. Hansen, A. Karlsson, S. Kolboe, M. Stöcker, Microporous Mesoporous Mater. 35–36 (2000) 327.
2122. F. Courivaud, E.W. Hansen, A. Karlsson, S. Kolboe, M. Stöcker, Microporous Mesoporous Mater. 37 (2000) 223.
2123. F. Stallmach, A. Gräser, J. Kärger, C. Krause, M. Jeschke, U. Oberhagemann, S. Spange, Microporous Mesoporous Mater. 44–45 (2001) 745.
2124. L. Gjerdaker, D.W. Aksnes, G.H.S. Sørland, M. Stöcker, Microporous Mesoporous Mater. 42 (2001) 89.
2125. W.-H. Chen, Q. Zhao, H.-P. Lin, Y.-S. Yang, C.-Y. Mou, S.-B. Liu, Microporous Mesoporous Mater. 66 (2003) 209.
2126. G.H. Kuehl, US Patent 5,583,277 (1996).
2127. M. Ghiaci, A. Abbaspur, R. Kia, F. Seyedeyn-Azad, Sep. Purif. Technol. 40 (2004) 217.
2128. E. Furuya, S. Kato, T. Kataoka, T. Horiguchi, Y. Otake, Sep. Purif. Technol. 39 (2004) 73.
2129. V.R. Choudhary, K. Mantri, Langmuir 16 (2000) 7031.
2130. V.R. Choudhary, K. Mantri, Langmuir 16 (2000) 8024.

2131. J. Aguado, D.P. Serrano, J.M. Escola, Microporous Mesoporous. Mater. 34 (2000) 43.
2132. R. Mokaya, W. Jones, J. Catal. 172 (1997) 211.
2133. H.A. Benesi, J. Phys. Chem. 61 (1957) 970.
2134. H.A. Benesi, J. Am. Chem. Soc. 78 (1956) 5490.
2135. C.H. Rhee, J.S. Lee, Catal. Today 38 (1997) 213.
2136. D. Barthomeuf, J. Phys. Chem. 83 (1979) 766.
2137. A.K. Ghosh, G. Curtýós, J. Chem. Soc., Faraday Trans. 1 79 (1983) 147.
2138. D. Fuentes-Perujo, J. Santamarýa-González, J. Mérida-Robles, E. Rodrýguez-Castellón, A. Jiménez-López, P. Maireles-Torres, R. Moreno-Tost, R. Mariscal, J. Solid State Chem. 179 (2006) 2182.
2139. M.J. Remy, M.J. Genet, G. Pocelet, P.F. Lardinois, P.P. Notté, J. Phys. Chem. 96 (1992) 2614.
2140. J.M. Mérida-Robles, P. Olivera-Pastor, A. Jiménez-López, E. Rodrýguez-Castellón, J. Phys. Chem. 100 (1996) 14726.
2141. M.E. Davis, Chem. Eng. Res. 30 (1991) 1675.
2142. V.R. Choudhary, K. Mantri, Microporous Mesoporous Mater. 40 (2000) 127.
2143. A.M. Mastral, T. Garcýá, M.S. Callén, R. Murillo, J.M. Lopez, M.V. Navarro, Fuel Proc. Technol. 77–78 (2002) 365.
2144. A.M. Mastral, T. Garcýa, M.S. Callén, R. Murillo, J.M. Lopez, M.V. Navarro, Fuel Proc. Technol. 77–78 (2002) 373.
2145. W.M. Meier, K. Siegmann, Microporous Mesopor. Mater. 33 (1999) 307.
2146. R.G. Harvey, Polycyclic Aromatic Hydrocarbons: Chemistry and Carcinogenicity, Cambridge University Press, Cambridge, 1991.
2147. T. Boger, R. Roesky, R. Glässer, G. Eisenberg, P.J. Weitkam, Microporous Mater. 8 (1997) 79.
2148. S.Z. Qiao, S.K. Bhatia, Ind. Eng. Chem. Res. 44 (2005) 6477.
2149. U. Kreibig, M. Vollmer, Optical Properties of Metal Clusters, Springer, Berlin, 1995.
2150. C. Estournes, T. Lutz, J. Happich, T. Quaranta, P. Wissler, J.L. Guille, J. Magn. Magn. Mater. 173 (1997) 83.
2151. N. Serpone, R.F. Khairutdinov, Stud. Surf. Sci. Catal. 103 (1997) 417.
2152. Y.G. Sun, Y.N. Xia, Science 298 (2002) 2176–2179.
2153. X.L. Feng, H.Y. Ma, S.X. Huang, W. Pan, X.K. Zhang, F. Tian, C.X. Gao, Y.W. Cheng, J.L. Luo, J. Phys. Chem. B 110 (2006) 12311.
2154. Y.G. Sun, B. Mayers, Y.N. Xia, Nano Lett. 3 (2003) 675.
2155. J.L. Gu, J.L. Shi, G.J. You, L.M. Xiong, S.X. Qian, Z.L. Hua, H.R. Chen, Adv. Mater. 17 (2005) 557.
2156. Q.F. Zhang, W.M. Liu, Z.Q. Xue, J.L. Wu, S.F. Wang, D.L. Wang, Q.H. Gong, Appl. Phys. Lett. 82 (2003) 958.
2157. P. Zhou, G.J. You, Y.G. Li, T. Han, J. Li, S.Y. Wang, L.Y. Chen, Y. Liu, S.X. Qian, Appl. Phys. Lett. 83 (2003) 3876.
2158. C.F. Zhang, Y. Liu, G.J. You, B. Li, J.L. Shi, S.X. Qian, Phys. B 357 (2005) 334.
2159. J. Sort, S. Surinach, M.D. Baro, D. Muraviev, G.I. Dzhardimalieva, N.D. Golubeva, S.I. Pomogailo, A.D. Pomogailo, W.A.A. Macedo, D. Weller, V. Skumryev, J. Nogues, Adv. Mater. 18 (2006) 466.
2160. N.D. Fatti, F. Vallee, C. Flytzanis, Y. Hamanaka, A. Nakamura, Chem. Phys. 251 (2000) 215.
2161. N.D. Fatti, F. Vallee, Appl. Phys. B 68 (1999) 433–437.
2162. A. Meldrum, R.F. Haglund, L.A. Boatner, C.W. White, Adv. Mater. 13 (2001) 1431–1444.
2163. Y.S. Shin, J. Lin, J.H. Chan, Z.M. Nie, G.J. Exarhos, Adv. Mater. 13 (2001) 728–732.
2164. M.W. Anderson, S.M. Holmes, N. Hanif, C.S. Cundy, Angew. Chem. Int. Ed. 39 (2000) 2707.
2165. A. Zampieri, H. Sieber, T. Selvam, G.T.P. Mabande, W. Schwieger, F. Sche.er, M. Scheer, P. Greil, Adv. Mater. 17 (2005) 344.
2166. T.X. Fan, T. Hirose, T. Okabe, D. Zhang, J. Porous Mater. 8 (2001) 211.
2167. T.X. Fan, T. Hirose, T. Okabe, D. Zhang, R. Teranisi, M. Yoshimura, J. Porous Mater. 9 (2002) 35.

2168. A.G. Dong, Y.J. Wang, Y. Tang, N. Ren, Y.H. Zhang, Y.H. Yue, Z. Gao, Adv. Mater. 14 (2002) 926.
2169. H. Sieber, C.R. Rambo, J. Cao, Key Eng. Mater. 206 (2002) 2009.
2170. T.X. Fan, B.H. Sun, J.J. Gu, D. Zhang, W.M.L. Leo, Scripta Mater. 53 (2005) 893.
2171. C.R. Rambo, H. Sieber, Adv. Mater. 17 (2005) 1088.
2172. B.H. Sun, T.X. Fan, J.Q. Xu, D. Zhang, Mater. Lett. 59 (2005) 2325.
2173. W. Zhang, D. Zhang, T.X. Fan, J. Ding, Q.X. Guo, O. Hiroshi, Nanotechnology 17 (2006) 840.
2174. T.X. Fan, X.F. Li, Z.T. Liu, J.J. Gu, D. Zhang, J. Am. Ceram. Soc. 89 (2006) 3511.
2175. X.F. Li, T.X. Fan, Z.T. Liu, J. Ding, Q.X. Guo, D. Zhang, J. Eur. Ceram. Soc. 26 (2006) 3657.
2176. Z.T. Liu, T.X. Fan, D. Zhang, J. Am. Ceram. Soc. 82 (2006) 662.
2177. R.A. Caruso, Angew. Chem. Int. Ed. 43 (2004) 2746.
2178. N. Popovska, D.A. Streitwieser, C. Xu, H. Gerhard, J. Eur. Ceram. Soc. 25 (2005) 829.
2179. Z.T. Liu, T.X. Fan, W. Zhang, D. Zhang, Microporous Mesoporous Mater. 85 (2005) 82.
2180. C.N. Sattereld, Heterogeneous Catalysts in Practice, McGraw-Hill, New York, 1980.
2181. R. Prins, Adv. Catal. 46 (2001) 399–464.
2182. R.A. Caruso, M. Antonietti, Chem. Mater. 13 (2001) 3272.
2183. G. Paglia, E.S. Bozin, S.J.L. Billinge, Chem. Mater. 18 (2006) 3242.
2184. N.A. Seaton, Chem. Eng. Sci. 46 (1991) 1895.
2185. R.R. Xu, W.Q. Pang, Chemistry-Zeolites and Porous Materials, Science Press, Beijing, 2004.
2186. J.H. He, K. Toyoki, N. Aiko, Chem. Mater. 15 (2003) 4401.
2187. J.H. He, K. Toyoki, Chem. Mater. 16 (2004) 2656.
2188. A.L. Pan, Z.P. Yang, H.G. Zheng, F.X. Liu, Y.C. Zhu, X.B. Su, Z.J. Ding, Appl. Surf. Sci. 205 (2003) 323.
2189. E. Dujardin, C. Peet, G. Stubbs, J.N. Culver, S. Mann, Nano Lett. 3 (2003) 413.
2190. J. Chen, L. Eberlein, C.H. Langford, J. Photochem. Photobiol. A: Chem. 148 (2002) 183.
2191. I. Othman, R. Mohamady, I.A. Ibraheem, M.M. Mohamed, Appl. Catal. A 299 (2006) 95.
2192. S. Hashimoto, J. Photochem. Photobiol. C: Photochem. Rev. 4 (2003) 19.
2193. V. Ramamurthy, J. Photochem. Photobiol. C 1 (2000) 145.
2194. M. Iwamoto, H. Yahiro, N. Mizuno, W.X. Zhang, Y. Mine, H. FuruKawa, S. Kagawa, J. Phys. Chem. 96 (1992) 9360.
2195. M. Anpo, M. Matsuoka, Y. Shioya, H. Yamashito, E. Giamello, C. Morterra, M. Che, H.H. Patterson, S. Webber, S. Ouellettl, M.A. Fox, J. Phys. Chem. 98 (1994) 5744.
2196. M. Anpo, H. Yamashta, Y. Fchihashi, Y. Fujii, M. Monda, J. Phys.Chem. B 101 (1997) 2632.
2197. J. Dedecek, D. Kaucky, B. Wichterlova, Microporous Mesoporous Mater. 35 (2000) 483.
2198. H. Katsuki, S. Furuta, T. Watari, S. Komarneni, Microporous Mesoporous Mater. 86 (2005) 145.
2199. J. James, M.S. Rao, Thermochim. Acta 97 (1986) 329.
2200. A. Chakraverty, P. Mishra, H.D. Banerjee, J. Mater. Sci. 23 (1988) 21.
2201. F.W. Chang, T.J. Hsiao, S.W. Chung, J.J. Lo, Appl. Catal. A 164 (1997) 225.
2202. C. Real, M.D. Alcala, J.M. Criado, J. Am. Ceram. Soc. 79 (8) (1996) 2012.
2203. I.A. Rahman, F.L. Riley, J. Eur. Ceram. Soc. 5 (1989) 11.
2204. R.V. Krishnarao, J. Eur. Ceram. Soc. 12 (1993) 395.
2205. P.K. Bajpai, M.S. Rao, Ind. Eng. Chem. Prod. Res. Dev. 20 (1981) 721.
2206. A.K. Dalai, M.S. Rao, K.V.G.K. Gokhale, Ind. Eng. Chem. Prod. Res. Dev. 24 (1985) 465.
2207. H. Nur, Indones. J. Agric. Sci. 1 (2001) 40.
2208. M.M. Mohamed, F.I. Zidan, M.H. Fodail, J. Mater. Sci. 42 (2007) 721.
2209. M.M. Mohamed, T.M. Salama, I. Othman, I. Abd-Ellah, Microporous Mesoporous Mater. 84 (2005) 84.
2210. M.C. Dalconi, A. Alberti, G. Cruciani, P. Ciambelli, E. Fonda, Microporous Mesoporous Mater. 62 (2003) 191.
2211. X. Wang, H-Y. Chen, W-M.H. Saoch Her, Appl. Catal. B 26 (2000) 1227.

2212. H. Hamdan, M.N.M. Muhid, S. Endud, E. Lostiorini, Z. Ramli, J. Non-crys. solids 211 (1997) 126.
2213. V. Sundraramurthy, N. Lingappan, J. Mol. Catal. A 160 (2000) 367.
2214. A.W. Chester, R.-M. Dessan, L.B. Alemani, G.L. Woolery, Zeolites 6 (1986) 14.
2215. M.M. Mohamed, T.M. Salama, R. Ohnishi, M. Ichikawa, Langmuir 17 (2001) 5678.
2216. C. Morterra, A. Chiorino, G. Ghiotti, E. Garrone, J.C.S. Faraday I 74 (1978) 271.
2217. K.B. Yoon, J.K. Kochi, J. Phys. Chem. 95 (1991) 3780.
2218. H.-P. Boehm, H. Knozinger, in: J.R. Anderson, M. Boudart (Eds.), Catalytic Science Technology, vol. 4, Springer, Berlin, 1983.
2219. Z. Sobalik, J. Dedecek, I. Ikonnikov, B. Wichterlova, Microporous Mesoporous 21 (1998) 525.
2220. A.M. Daifullah, M.M. Mohamed, J. Chem. Technol. Biotechnol. 79 (2004) 468.
2221. G. Yan, X. Wang, X. Fu, D. Li, Catal. Today 93 (2004) 851.
2222. S. Hashimoto, M. Hagiri, N. Matsabara, S. Tobita, Phys. Chem. Chem. Phys. 3 (2001) 5043.
2223. S. Dzwigaj, M. Matsuoka, M. Anpo, M. Che, J. Phys. Chem. B 104 (2000) 6012.
2224. L. Sun, H. Zhang, C. Peng, J. Yu, Q. Meng, L. Fu, F. Liu, X. Guo, J. Phys. Chem. B 110 (2006) 7249.
2225. P. Kustrowski, L. Chmielarz, J. Surman, E. Bidzinska, R. Dziembaj, P. Cool, E.F. Vansant, J. Phys. Chem. A 109 (2005) 9808.
2226. A. Sayari, S. Hamoudi, Y. Yang, Chem. Mater. 17 (2005) 212.
2227. Y. Yamada, H.S. Zhou, H. Uchida, M. Tomita, Y. Ueno, T. Ichino, I. Honmu, K. Asai, T. Katsube, Adv. Mater. 14 (2004) 812.
2228. G. Wimsberger, P. Yang, B.J. Scott, B.F. Chmelka, G.D. Stucky, Spectrochim. Acta. A 57 (2001) 2049.
2229. Y. Gucbilmez, T. Dogu, Ind. Eng. Chem. Res. 45 (2006) 3496.
2230. P. Liu, S.H. Lee, C.E. Tracy, Y. Yan, J.A. Turner, Adv. Mater. 14 (2002) 27.
2231. F. Schüth, W. Schmidt, Adv. Eng. Mater. 4 (2002) 269.
2232. J. Xi, S. Miao, X. Tang, Macromolecules 37 (2004) 8592.
2233. B.G. Trewyn, C.M. Whitman, V.S.-Y. Lin, Nano Lett. 4 (2004) 2139.
2234. D. Zhao, P. Yang, Q. Huo, B.F. Chmelka, G.D. Stucky, Curr. Opin. Colloid Interface Sci. 3 (1998) 174.
2235. M. Kruk, M. Jaroniec, C.H. Ko, R. Ryoo, Chem. Mater. 12 (2000) 1961.
2236. Y. Sakamoto, I. Daiz, O. Terasaki, D. Zhao, J. Perez-Pariente, J.M. Kim, G.D. Stucky, J. Phys. Chem. B 106 (2002) 3118.
2237. Q. Huo, D.I. Margolese, U. Ciesla, P. Feng, T.E. Gier, P. Sieger, R. Leon, P.M. Petro., F. Schuth, G.D. Stucky, Nature 368 (1994) 317.
2238. Y. Sakamoto, M. Kaneda, O. Terasaki, D. Zhao, J.M. Kim, G.D. Stucky, H.J. Shin, R. Ryoo, Nature 408 (2000) 449.
2239. M.J. Kim, R. Ryoo, Chem. Mater. 11 (1999) 487.
2240. L.X. Dai, K. Tabata, E. Suzuki, Chem. Mater. 13 (2001) 208.
2241. C. Yu, B. Tian, J. Fan, G.D. Stucky, D. Zhao, J. Am. Chem. Soc. 124 (2002) 4556.
2242. J. Fan, C. Yu, F. Gao, J. Lei, B. Tian, L. Wang, Q. Luo, B. Tu, W. Zhou, D. Zhao, Angew. Chem. Int. Edit. 42 (2003) 3146.
2243. L. Wang, J. Fan, B. Tian, H. Yang, C. Yu, B. Tu, D. Zhao, Microporous Mesoporous Mater. 67 (2004) 135.
2244. Y.K. Hwang, J.S. Chang, Y.U. Kwon, S.E. Park, Microporous Mesoporous Mater. 68 (2004) 21.
2245. T.W. Kim, R. Ryoo, M. Kruk, K.P. Gierszal, M. Jaroniec, S. Kamiya, O. Terasaki, J. Phys. Chem. B 108 (2004) 11480.
2246. F. Kleitz, T.W. Kim, R. Ryoo, Langmuir 22 (2006) 440.
2247. L. Mercier, T.J. Pinnavaia, Adv. Mater. 9 (1997) 500.
2248. G.J.A.A. Soler-Illia, C. Sanchez, B. Lebeau, J. Patarin, Chem. Rev. 102 (2002) 4093.
2249. J.R. Matos, L.P. Mercuri, M. Kruk, M. Jaroniec, Langmuir 18 (2002) 884.
2250. C.F. Cheng, Y.C. Lin, H.H. Cheng, Y.C. Chen, Chem. Phys. Lett. 382 (2003) 496.

2251. D. Zhao, P. Yang, N. Melosh, J. Feng, B.F. Chmelka, G.D. Stucky, Adv. Mater. 10 (1998) 1380.
2252. C. Boissiere, M.A.U. Martines, M. Tokumoto, A. Larbot, E. Prouzet, Chem. Mater. 15 (2003) 509.
2253. C. Yu, J. Fan, B. Tian, D. Zhao, Chem. Mater. 16 (2004) 889.
2254. M. Kruk, M. Jaroniec, Chem. Mater. 15 (2003) 2942.
2255. J.M. Kim, Y. Sakamoto, Y.K. Hwang, Y.U. Kwon, O. Terasaki, S.E. Park, G.D. Stucky, J. Phys. Chem. B 106 (2002) 2552.
2256. J.N. Israelachvili, D.J. Mitchell, B.W. Niham, J. Chem. Soc., Faraday Trans. 272 (1976) 1525.
2257. H. Lin, S. Cheng, C. Mou, Microporous Mesoroporous Mater. 10 (1997) 111.
2258. G. Villemuret, C. Detellier, Langmuir 7 (1991) 1215.
2259. P. Komadel, M. Janek, J. Madejova, A. Weekes, C. Breen, J. Chem. Soc., Faraday Trans. 93 (1997) 4207.
2260. S.T. Frey, B.M. Hutchins, B.J. Anderson, T.K. Schreiber, M.E. Hagerman, Langmuir 19 (2003) 2188.
2261. D. Bonn, H. Kellay, H. Tanaka, G. Wegdam, J. Meunier, Langmuir 15 (1999) 7534.
2262. K.A. Carrado, R. Csencsits, P. Thiyagarajan, S. Seifert, S.M. Macha, J.S. Harwood, J. Mater. Chem. 12 (2002) 3228.
2263. J.S. Yadav, B.V. Subba Reddy, C. Madan, New J. Chem. 25 (2001) 1114.
2264. J. Farkas, S. Békássy, J. Madarász, F. Figueras, New J. Chem. 26 (2002) 750.
2265. L. Lami, B. Casal, L. Cuadra, J. Merino, A. Alvarez, E. Ruiz-Hitzky, Green Chem. (1999) 199.
2266. G.W. Brindley, R.E. Semples, Clay Miner. 12 (1977) 229.
2267. R. Burch (Ed.), Catalysis Today, 2, Elsevier, New York, 1988.
2268. S. Yamanaka, T. Nishihara, M. Hattori, Y. Suzuki, Mater. Chem. Phys. 17 (1987) 87.
2269. S. Yamanaka, Y. Inoue, M. Hattori, F. Okumura, M. Yoshikawa, Bull. Chem. Soc. Jpn. 65 (1992) 2494.
2270. H.Y. Zhu, J.A. Orthman, J.Y. Li, J.C. Zhao, G.J. Churchman, E.F. Vansant, Chem. Mater. 14 (2002) 5037.
2271. R. Toranzo, M.A. Vicente, M.A. Bañares-Muñoz, Chem. Mater. 9 (1997) 1829.
2272. K.J. Balkus Jr., J. Shi, J. Phys. Chem. 100 (1996) 16429.
2273. M. Polverejan, Y. Liu, T.J. Pinnavaia, Chem. Mater. 14 (2002) 2283.
2274. N. Aihara, K. Torigoe, K. Esumi, Langmuir 14 (1998) 4945.
2275. M.E. Hagerman, S.J. Salamone, R.W. Herbst, A.L. Payeur, Chem. Mater. 15 (2003) 443.
2276. A.P. Carvalho, C. Castanheira, B. Cardoso, J. Pires, A.R. Silva, C. Freire, B. de Castro, M.B. de Carvalhoa, J. Mater. Chem. 14 (2004) 374.
2277. Q. Sun, Y. Deng, Z.L. Wang, Macromol. Mater. Eng. 289 (2004) 288.
2278. D.W. Kim, A. Blumstein, J. Kumar, L.A. Samuelson, B. Kang, C. Sung, Chem. Mater. 14 (2002) 3925.
2279. R.A. Caruso, A. Susha, F. Caruso, Chem. Mater. 13 (2001) 400.
2280. M.A. Vicente, J.F. Lambert, Phys. Chem. Chem. Phys. 3 (2001) 4843.
2281. R. Burch, C.I. Warburton, Appl. Catal. 33 (1987) 395.
2282. C.I. Warburton, Catal. Today 2 (1988) 271.
2283. M. Sychev, V.H.J. San de Beer, A. Kodentsov, E.M. van Oers, R.A. van Santen, J. Catal. 168 (1997) 245.
2284. Z. Han, H. Zhu, S.R. Bulcock, S.P. Ringer, J. Phys. Chem. B 109 (2005) 2673.
2285. Z. Han, H. Zhu, J. Shi, G.Q. Lu, Mater. Lett. 60 (2006) 2309.
2286. F. Kooli, W. Jones, Chem. Mater. 9 (1997) 2913.
2287. O. Enea, A.J. Bard, J. Phys. Chem. 90 (1986) 301.
2288. T.J. Pinnavaia, Science 220 (1983) 365.
2289. D.R. Lide, H.P.R. Frederikse, CRC Handbook of Chemistry and Physics, CRC Press, Inc., Boca Raton, 1995.
2290. M.K. Arora, N. Sahu, S.N. Upadhyay, A.S.K. Sinha, Ind. Eng. Chem. Res. 38 (1999) 2659.
2291. C.E. Harland, Ion Exchange: Theory and Practice, second ed., The Royal Society of Chemistry, Cambridge, 1994.
2292. W.H.R. Shaw, D.G. Walker, J. Am. Chem. Soc. 78 (1956) 5769.

2293. P.C. Rieke, S.B. Bentjen, Chem. Mater. 5 (1993) 43.
2294. R. Ortega-Borges, D. Lincot, J. Electrochem. Soc. 140 (1993) 3464.
2295. F. Gao, Q. Lu, D. Zhao, Adv. Mater. 15 (2003) 739.
2296. D. Kashchiev, G.M. van Rosmalen, Cryst. Res. Technol. 38 (2003) 555.
2297. S. Licht, J. Electrochem. Soc. 135 (1988) 2971.
2298. K. Norrish, Nature 173 (1954) 256.
2299. K. Norrish, Discuss Faraday Soc. 18 (1954) 120.
2300. F. Huang, H. Zhang, J.F. Baneld, Nano Lett. 3 (2003) 373.
2301. R.L. Penn, J.F. Baneld, Science 281 (1998) 969.
2302. J.F. Baneld, S.A. Welch, H. Zhang, T.T. Ebert, R.L. Penn, Science 289 (2000) 751.
2303. S.J. Gregg, K.S.W. Sing, Adsorption, Surface Area and Porosity, second ed., Academic Press, New York, 1982.
2304. H.Y. Zhu, P. Cool, E.F. Vansant, B.L. Su, X.P. Gao, Langmuir 20 (2004) 10115.
2305. H. He, L. Zhang, J. Klinowski, J. Phys. Chem. 99 (1995) 6980.
2306. L. Spanhel, M. Haase, H. Weller, A. Henglein, J. Am. Chem. Soc. 109 (1987) 5649.
2307. G. Fellenberg, The Chemistry of Pollution, J. Wiley, Chichester, 2000.
2308. D.L. Ford, Toxicity Reduction Evaluation and Control, vol. 3, Technomic Publishing AG, Lancaster, PA, 1992.
2309. Toxicological Review of Hexavalent Chromium, U.S. Environmental Protection Agency, National Center for Environmental Assessment, Office of Research and Development, Washington, DC, 1998.
2310. Toxicological Profile for Chromium, Agency for Toxic Substances and Disease Registry (ATSDR), U.S. Public Health Service, U.S. Department of Health and Human Services, Atlanta, GA, 1998.
2311. F. Cadena, R. Rizni, R.W. Peters, in: Hazardous and Industrial Wastes, Proceedings of the 22nd Mid-Atlantic Industrial Waste Conference, Technomic Publ. Co., Lancaster PA, 1990.
2312. D.C. Sharma, C.F. Forster, Water Res. 27 (1993) 1201.
2313. D.C. Sharma, C.F. Forster, Bioresour. Technol. 49 (1994) 31.
2314. C.P. Huang, M.H. Wu, Water Res. 11 (1977) 673.
2315. M.P. Candela, M. Jose, M. Martinez, R.T. Macia, Water Res. 29 (1995) 2174.
2316. P.B. Orhan, H. Büjükgüngor, Water Sci. Technol. 28 (1993) 247.
2317. R.S. Bowman, G.M. Haggerty, R.G. Huddleston, D. Neel, M.M. Flynn, in: D.A. Sabatini, R.C. Knox, J.H. Harwell (Eds.), Surfactant-Enhanced Subsurface Remediation: Emerging Technologies, American Chemical Society Symposium Series, vol. 594, Oxford University Press, New York, 1995.
2318. A.D. Vujakocic, M. Tomasevic-Canovic, A. Dakovic, V.T. Dondur, Appl. Clay Sci. 17 (2000) 265.
2319. G.M. Haggerty, R.S. Bowman, Environ. Sci. Technol. 28 (1994) 452.
2320. M. Ghiaci, R. Kia, A. Abbaspur, F. Seyedeyn-Azad, Sep. Purif. Technol. 40 (2004) 285.
2321. H. Faghihian, R.S. Bowman, Water Res. 39 (2005) 1099.
2322. J. Warchol, P. Misaelides, R. Petrus, D. Zamboulis, J. Hazard. Mater. 137 (2006) 410.
2323. R.S. Bowman, Microporous Mesoporous Mater. 61 (2003) 43.
2324. Z. Li, T. Burt, R. Bowman, Environ. Sci. Technol. 34 (2000) 3756.
2325. F. Cadena, E. Cazares, in: B. Sahwney (Ed.), Organic Pollutants in the Environment, CMS Workshop, vol. 8, The Clay Minerals Society, Boulder, 1996.
2326. A. Dakovic, M. Tomasevic-Canovic, V. Dondur, G.E. Rottinghaus, V. Medakovic, S. Zaric, Colloids Surf. B 46 (2005) 20.
2327. G. Limousin, J.-P. Gaudet, L. Charlet, S. Szenknect, V. Barthes, M. Krimissa, Appl. Geochem. 22 (2007) 249.
2328. A. Velayudhan, C. Horvath, J. Chromatogr. A 663 (1994) 1.
2329. K.V. Kumar, S. Sivanesan, J. Hazard. Mater. B 134 (2006) 237.
2330. O. Altin, O. Ozbelge, T. Dogu, J. Colloid Interface Sci. 198 (1998) 130.
2331. S. Balci, Water Res. 38 (2004) 1129.
2332. A. Tsirambides, A. Filippidis, A. Kassoli-Fournaraki, Appl. Clay Sci. 7 (1993) 509.

2333. T.W. Batiashvili, R.A. Achvlediani, G.V. Gvachariya, in: Clinoptilolite: Symposium on the Investigation and Utilization of Natural Zeolites, Mezniereba, Tbilisi, 1977.
2334. V.A. Nikashina, B.F. Myasoedov, in: P. Misaelides, F. Macasek, T.J. Pinnavaia, C. Colella (Eds.), Natural Microporous Materials in Environmental Technology, NATO Science Series, vol. E-362, Kluwer Academic Publishers, Dordrecht, 1999.
2335. V.A. Nikashina, P.A. Gembitskii, E.M. Kats, L.F. Boksha, A.K. Galuzinskaya, Russ. Chem. Bull. 43 (1994) 1462.
2336. P. Misaelides, V.A. Nikashina, A. Godelitsas, P.A. Gembitskii, E.M. Kats, J. Radioanal. Nucl. Chem. 227 (1998) 183.
2337. C.F. Baes Jr., R.E. Messmer, Hydrolysis of Cations, Krieger Pub. Co. Inc., Malabar, 1986.
2338. D.W. Marquardt, J. Soc. Ind. Appl. Math. 11 (1963) 431.
2339. R.M. Barrer, Hydrothermal Chemistry of Zeolites, Academic Press, London, 1978.
2340. S. Chandrasekhar, Met. Mater. Process. 13 (2001) 109.
2341. L.V.C. Rees, S. Chandrasekhar, Zeolites 13 (1993) 524.
2342. S. Chandrasekhar, Clay Miner. 31 (1996) 253.
2343. P. Chu, F.G. Dwyer, C.J. Vartuli, US Patent 4,778,666 (1988).
2344. C.S. Cundy, Collect. Czech. Chem. Commun. 63 (1998) 1699.
2345. P.M. Slangen, J.C. Jansen, H. van Bekkum, Microporous Mater. 9 (1997) 259.
2346. L. Bonaccorsi, E. Proverbio, J. Cryst. Growth 247 (2003) 333.
2347. M.D. Romero, J.M. Gomez, G. Ovejero, A. Rodriguez, Mater. Res. Bull. 39 (2004) 389.
2348. H. Katsuki, S. Furuta, J. Porous Mater. 8 (2001) 5.
2349. O.G. Somani, A.L. Choudhari, B.S. Rao, S.P. Mirajkar, Mater. Chem. Phys. 82 (2003) 538.
2350. M. Inada, H. Tsujimoto, Y. Eguchi, N. Enomoto, J. Hojo, Fuel 84 (2005) 1482.
2351. R.V. Ballmoos, Collection of Simulated XRD Powder Patterns for Zeolites, Butterworth Scientic Ltd., 1984, On Behalf of the Structure Commission of the International Zeolite Association, UK.
2352. E.M. Flanigen, H. Khatami, H.A. Szymansky, in: IR Spectral Studies of Zeolite Frameworks, Molecular Sieve Zeolites I., Advances in Chemistry Series, American Chemical Society, Washington DC, 1971.
2353. H. Bennet, R.A. Reed, Chemical Methods of Silicate Analysis – A Hand Book, Academic Press Ltd., London, 1971.
2354. American Standards for Testing Materials, C 373, 1972.
2355. C.G. Pope, Microporous Mesoporous Mater. 21 (1998) 333.
2356. K.J.D. MacKenzie, I.M.N. Brown, M.E. Bowden, J. Am. Ceram. Soc. 68 (1985) 293.
2357. N.A. Eltekova, D. Berek, I. Novak, Carbon 38 (3) (2000) 373.
2358. O. Olkhovyk, M. Jaroniec, J. Int. Adsorp. Soc. 11 (2005) 205.
2359. M.A. Abu-Daabes, N.G. Pinto, Chem. Eng. Sci. 60 (2005) 1901.
2360. E. Sipos, G. Fogassy, A. Tungler, P.V. Samant, J.L. Figueiredo, J. Mol. Catal. A: Chem. 212 (2004) 245.
2361. F.B. Su, J.H. Zeng, X.Y. Bao, Y.S. Yu, J.Y. Lee, X.S. Zhao, Chem. Mater. 17 (2005) 3960.
2362. P.V. Samant, C.M. Rangel, M.H. Romero, J.B. Fernandes, J.L. Figueiredo, J. Power Sources 151 (2005) 79.
2363. I. Moriguchi, F. Nakahara, H. Furukawa, H. Yamada, T. Kudo, Electrochem. Solid-State Lett. 7 (8) (2004) A221.
2364. S. Yoon, J.W. Lee, T. Hyeon, S.M. Oh, J. Electrochem. Soc. 147 (7) (2000) 2507.
2365. J. Lee, S. Yoon, T. Hyeon, S.M. Oh, K.B. Kim, Chem. Commun. 21 (1999) 2177.
2366. S.J. Han, Y.K. Yun, K.W. Park, Y.E. Sung, T. Hyeon, Adv. Mater. 15 (2003) 1922.
2367. J. Fan, T. Wang, C.Z. Yu, B. Tu, Z.Y. Jiang, D.Y. Zhao, Adv. Mater. 16 (2004) 1432.
2368. I. Grigoriants, L. Sominski, H.L. Li, I. Ifargan, D. Aurbach, A. Gedanken, Chem. Commun. 7 (2005) 921.
2369. R. Hurt, G. Krammer, G. Crawford, K.Q. Jian, C. Rulison, Chem. Mater. 14 (2002) 4558.
2370. C. Chan, G.P. Crawford, Y. Gao, R.H. Hurt, K.Q. Jian, H. Li, B. Sheldon, M. Sousa, N. Yang, Carbon 43 (2005) 2431.
2371. K.Q. Jian, H.S. Shim, A. Schwartzman, G.P. Crawford, R.H. Hurt, Adv. Mater. 15 (2003) 164.

2372. K.Q. Jian, A. Yan, I. Kulaots, G.P. Crawford, R.H. Hurt, Carbon 44 (2006) 2102.
2373. H. Konno, S. Sato, H. Habazaki, M. Inagaki, Carbon 42 (2004) 2756.
2374. L.J. Zhi, T. Gorelik, J.S. Wu, U. Kolb, K. Mullen, J. Am. Chem. Soc. 127 (2005) 12792.
2375. L.J. Zhi, J.S. Wu, J.X. Li, U. Kolb, K. Mullen, Angew. Chem. -Int. Ed. 44 (2005) 2120.
2376. K.Q. Jian, H.Q. Xianyu, J. Eakin, Y.M. Gao, G.P. Crawford, R.H. Hurt, Carbon 43 (2005) 407.
2377. T.W. Kim, I.S. Park, R. Ryoo, Angew. Chem. Int. Ed. 42 (2003) 4375.
2378. N. Yang, K.Q. Jian, I. Kulaots, G.P. Crawford, R.H. Hurt, J. Nanosci. Nanotechnol. 3 (2003) 386.
2379. K.P. Gierszal, M. Jaroniec, Chem. Commun. 22 (2004) 2576.
2380. K.P. Gierszal, T.W. Kim, R. Ryoo, M. Jaroniec, J. Phys. Chem. B 109 (2005) 23263.
2381. S.B. Yoon, G.S. Chai, S.K. Kang, J.S. Yu, K.P. Gierszal, M. Jaroniec, J. Am. Chem. Soc. 127 (2005) 4188.
2382. K.Q. Jian, H.S. Shim, D. Tuhus-Dubrow, S. Bernstein, C. Woodward, M. Pfeer, D. Steingart, T. Gournay, S. Sachsmann, G.P. Crawford, R.H. Hurt, Carbon 41 (2003) 2073.
2383. J.D. Brooks, G.H. Taylor, Carbon 3 (1965) 185.
2384. G.R. Hennig, in: P.L. Walker, Jr. (Ed.), Chemistry and Physics of Carbon, vol. 2, Marcel Dekker, New York, 1966.
2385. J.M. Thomas, in: P.L. Walker Jr. (Ed.), Chemistry and Physics of Carbon, vol. 2, Marcel Dekker, New York, 1965.
2386. R. Phillips, F.J. Vastola, P.L. Walker Jr., Carbon 8 (1970) 197.
2387. H. Ye, N. Naguib, Y. Gogotsi, A.G. Yazicioglu, C.M. Megaridis, Nanotechnology 15 (2004) 232.
2388. H.P. Boehm, Carbon 32 (5) (1994) 759.
2389. H.P. Boehm, Carbon 40 (2002) 145.
2390. C.U. Pittman, G.R. He, B. Wu, S.D. Gardner, Carbon 35 (1997) 317.
2391. N.R. Laine, F.J. Vastola, P.L. Walker Jr., J. Phys. Chem. 67 (1963) 2030.
2392. W.P. Hoffman, F.J. Vastola, P.L. Walker Jr., Carbon 22 (1984) 585.
2393. W.P. Hoffman, Carbon 29 (1991) 769.
2394. W.P. Hoffman, in: The Science of Carbon Materials, University of Alicanti Press, 2000.
2395. I.I. Salame, T.J. Bandosz1, J. Colloid Interface Sci. 240 (2001) 252.
2396. S.V. Rotkin, Y. Gogotsi, Mater. Res. Innovations 5 (2002) 191.
2397. S. Lim, S.H. Yoon, I. Mochida, J.H. Chi, J. Phys. Chem. B 108 (2004) 1533.
2398. J.Y. Li, L. Ma, X.N. Li, C.S. Lu, H.Z. Liu, Ind. Eng. Chem. Res. 44 (2005) 5478.
2399. B.C, Satishkumar, A. Govindaraj, J. Mofokeng, G.N. Subbanna, C.N.R. Raoy, J. Phys. B: At., Mol. Opt. Phys. 29 (1996) 4925.
2400. H. Hu, P. Bhowmik, B. Zhao, M.A. Hamon, M.E. Itkis, R.C. Haddon, Chem. Phys. Lett. 345 (2001) 25.
2401. P. Ehrburger, N. Pusset, P. Dziedzinl, Carbon 30 (1992) 1105.
2402. W.P. Hoffman, J.A. Lowry, in: Proceedings of the 16th Biennial Conference on Carbon, San Diego, 1985.
2403. E. Piera, C.A.M. Brenninkmeijer, J. Santamaria, J. Coronas, J. Membr. Sci. 201 (2002) 229.
2404. Y. Morigami, M. Kondo, J. Abe, H. Kita, K. Okamoto, Sep. Purif. Tech. 25 (2001) 251.
2405. J. Zhang, J. Dong, M. Luo, H. Xiao, S. Murad, R.A. Normann, Langmuir 21 (2005) 8609.
2406. D.P. Mann, T. Paraskeva, K.F.E. Pratt, I.P. Parkin, D.E. Williams, Meas. Sci. Technol. 16 (2005) 1193.
2407. M. Grahn, Z. Wang, M. Lidstrom-Larsson, A. Holmgren, J. Hedlund, J. Sterte, Microporous Mesoporous Mater. 81 (2005) 357.
2408. A. Mitra, T.G. Cao, H.T. Wang, Z.B. Wang, L. Huang, S. Li, Z.J. Li, Y.S. Yan, Ind. Eng. Chem. Res. 43 (2004) 2946.
2409. J. Caro, M. Noack, P. Koelsch, R. Schaefer, Microporous Mesoporous Mater. 38 (2000) 3.
2410. E. Mateo, R. Lahoz, G.F. De La Fuente, A. Paniagua, J. Coronas, J. Santamaria, Chem. Mater. 16 (2004) 4847.
2411. J. Hedlund, F. Jareman, A.-J. Bons, M. Anthonis, J. Membr. Sci. 222 (2003) 163.
2412. C.J. Gump, X. Lin, J.L. Falconer, R.D. Noble, J. Membr. Sci. 173 (2000) 35.

2413. W.J.W. Bakker, F. Kapteijn, J. Poppe, J.A. Moulijn, J. Membr. Sci. 117 (1996) 57.
2414. S. Gopalakrishnan, T. Yamaguchi, S.-I. Nakao, J. Membr. Sci. 274 (2006) 102.
2415. J.L. Falconer, M. Arruebo, R.D. Noble, J. Membr. Sci. 269 (2006) 171.
2416. Z. Lai, G. Bonilla, I. Diaz, J.G. Nery, K. Sujaoti, M.A. Amat, E. Kokkoli, O. Terasaki, R.W. Thompson, M. Tsapatsis, D.G. Vlachos, Science 300 (2003) 456.
2417. J. Hedlund, M. Noack, P. Kolsch, D. Creaser, J. Caro, J. Sterte, J. Membr. Sci. 159 (1999) 263.
2418. M.A. Salomon, J. Coronas, M. Menendez, J. Santamaria, Appl. Catal. Gen. 200 (2000) 201.
2419. M. Lassinantti, J. Hedlund, J. Sterte, Microporous Mesoporous Mater. 38 (1) (2000) 25.
2420. H. Kita, T. Inoue, H. Asamura, A. Murata, K. Tanaka, K. Okamoto, Chem. Commun. (1997) 4.
2421. S. Nair, M. Tsapatsis, in: S.M. Auerbach, K.A. Carrado, P.K. Dutta (Eds.), Handbook of Zeolite Science and Technology, Marcel Dekker, New York, 2003.
2422. M.P. Rohde, D. Unruh, G. Schaub, Membrane application in Fischer–Tropsch synthesis reactors – overview of concepts, in: International Conference on Gas–Fuel 05, Elsevier, Amsterdam, 2005.
2423. M.P. Rohde, D. Unruh, G. Schaub, Ind. Eng. Chem. Res. 44 (2005) 9653.
2424. O. Pachtova, I. Kumakiri, M. Kocirik, S. Miachon, J.-A. Dalmon, J. Membr. Sci. 226 (2003) 101.
2425. J. Hedlund, J. Sterte, M. Anthonis, A.-J. Bons, B. Carstensen, N. Corcoran, D. Cox, H. Deckman, W. De Gijnst, P.-P. de Moor, F. Lai, J. McHenry, W. Mortier, J. Reinoso, J. Peters, Microporous Mesoporous Mater. 52 (2002) 179.
2426. D.M. Bibby, L.P. Aldridge, N.B. Milestone, J. Catal. 72 (1981) 373.
2427. F. Jareman, J. Hedlund, D. Creaser, J. Sterte, J. Membr. Sci. 236 (2004) 81.
2428. F. Jareman, J. Hedlund, Microporous Mesoporous Mater. 82 (2005) 201.
2429. J. Hedlund, F. Jareman, C. Andersson, Factors affecting the performance of MFI membranes, in: 14th International Zeolite Conference, Cape Town, South Africa, 2004.
2430. M.P. Bernal, E. Piera, J. Coronas, M. Menendez, J. Santamaria, Catal. Today 56 (2000) 221.
2431. A. Walcarius, C. Delacote, Chem. Mater. 15 (2003) 4181.
2432. R. Hoppe, A. Ortlam, J. Rathousky, G. Schulz, E.A. Zukel, Microporous Mater. 8 (1997) 267.
2433. S.A. Davis, M. Breulmann, K.H. Rhodes, B. Zhang, S. Mann, Chem. Mater. 13 (2001) 3218.
2434. G.S. Attard, J.C. Glyde, C.G. Göltner, Nature 378 (1995) 366.
2435. C.G. Göltner, S. Henke, M.C. Weisenherger, M. Antonietti, Angew. Chem. Int. Ed. 37 (1998) 613.
2436. S.A. Bagshaw, E. Prouzet, J. Thomas, Science 269 (1995) 1242.
2437. C.E. Fowler, D. Khushalani, S. Mann, Chem. Commun. (2001) 2028.
2438. D. Walsh, J.D. Hopwood, S. Mann, Science 264 (1994) 1576.
2439. D. Volkmer, S. Tugulu, M. Fricke, T. Nielsen, Angew. Chem. Int. Ed. 42 (2003) 58.
2440. S.R. Hall, H. Bolger, S. Mann, Chem. Commun. (2003) 2784.
2441. S.A. Davis, S.L. Burkett, N.H. Mendelson, S. Mann, Nature 385 (1997) 420.
2442. H. Yang, N. Coombs, G.A. Ozin, Nature 386 (1997) 692.
2443. R.K. Rana, Y. Mastai, A. Gedanken, Adv. Mater. 14 (2002) 1414.
2444. Z.Y. Yuan, J.Q. Liu, L.M. Peng, B.L. Su, Langmuir 18 (2002) 2450.
2445. H.P. Lin, Y. Cheng, C.Y. Mou, Chem. Mater. 10 (1998) 3772.
2446. B. Jönsson, B. Lindman, K. Holmberg, B. Kronberg, Surfactants and Polymers in Aqueous Solution, John Wiley & Sons, New York, 1998.
2447. S.A. Bagshaw, Chem. Commun. (1999) 767.
2448. Q.Y. Sun, P.J. Kooyman, J.G. Grossmann, P.H.H. Bomans, P.M. Frederik, P.C.M.M. Magusin, T.P.M. Beelen, R.A. Van Santen, N.A.M. Sommerdijk, Adv. Mater. 15 (2003) 1097.
2449. E. Muthusamy, D. Walsh, S. Mann, Adv. Mater. 14 (2002) 969.
2450. S.S. Kim, W.Z. Zhang, T.J. Pinnavaia, Science 282 (1998) 1302.
2451. Q. Cai, Z. Luo, W. Pang, Y. Fan, X. Chen, F. Cui, Chem. Mater. 13 (2001) 258.
2452. S.H. Tolbert, C.C. Landry, G.D. Stucky, B.F. Chmelka, P. Norby, J.C. Hanson, A. Monnier, Chem. Mater. 13 (7) (2001) 2247.
2453. G.A. Ozin, Acc. Chem. Res. 30 (1997) 17.

2454. K. Cui, Q. Cai, X.H. Chen, Q.L. Feng, H.D. Li, Microporous Mesoporous Mater. 68 (2004) 61.
2455. S.J. Greg, K.S.W. Sing, Adsorption, Surface Area and Porosity, second ed., Academic Press, London, 1995.
2456. H. Stephen, T. Stephen, Solubilities of Inorganic and Organic Compounds, first ed., Pergamon Press, Oxford, 1963.
2457. H.S. Jeon, A.I. Nakatani, E.K. Hobbie, C.C. Han, Langmuir 17 (10) (2001) 3087.
2458. I. Prigogine, Non-Equilibrium Statistical Mechanics, Interscience Publishers, New York, 1962.
2459. P.T. Tanev, Y. Liang, T.J.J. Pinnavaia, Am. Chem. Soc. 119 (1997) 8616.
2460. D.H.W. Hubert, M. Jung, P.M. Frederik, P.H.H. Bmans, J. Meuldijk, A.L. German, Adv. Mater. 12 (2000) 1286.
2461. Q. Huo, J. Feng, F. Schuth, G.D. Stucky, Chem. Mater. 9 (1997) 14.
2462. Q. Huo, R. Leon, P.M. Petro., G.D. Stucky, Science 268 (1995) 1324.
2463. J.N. Israelachvili, D.J. Mitchell, B.W. Ninham, J. Chem. Soc., Faraday Trans. II 72 (1976) 1525.
2464. J.N. Israelachvili, D.J. Mitchell, B.W. Ninham, Biochem. Biophys. Acta 470 (1977) 185.
2465. A. Shioi, T.A. Hatton, Langmuir 18 (2002) 7341.
2466. M. Antonietti, S. Förster, Adv. Mater. 15 (2003) 1323.
2467. Q. Sun, P.C.M.M. Magusin, B. Mezari, P. Panine, R.A. van Santen, N.A.J.M. Sommerdijk, J. Mater. Chem. 15 (2) (2005) 256.
2468. J.P. Holman, Thermodynamics, second ed., McGraw-Hill, New York, 1974.
2469. P. Tarazona, U. Marini Bettolo Marconi, R. Evans, Mol. Phys. 60 (1987) 573.
2470. C. Lastoskie, K.E. Gubbins, N. Quirke, J. Phys. Chem. 97 (1993) 4786.
2471. W.A. Steele, The Interaction of Gases with Solid Surfaces, Pergamon, Oxford, 1974.
2472. P.I. Ravikovitch, A. Vishnyakov, R. Russo, A.V. Neimark, Langmuir 16 (2000) 2311.
2473. G. Stan, M.W. Cole, J. Low Temp. Phys. 110 (1998) 539.
2474. R.P. Feynman, Statistical Mechanics, Benjamin, New York, 1972.
2475. J. Jagiello, M. Thommes, A. Linares-Solano, D. Cazorla-Amorós, D. Lozano-Castelló, Extended Abstracts Carbon'04, Providence, RI, 2004.
2476. D.H. Everett, J.C. Powl, J. Chem. Soc., Faraday Trans. 1 (1976) 619.
2477. J. Jagiello, Langmuir 10 (1994) 2778.
2478. J. Jagiello, D. Tolles, in: F. Meunier (Ed.), Fundamentals of Adsorption, Elsevier, Paris, 1998.
2479. M. Konstantakou, S. Samios, Th.A. Steriotis, M. Kainourgiakis, G.K. Papadopoulos, E.S. Kikkinides, A.K. Stubos, in: P.L. Llewellyn, F. Rodriguez-Reinoso, J. Rouquerol, N. Seaton (Eds.), Characterization of Porous Solids VII, Studies in Surface Science and Catalysis, vol. 160, Elsevier, Amsterdam, 2007.
2480. V. Lopez-Ramon, J. Jagiello, T.J. Bandosz, N.A. Seaton, Langmuir 13 (1997) 4435.
2481. J.B. Parra, C.O. Ania, A. Arenillas, F. Rubiera, J.M. Palacios, J.J. Pis, J. Alloys Compd. 379 (2004) 280.
2482. W. Betz, S. Lambiase, J. Chromatogr. 556 (1991) 433.
2483. W. Betz, M. Keeler, Extended Abstracts Carbon'99, Lexington, KY, 1999.
2484. D.F. Quinn, Carbon 40 (2002) 2767.
2485. M.G. Nijkamp, J.E.M.J. Raaymakers, A.J. van Dillen, K.P. de Jong, Appl. Phys. A 72 (2001) 619.
2486. G. Yushin, R. Dash, J. Jagiello, J.E. Fischer, Y. Gogotsi, Adv. Funct. Mater. 16 (2006) 2288.
2487. C. Valenzuela-Calahorro, A. Navarrete-Guijosa, M. Stitou, E. Cuerda-Correa, Colloids Surf. A: Physicochem. Eng. Asp. 224 (2003) 135.
2488. C. Aguzzi, P. Cerezo, C. Viseras, C. Caramella, Appl. Clay Sci. 36 (2007) 22.
2489. T.F. Bates, F.A. Hildebrand, A. Swinefor, Am. Miner. 35 (1950) 463.
2490. S.R. Levis, P.B. Deasy, Int. J. Pharm. 253 (2003) 145.
2491. R.S. Byrne, P.B. Deasy, J. Microencapsulation 22 (4) (2005) 423.
2492. Y. Lvov, K. Ariga, I. Ichinose, T. Kunitake, Thin Solid Films 284/285 (1996) 797.
2493. A.K. Dash, H.G. Brittain, in: H.G. Brittain (Ed.), Analytical Profiles of Drugs Substances and Excipients, vol. 25, Academic Press, New Jersey, 1998.
2494. S.B. Hanauer, in: M. Peppercorn (Ed.), The Management of Inflammatory Bowel Disease: New Medical and Surgical Approaches, Marcel Dekker, Philadelphia, 1990.

2495. M.C.M. Rijk, A.V. Schaik, J.H.M.V. Tongeren, Scand. J. Gastroenterol. 23 (1998) 107.
2496. E. Schacht, A. Gevaert, E. Refaie-Kenawy, K. Molly, W. Verstraete, P. Adriaensens, R. Carleer, J. Gelan, J. Control. Release 39 (1996) 327.
2497. V.K. Gupta, T.E. Beckert, J.C. Price, Int. J. Pharm. 213 (2001) 83.
2498. V.K. Gupta, M.W. Assmus, T.E. Beckert, J.C. Price, Int. J. Pharm. 213 (2001) 93.
2499. H. Tozaki, T. Odoriba, N. Okada, T. Fujita, A. Terabe, T. Suzuki, S. Okabe, S. Muranishi, A. Yamamoto, J. Control. Release 82 (2002) 51.
2500. R. Wiwattanapatapee, L. Lomlim, K. Saramunee, J. Controlled Release 88 (2003) 1.
2501. M.K. Chourasia, S.K. Jain, J. Pharm. Pharm. Sci. 6 (1) (2003) 33.
2502. C. Valenzuela-Calahorro, E. Cuerda-Correa, A. Navarrete-Guijosa, E.G. Pradas, J. Colloid Interface Sci. 248 (2002) 33.
2503. C. Valenzuela-Calahorro, E. Cuerda-Correa, A. Navarrete-Guijosa, E.G. Pradas, J. Colloid Interface Sci. 250 (2002) 67.
2504. C. Valenzuela-Calahorro, A. Navarrete-Guijosa, M. Stitou, E. Cuerda-Correa, Colloids Surf. A: Physicochem. Eng. Asp. 237 (2004) 7.
2505. W. Lund, in: W. Lund (Ed.), The pharmaceutical Codex, Principles and Practice of Pharmaceutics, twenty fifth ed., The Pharmaceutical Press, London, 1994.
2506. C. Hinz, Geoderma 99 (2001) 225.
2507. M.L. Poutsma, in: J.A. Rabo (Ed.), Zeolite Chemistry and Catalysis, American Chemistry Society, Washington, DC, 1976.
2508. H. Knözinger, in: G. Ertl, H. Knözinger, J. Weitkamp (Eds.), Handbook of Heterogeneous Catalysis, vol. 2, Wiley-VCH, Weinheim, 1997.
2509. K. Pitchumani, D.R. Corbin, V.J. Ramamurthy, J. Am. Chem. Soc. 118 (1996) 8152.
2510. K.B. Yoon, Chem. Rev. 93 (1993) 321.
2511. D.K. Murray, J.W. Chang, J.F. Haw, J. Am. Chem. Soc. 115 (1993) 4732.
2512. D.K. Murray, T. Howard, P.W. Goguen, T.R. Krawietz, J.F. Haw, J. Am. Chem. Soc. 116 (1994) 6354.
2513. R.J. Correa, C.J.A. Mota, Appl. Catal. A 255 (2003) 255.
2514. J.F. Haw, J.B. Nicholas, T. Xu, L.W. Beck, D.B. Ferguson, Acc. Chem. Res. 29 (1996) 259.
2515. R.J. Correa, E.F. Sousa-Aguiar, A. Ramirez-Solis, C. Zicovich- Wilson, C.J.A. Mota, J. Phys. Chem. B. 108 (2004) 10658.
2516. V. Bosacek, M. Dlouha, S. Vratislav, Physica B 214–243 (1998) 400.
2517. V. Bosacek, M. Dlouha, S. Vratislav, Physica B 276–278 (2000) 400.
2518. T.P. Beebe, J.E. Crowell, J.T. Yates, J. Phys. Chem. 92 (1988) 1296.
2519. J.F. Haw, B.R. Richardson, I.S. Oshiro, N.D. Lazo, J.A. Speed, J. Am. Chem. Soc. 111 (1989) 2052.
2520. N.D. Lazo, B.R. Richardson, P.D. Schettler, J.L. White, E.J. Munson, J.F. Haw, J. Phys. Chem. 95 (1991) 9420.
2521. M.T. Aronson, R.J. Gorte, W.E. Farneth, D. White, J. Am. Chem. Soc. 111 (1989) 840.
2522. A.G. Stepanov, K.I. Zamaraev, J.M. Thomas, Catal. Lett. 13 (1992) 407.
2523. J.F. Haw, D.K. Murray, J.W. Chang, J. Am. Chem. Soc. 115 (1993) 4732.
2524. J.F. Haw, D.K. Murray, T. Howard, P.W. Goguen, T.R. Krawietz, J. Am. Chem. Soc. 116 (1994) 6354.
2525. V. Bocasek, J. Phys. Chem. 97 (1993) 10732.
2526. T.R. Krawietz, P.W. Goguen, J.F. Haw, Catal. Lett. 42 (1996) 41.
2527. W. Kladnig, H. Noller, J. Catal. 29 (1973) 385.
2528. C. Pistarino, E. Finocchio, G. Romenzzano, F. Brichese, R. Felice, G. Busca, M. Baldi, Ind. Eng. Chem. Res. 39 (2000) 2752.
2529. N. Rosenbach, C.J.A. Mota, J. Mol. Struct.: Theochem. 731 (2005) 157.
2530. C. Pistarino, E. Finocchio, M.A. Larubia, B. Serra, S.I. Braggio, G. Busca, Ind. Eng. Chem. Res. 40 (2001) 3262.
2531. N. He, S. Bao, Q. Xu, Appl. Catal. A 169 (1998) 29.
2532. A.M.F. Bidart, A.P. Borges, L. Nogueira, E.R. Lachter, C.J.A. Mota, Catal. Lett. 75 (2001) 155.
2533. R.J. Corrêa, C.J.A. Mota, Phys. Chem. Chem. Phys. 4 (2002) 4268.

2534. C.W. Kanyi, D.C. Doetschman, J.T. Schulte, K. Yan, R.E. Wilson, B.R. Jones, C.O. Kowenje, Microporous Mesoporous Mater. 92 (2006) 292.
2535. A.G. Stepanov, M.V. Luzgin, S.S. Arzumanov, H. Ernst, D. Freude, J. Catal. 211 (2002) 165.
2536. P.Y. Bruice, Organic Chemistry, third ed., Prentice Hall, New Jersey, 2001.
2537. E. Finocchio, C. Pistarino, S. Dellepiane, B. Serra, S. Braggio, M. Baldi, G. Busca, Catal. Today 75 (2002) 263.
2538. K. Góra-Marek, M. Derewinski, P. Sarv, J. Datka, Catal. Today 101 (2005) 131.
2539. J. Liu, D. Yin, D. Yin, Z. Fu, Q. Li, G. Lu, J. Mol. Catal. A Chem. 209 (2004) 171.
2540. J.F. Bunnet, Angew. Chem. Int. Ed. 1 (1962) 225.
2541. W. Ortiz, F.L. Cozens, N.P. Shepp, Org. Lett. 1 (1999) 531.
2542. D.H. Olson, Zeolites 15 (1995) 439.
2543. M. Feuerstein, M. Hunger, G. Engelhardt, J.P. Amoureux, Solid State Nucl. Magn. Reson. 7 (1996) 95.
2544. J.D. Fox, A. Meenakshi, J. Phys. Chem. B 109 (2005) 9917.
2545. V. Nieminen, M. Sierka, D.Y. Murzin, J. Sauer, J. Catal. 231 (2005) 393.
2546. F.A. Carey, R.J. Sundberg, Advanced Organic Chemistry, fourth ed., Kluwer Academic/ Plenum Publishers, New York, 2000.
2547. R.A. Bartsch, B.M. Pruss, B.A. Bushaw, K.E. Wiegers, J. Am. Chem. Soc. 95 (1973) 3405.
2548. D.L. Griffith, D.L. Meges, H.C. Brown, J. Chem. Soc., Chem. Commun. (1968) 90.
2549. M.L. Dhar, E.D. Hughes, C.K. Ingold, J. Chem. Soc. (1948) 2058.
2550. R. Schollhorn, Chem. Mater. 8 (1996) 174.
2551. A.M. Fogg, V.M. Green, D. O'Hare, Chem. Mater. 11 (1999) 216.
2552. D. Lunau, Curr. Opinion Solid State Mater. Sci. 5 (2001) 123.
2553. K.R. Dunbar, R.A. Heintz, in: K.D. Karlin (Ed.), Progress in Inorganic Chemistry, vol. 45, John-Wiley, 1997.
2554. A. Leaustic, J.P. Audiere, P.G. Lacroix, R. Clement, Chem. Mater. 7 (1996) 1103.
2555. Y. Tsunoda, W. Sugimoto, Y. Sugahara, Chem. Mater. 15 (2003) 632.
2556. A. Zak, Y. Feldman, V. Lyakhovitskaya, G. Leitus, R. Popovitz-Biro, E. Wachtel, H. Cohen, S. Reich, R. Tenne, J. Am. Chem. Soc. 124 (2002) 4747.
2557. A. Leaustic, E. Riviere, R. Clement, Chem. Mater. 15 (2003) 4785.
2558. R. Clement, L. Lomas, J.P. Audiere, Chem. Mater. 2 (1990) 641.
2559. T. Yi, R. Clement, C. Haut, L. Catala, T. Gacoin, N. Tancrez, I. Ledoux, J. Ziss, Adv. Mater. 17 (2005) 335.
2560. I. Lagadic, P.G. Lacroix, R. Clement, Chem. Mater. 9 (1997) 2004.
2561. V. Laget, C. Hornick, P. Rabu, M. Drillon, R. Ziessel, Coord. Chem. Rev. 180 (1998) 1533.
2562. S. Bhattacharjee, J.A. Anderson, Chem. Commun. 554 (2004).
2563. B.M. Choudary, S. Madhi, N.S. Chowdari, M.L. Kantam, B. Sreedhar, J. Am. Chem. Soc. 124 (2002) 14127.
2564. B.M. Lira-Cantu, P. Gomez-Romero, J. Electrochem. Soc. 146 (1999) 2029.
2565. P.G. Lacroix, R. Clement, K. Nakatani, J. Zyss, I. Ledoux, Science 263 (1994) 558.
2566. D. O'Hare, in: D.W. Bruce, D. O'Hare (Eds.), Inorganic Materials, Wiley, Chichester, UK, 1992.
2567. W. Kliggen, R. Ott, H. Hahn, Z. Anorg. Allg. Chem. 396 (1973) 271.
2568. S. Soled, A. World, Mater. Res. Bull. 11 (1976) 657.
2569. R. Brec, D.M. Schleich, G. Ouvrard, J. Rouxel, Inorg. Chem. 18 (1979) 1814.
2570. R. Brec, Solid State Ionics 22 (1986) 3.
2571. R. Clement, I. Lagadic, A. Leaustic, J.P. Audiere, L. Lomas, in: P. Bernier et al., (Eds.), Chemical Physics of Intercalation II, Plenum Press, New York, 1993.
2572. A. Leaustic, J.P. Audiere, D. Cointereau, R. Clement, Chem. Mater. 8 (1996) 1954.
2573. A. Leaustic, A. Sour, E. Riviere, R. Clement, C. R. Acad. Sci. Ser. II Fascicule C-Chimie 4 (2001) 91.
2574. S. Benard, A. Leaustic, E. Riviere, P. Yu, R. Clement, Chem. Mater. 13 (2001) 3709.
2575. Y. Mathey, A. Michalowicz, P. Toffoli, G. Vlaic, Inorg. Chem. 23 (1984) 897.
2576. V. Manriquez, P. Barahona, O. Pena, M. Mouallem-Bahout, R.E. Avila, J. Alloys Compd. 329 (2001) 92.

2577. J.P. Odile, J.J. Steger, A. Wold, Inorg. Chem. 14 (1975) 2400.
2578. D.J. Goossens, T.J. Hicks, J. Magn. Magn. Mater. 177–181 (1998) 721.
2579. V. Manriquez, P. Barahona, O. Pena, Mater. Res. Bull. 35 (2000) 1889.
2580. S. Lee, P. Colombet, G. Ouvrard, R. Brec, Inorg. Chem. 27 (1988) 1291.
2581. P. Colombet, A. Leblanc, M. Danot, J. Rouxel, J. Solid State Chem. 41 (1982) 174.
2582. A. Leaustic, E. Riviere, R. Clement, E. Manova, I. Mitov, J. Phys. Chem. B 103 (23) (1999) 4833.
2583. J.S.O. Evans, D. O'Hare, R. Clement, A. Leaustic, P. Thuery, Adv. Mater. 7 (1995) 735.
2584. N. Chandrasekharan, S. Vasudevan, Phys. Rev. B – Condens. Matter 54 (1996) 14903.
2585. P.A. Joy, S. Vasudevan, J. Am. Chem. Soc. 14 (1992) 7792.
2586. P.A. Joy, S. Vasudevan, J. Chem. Phys. 99 (1993) 4411.
2587. R. Clement, O. Gamier, J. Jegoudez, Inorg. Chem 25 (1986) 1404.
2588. J. Qin, C. Yang, K. Yakushi, Y. Nakazawa, K. Ichimura, Solid State Commun. 100 (1996) 427.
2589. X. Chen, C. Yang, J. Qin, K. Yakushi, Y. Nakazawa, K. Ichimura, J. Solid State Chem. 150 (2000) 258.
2590. J.F. Diaz, K.J. Balkus Jr., J. Mol. Catal. B: Enzym. 2 (1996) 115.
2591. T.K. Jain, I. Roy, T.K. De, A. Maitra, J. Am. Chem. Soc. 120 (1998) 11092.
2592. S. Inagaki, S. Guan, T. Ohsuna, O. Terasaki, Nature 416 (2002) 304.
2593. D. Coutinho, R.A. Orozio-Tevan, R.F. Reidy, K.J. Balkus Jr., Microporous Mesoporous Mater. 54 (2002) 229.
2594. S. Madhugiri, W. Zhou, J.P. Ferraris, K.J. Balkus Jr., Microporous Mesoporous Mater. 63 (2003) 75.
2595. B. Hatton, K. Landskron, W. Whitnall, D. Perovic, G.A. Ozin, Acc. Chem. Res. 38 (2005) 305.
2596. W.J. Hunks, G.A. Ozin, Chem. Commun. (2004)2426.
2597. J.R. Matos, M. Kruk, L.P. Mercuri, M. Jaroniek, T. Asefa, N. Coombs, G.A. Ozin, T. Kamiyama, O. Terasaki, Chem. Mater. 14 (2002) 1903.
2598. W.J. Hunks, G.A. Ozin, Chem. Mater. 16 (2004) 5465.
2599. W. Wang, S. Xie, W. Zhou, A. Sayari, Chem. Mater. 16 (2004) 1756.
2600. M. Kapoor, S. Inagaki, Chem. Lett. 33 (2004) 88.
2601. W. Wang, W. Zhou, A. Sayari, Chem. Mater. 15 (2003) 4886.
2602. Y. Goto, S. Inagaki, Chem. Commun. (2002)2410.
2603. S. Dai, M.C. Burleigh, Y. Shin, C.C. Morrow, C.E. Barnes, Z. Xue, Angew. Chem. Int. Ed. 38 (1999) 1235.
2604. A. Walcarius, N. Luthi, J.-L. Blin, B.-L. Su, L. Lamberts, Electrochim. Acta 44 (1999) 4601.
2605. J.F. Diaz, K.J. Balkus Jr., F. Bedioui, V. Kurshev, L. Kevan, Chem. Mater. 9 (1997) 61.
2606. V. Antochshuuk, M. Jaroniec, Chem. Mater. 12 (2000) 2496.
2607. L. Mercier, T.J. Pinnavaia, Environ. Sci. Technol. 32 (1998) 2749.
2608. J.H. Clark, D.J. Macquarrie, E.B. Mubofu, Green Chem. (2000) 53.
2609. M.H. Lin, A. Stein, Chem. Mater. 11 (1999) 3285.
2610. M.H. Lim, C.F. Blanford, A. Stein, Chem. Mater. 10 (1998) 467.
2611. C.M. Bambrough, R.C.T. Slade, R.T. Williams, J. Mater. Chem. 8 (1998) 569.
2612. C.E. Fowler, S.L. Burkett, S. Mann, Chem. Commun. (1997)1769.
2613. S.R. Hall, C.E. Fowler, B. Lebeau, S. Mann, Chem. Commun. (1999) 201.
2614. F. Babonneau, L. Leite, S. Fontlupt, J. Mater. Chem. 9 (1999) 175.
2615. C.E. Fowler, B. Lebeau, S. Mann, Chem. Commun. (1998)1825.
2616. R. Richer, L. Mercier, Chem. Commun. (1998)1775.
2617. L. Mercier, T.J. Pinnavaia, Chem. Mater. 12 (2000) 188.
2618. W. Van Rhijn, D. De Vos, W. Bossaert, B. Wouters, P. Groubet, P. Jacobs, Stud. Surf. Sci. Catal. 117 (1998) 183.
2619. M.A. Markowitz, J. Klaehn, R.A. Hendel, S.B. Qadrip, S.L. Golledge, D.G. Castner, B.P. Gaber, J. Phys. Chem. B 104 (2000) 10820.
2620. D.J. Macquarrie, D.B. Jackson, S. Tailland, K.A. Utting, J. Mater. Chem. 11 (2001) 1843.
2621. D.J. Macquarrie, Green Chem. (1999) 195.

2622. H.H.P. Yiu, C.H. Botting, N.P. Botting, P.A. Wright, Phys. Chem. Chem. Phys. 3 (2001) 2983.
2623. V.S.-Y. Lin, D.R. Radu, M.-K. Han, W. Deng, S. Kuroki, B.H. Shanks, M. Pruski, J. Am. Chem. Soc. 124 (2002) 9040.
2624. I.K. Mbaraka, D.R. Radu, V.S.-Y. Lin, B.H. Shanks, J. Catal. 219 (2003) 329.
2625. S. Huh, J.W. Wiench, J.-C. Yoo, M. Pruski, V.S.-Y. Lin, Chem. Mater. 15 (2003) 4247.
2626. D. Coutinho, Z. Yang, J.P. Ferraris, K.J. Balkus Jr., Microporous Mesoporous Mater. 81 (2005) 321.
2627. D. Coutinho, S. Madhugiri, K.J. Balkus Jr., J. Porous Mater. 11 (2004) 239.
2628. X. Wang, K.S.K. Lin, J.C.C. Chan, S. Cheng, J. Phys. Chem. B 109 (2005) 1763.
2629. R.J.P. Corriu, E. Lancelle-Beltran, A. Mehdi, C. Reye, S. Brandes, R. Guilard, J. Mater. Chem. 12 (2002) 1355.
2630. G. Deng, M.A. Markowitz, P.R. Kust, B.P. Gaber, Mater. Sci. Eng. C 11 (2000) 165.
2631. T. Kang, Y. Park, K. Choi, J.S. Lee, J. Yi, J. Mater. Chem. 14 (2004) 1043.
2632. M.C. Burleigh, M.A. Markowitz, M.S. Spector, B.P. Gaber, J. Phys. Chem. B 105 (2001) 9935.
2633. M.C. Burleigh, M.A. Markowitz, M.S. Spector, B.P. Gaber, Chem. Mater. 13 (2001) 4760.
2634. M.A. Wahab, I. Kim, C.S. Ha, J. Solid State Chem. 177 (2004) 3439.
2635. Q. Yang, M.P. Kapoor, S. Inagaki, J. Am. Chem. Soc. 124 (2002) 9694.
2636. M. Kruk, M. Jaroniec, Chem. Mater. 13 (2001) 3169.
2637. Y. Yan, H. Wang, Encyclopedia of Nanoscience and Nanotechnology, vol. 7, American Scientific Publishers, Los Angeles, 2004, p. 763.
2638. M.C. Lovallo, M. Tsapatsis, T. Okubo, Chem. Mater. 8 (1996) 1579.
2639. A.M.P. McDonnell, D. Beving, A.J. Wang, W. Chen, Y.S. Yan, Adv. Funct. Mater. 15 (2005) 336.
2640. R.A. Munoz, D. Beving, Y.S. Yan, Ind. Eng. Chem. Res. 44 (2005) 4310.
2641. C. O'Neill, D.E. Beving, W. Chen, Y.S. Yan, AIChe J. 52 (2006) 1157.
2642. G.E. Totten, D.S. Mackenzie (Eds.), Handbook of Aluminum, vol. 1, Marcel Dekker, New York, 2003.
2643. Y.S. Yan, D.E. Beving, High aluminum coatings on corrodible metal surfaces, US Patent 7,179,547 (2007).
2644. M. Smaihi, O. Barida, V. Valtchev, Eur. J. Inorg. Chem. 24 (2003) 4370.
2645. D-3359-02 cross-cut tape test for adhesion, American Society for Testing and Materials, ASTM, West Conshohocken, PA.
2646. M.M. Cowan, K.Z. Abshire, S.L. Houk, S.M. Evans, J. Ind. Microbiol. Biotechnol. 30 (2003) 102.
2647. X.L. Cheng, Z.B. Wang, Y.S. Yan, Electrochem. Solid-State Lett. 4 (2001) B23.
2648. D.E. Beving, A.M.P. McDonnell, W. Yang, Y.S. Yan, J. Electrochem. Soc. 153 (2006) B325.
2649. A. Mitra, Z.B. Wang, T.G. Cao, H.T. Wang, L.M. Huang, Y.S. Yan, J. Electrochem. Soc. 149 (2002) B472.
2650. H. Ai, Y.M. Lvov, D.K. Mills, M. Jennings, J.S. Alexander, S.A. Jones, Cell Biochem. Biophys. 38 (2003) 103.
2651. X.S. Zhao, G.Q. Lu, G.J. Millar, Ind. Eng. Chem. Res. 35 (1996) 2075.
2652. C.G. Guizzard, A.C. Julbe, A. Ayral, J. Mater. Chem. 9 (1999) 55.
2653. U. Ciesla, F. Schüth, Microporous Mesoporous Mater. 27 (1999) 131.
2654. Q. Huo, D.I. Margolese, U. Ciesla, D.G. Demuth, P. Feng, T.E. Gier, P. Sieger, A. Firouzi, B.F. Chmekla, F. Schüth, G.D. Stucky, Chem. Mater. 6 (1994) 1176.
2655. Y. Zhou, M. Antonietti, Adv. Mater. 15 (2003) 1452.
2656. Y. Wei, J. Xu, H. Dong, J.H. Dong, K. Qiu, S.A. Jansen-Varnum, Chem. Mater. 11 (1999) 2023.
2657. C. Stubenrauch, Curr. Opin. Colloid Interface Sci. 6 (2001) 160.
2658. B. Hoffmann, G. Platz, Curr. Opin. Colloid Interface Sci. 6 (2001) 171.
2659. C. Pinel, P. Loisil, P. Gallezot, Adv. Mater. 9 (1997) 582.
2660. J.H. Jung, M. Amaike, S. Shinkai, Chem. Commun. (2000)2343.
2661. M.A. Markowitz, P.R. Kust, G. Deng, P.E. Schoen, J.S. Dordick, D.S. Clark, B.P. Gaber, Langmuir 16 (2000) 1759.

2662. P. Sakya, J.M. Seddon, R.H. Templer, J. Phys. II France 4 (1994) 1311.
2663. F. Nilsson, O. Söderman, Langmuir 12 (1996) 902.
2664. U.L. Stangar, M. Puchberger, N. Hüsing, Mater. Res. Soc. Symp. Proc. 726 (2002) 111.
2665. G. Persson, H. Edlund, H. Amenitsch, P. Laggner, G. Lindblom, Langmuir 19 (2003) 5813.
2666. A.B. Cortés, M. Valiente, Colloid Polym. Sci. 281 (2003) 319.
2667. P.T. Tanev, T.J. Pinnavaia, Science 10 (1995) 865.
2668. A. Firouzi, D. Kumar, L.M. Bull, T. Besier, P. Sieger, Q. Huo, S.A. Walker, J.A. Zasadzinski, C. Glinka, J. Nicol, D. Margolese, G.D. Stucky, B.F. Chmelka, Science 267 (1995) 1138.
2669. A. Monnier, F. Schüth, Q. Huo, D. Kumar, D. Margolese, R.S. Maxwell, G.D. Stucky, M. Krishnamurty, P. Petro., A. Firouzi, M. Janicke, B.F. Chmelka, Science 261 (1993) 1299.
2670. S.E. Friberg, C.C. Yang, J. Sjöblom, Langmuir 8 (1992) 372.
2671. C.G. Göltner, M. Antonietti, Adv. Mater. 9 (1997) 431.
2672. C.G. Spickermann, Top. Curr. Chem. 226 (2003) 29.
2673. Y. Zhou, M. Antonietti, Chem. Mater. 16 (2004) 544.
2674. C.J. Brinker, Y.F. Lu, A. Sellinger, H.Y. Fan, Adv. Mater. 11 (1999) 579.
2675. C.G. Göltner, B. Smarsly, B. Berton, M. Antonietti, Chem. Mater. 13 (2001) 1617.
2676. H.P. Hentze, E. Krämer, B. Berton, S. Förster, M. Antonietti, Macromolecules 32 (1999) 5803.
2677. S. Polarz, B. Smarsly, L. Bronsterin, M. Antonietti, Angew. Chem. Int. Ed. 40 (2001) 4417.
2678. M. Antonietti, B. Berton, C. Göltner, H. Hentze, Adv. Mater. 10 (1998) 154.
2679. M. Groenewolt, M. Antonietti, Langmuir 20 (2004) 7811.
2680. P. Feng, X. Bu, D.J. Pine, Langmuir 16 (2000) 5304.
2681. M.G. Song, J.Y. Kim, S.H. Cho, J.D. Kim, Langmuir 18 (2002) 6110.
2682. G.M. Wu, J. Wang, J. Shen, Q.Y. Zhang, B. Zhou, Z.S. Deng, B. Fan, D.P. Zhou, F.S. Zhang, J. Phys. D, Appl. Phys. 34 (2001) 1301.
2683. H.P. Lin, C.Y. Mou, S.B. Liu, C.Y. Tang, C.Y. Lin, Microporous Mesoporous Mater. 44–45 (2001) 129.
2684. Z.Y. Yuan, J.L. Blin, B.L. Su, Chem. Comm. (2002) 504.
2685. M. Kruk, M. Jaroniec, Langmuir 13 (1997) 6267.
2686. A. Sayari, P. Liu, M. Kruk, M. Jaroniec, Chem. Mater. 9 (1997) 2499.
2687. M. Jaroniec, M. Kruk, Langmuir 15 (1999) 5410.
2688. S. Polarz, M. Antonietti, Chem. Comm. (2002)2593.
2689. A.R. Hind, S.K. Bhargava, S.C. Grocott, Langmuir 13 (1997) 6255.
2690. D.C. Calabro, E.W. Valyocsik, F.X. Ryan, Microporous Mater. 7 (1996) 243.
2691. D.R. Scheuing, J.G. Weers, Colloids Surf. 55 (1991) 41.
2692. P.J. Innocenzi, Non-Cryst. Solids 316 (2003) 309.
2693. T.M. Parrill, J. Mater. Res. 7 (1992) 2230.
2694. N.R.B. Coleman, G.S. Attard, Microporous Mesoporous Mater. 44–45 (2001) 73.
2695. H.P. Lin, C.Y. Mou, S.B. Liu, Adv. Mater. 12 (2000) 103.
2696. D. Khushalani, A. Kuperman, G. Ozin, K. Tanaka, J. Garces, M. Olken, N. Coombs, Adv. Mater. 7 (1995) 842.
2697. K. Ishida, Bull. Chem. Soc. Jpn. 31 (1958) 143.
2698. G. Grygierczyk, W. Fischer, M. Sajewicz, P. Kuś, R. Wrzalik, M. Czaja, M. Dziadek, T.J. Kowalska, J. Planar Chromatogr. – Mod. TLC 15 (2002) 449.
2699. J.E. Hodge, J. Agric. Food Chem. 1 (1953) 928.
2700. C.G.A. Davies, B.L. Wedzicha, C. Gillard, Food Chem. 60 (1997) 323.
2701. Z.Y. Yuan, B.L. Su, Colloids Surf. A: Physicochem. Eng. Asp. 241 (2004) 95.
2702. C.K. Shu, B.M. Lawrence, in: Maillard Reactions in Chemistry, Food and Health, Royal Society of Chemistry, Cambridge, 1994.
2703. S.D. Kinrade, R.J. Hamilton, A.S. Schach, C.T.G. Knight, J. Chem. Soc., Dalton Trans. (2001)961.
2704. S. Inagaki, Y. Sakamoto, Y. Fukushima, O. Terasaki, Chem. Mater. 8 (1996) 2089.
2705. T.A. Zepeda, J.L.G. Fierro, B. Pawelec, R. Nava, T. Klimova, G.A. Fuentes, T. Halachev, Chem. Mater. 17 (2005) 4062.
2706. M. Onaka, N. Hashimoto, Y. Kitabata, R. Yamasaki, Appl. Catal. A 241 (2003) 307.
2707. M. Stockenhuber, M.J. Hudson, R.W. Joyner, J. Phys. Chem. B 104 (2000) 3370.

2708. S. Samanta, S. Giri, P.U. Sastry, N.K. Mal, A. Manna, A. Bhaumik, Ind. Eng. Chem. Res. 42 (2003) 3012.
2709. J.S. Choi, S.S. Yoon, S.H. Jang, W.S. Ahn, Catal. Today 111 (2006) 280.
2710. A. Tuel, S. Gontier, Chem. Mater. 8 (1996) 114.
2711. A. Tuel, I. Arcon, J.M.M. Millet, J. Chem. Soc., Faraday Trans. 94 (1998) 3501.
2712. K. Bachari, J.M.M. Millet, B. Benaichouba, O. Cherifi, F. Figueras, J. Catal. 221 (2004) 55.
2713. H. Liu, G.Z. Lu, Y.L. Guo, Y. Guo, J.S. Wang, Nanotechnology 17 (2006) 997.
2714. T.R. Pauly, Y. Liu, T.J. Pinnavaia, S.J.L. Billinge, T.P. Rieker, J. Am. Chem. Soc. 121 (1999) 8835.
2715. Y. Wang, Q. Zhang, T. Shishido, K. Takehira, J. Catal. 209 (2002) 186.
2716. S. Gontier, A. Tuel, Zeolites 15 (1995) 601.
2717. J.C. Vartuli, K.D. Schmitt, C.T. Kresge, W.J. Roth, M.E. Leonowicz, Chem. Mater. 6 (1994) 1816.
2718. A. Thangaraj, M.J. Eapen, S. Sivasanker, P. Ratnasamy, Zeolites 12 (1992) 943.
2719. P.T. Tanev, M. Chibwe, T.J. Pinnavaia, Nature 368 (1994) 321.
2720. F.D. Renzo, F. Testa, J.D. Chen, H. Cambon, A. Galarneau, D. Plee, F. Fajula, Microporous Mesoporous Mater. 28 (1999) 437.
2721. W. Zhao, Y.F. Luo, P. Deng, Q.Z. Li, Catal. Lett. 73 (2001) 199.
2722. Y. Han, X.J. Meng, H.B. Guan, Y. Yu, L. Zhao, X.Z. Xu, X.Y. Yang, S. Wu, N. Li, F.S. Xiao, Microporous Mesoporous Mater. 57 (2003) 191.
2723. R. Johnson, J. Pankow, D. Bender, C. Price, J. Zagorski, Environ. Sci. Technol. 34 (2000) 210A.
2724. H.W. Hung, T.F. Lin, C. Baus, F. Sacher, H.J. Brauch, Environ. Technol. 26 (2005) 1371.
2725. A.A. Michael, Environ. Sci. Technol. 34 (2000) 725.
2726. S.G. Li, V.A. Tuan, R.D. Noble, J.L. Falconer, Environ. Sci. Technol. 37 (2003) 4007.
2727. A.E. Senatalar, J.A. Bergendahl, A. Giaya, R.W. Thompson, Environ. Eng. Sci. 21 (2004) 722.
2728. Y.C. Tong, T.B. Zhao, F.Y. Li, Y. Wang, Chem. Mater. 18 (2006) 4218.
2729. D.J. Wang, Y.H. Zhang, A.G. Dong, Y. Tang, Y.J. Wang, J.C. Xia, N. Ren, Adv. Funct. Mater. 13 (2003) 563.
2730. Y.J. Wang, Y. Tang, A.G. Dong, X.D. Wang, N. Ren, Z. Gao, J. Mater. Chem. 12 (2002) 1812.
2731. A.E. Persson, B.J. Schoeman, J. Sterte, J.-E. Otterstedt, Zeolites 14 (1994) 557.
2732. P.V. Krivenko, in: First International Conference of Alkaline Cements and Concretes, State Technical University, Kiev, 1994.
2733. A. Fernández-Jiménez, A. Palomo, M. Criado, Mater. Construcc. 56 (2006) 51.
2734. H. Xu, J.S.J. van Deventer, Int. J. Miner. Process. 59 (2000) 247.
2735. V.F.F. Barbosa, K.J.D. MacKenzie, C. Thaumaturgo, Int. J. Inorg. Mater. 2 (2000) 309.
2736. A. Fernández-Jiménez, A. Palomo, M. Criado, Cem. Concr. Res. 35 (2005) 1204–1209.
2737. J. Davidovits, J. Therm. Anal. 37 (1991) 1633.
2738. P. Duxson, J.L. Provis, G.C. Lukey, F. Separovic, J.S.J. van Deventer, Langmuir 21 (2005) 3028.
2739. M.L. Granizo, M.T. Blanco-Varela, A. Palomo, J. Mater. Sci. 35 (2000) 6309.
2740. A. Palomo, M.T. Blanco Varela, M.S. Granizo, F. Puertas, T. Vázquez, M.W. Grutzeck, Cem. Concr. Res. 29 (1999) 997.
2741. A. Palomo, F.P. Glasser, Br. Ceram. Trans. 91 (1992) 107.
2742. P. Duxson, S.W. Mallicoat, G.C. Lukey, W.M. Kriven, J.S.J. van Deventer, Colloids Surf. A: Physicochem. Eng. Asp. 292 (2007) 8.
2743. H. Wang, H. Li, F. Yan, Colloids Surf. A: Physicochem. Eng. Asp. 268 (2005) 1.
2744. A. Fernández-Jimánez, A. Palomo, Fuel 82 (2003) 2259.
2745. A. Fernández-Jiménez, A. Palomo, C. Lopez-Hombrados, ACI Mater. J. 103 (2006) 106.
2746. J.S.G. van Jaarsveld, J.S.J. van Deventer, G.C. Lukey, Mater. Lett. 57 (2003) 1272.
2747. A. Buchwald, M. Schulz, Cem. Concr. Res. 35 (2005) 968.
2748. A. Palomo, M.W. Grutzeck, M.T. Blanco, Cem. Concr. Res. 19 (1999) 1323.
2749. A. Madani, A. Aznar, J. Sanz, J.M. Serratosa, J. Phys. Chem. 94 (1990) 760.
2750. M. Alkan, C. Hopa, Z. Yilmaz, H. Guler, Microporous Mesoporous Mater. 86 (2005) 176.

2751. P. Duxson, J.L. Provis, G.C. Lukey, S.W. Mallicoat, W.M. Kriven, J.S.J. van Deventer, Colloids Surf. A: Physicochem. Eng. Asp. 269 (2005) 47.
2752. A. Palomo, S. Alonso, A. Fernández-Jiménez, I. Sobrados, J. Sanz, J. Am. Ceram. Soc. 87 (2004) 1141.
2753. A. Fernández-Jiménez, A. Palomo, I. Sobrados, J. Sanz, Microporous Mesoporous Mater. 91 (2006) 111–119.
2754. F. Puertas, A. Palomo, A. Fernández-Jiménez, J.D. Izquierdo, M.L. Granizo, Adv. Cem. Res. 15 (2003) 23.
2755. A. Fernández-Jiménez, G. de la Torre, A. Palomo, G. López-Olmo, M.M. Alonso, M.A.G. Aranda, Fuel 85 (2006) 1960.
2756. V.F.F. Barbosa, K.J.D. MacKenzie, Mater. Res. Bull. 38 (2003) 319.
2757. A. Fernández-Jiménez, A. Palomo, Microporous Mesoporous Mater. 86 (2005) 207.
2758. W. Mozgawa, M. Sitarz, J. Mol. Struct. 614 (2002) 273.
2759. G. Engelhardt, D. Michel, High Resolution Solid State NMR of Silicates and Zeolites, Wiley, London, 1987.
2760. J. Klinowski, Prog. NMR Spectrosc. 16 (1984) 237–309.
2761. A. Fernández-Jiménez, A. Palomo, Cem. Concr. Res. 35 (2005) 1984.
2762. G. Kovalchuk, A. Fernández-Jiménez, A. Palomo, Fuel 86 (2007) 315.
2763. J.H. Clark, D.J. Macquarrie, Chem. Commun. (1998)853.
2764. S. Inagaki, S. Guan, Y. Fukushima, T. Ohsuna, O. Terasaki, J. Am. Chem. Soc. 121 (1999) 9611.
2765. M.H. Lim, A. Stein, Chem. Mater. 11 (1999) 3285.
2766. A. Sayari, S. Hamoudi, Chem. Mater. 13 (2001) 3151.
2767. X. Feng, G.E. Fryxell, L.Q. Wang, A.Y. Kim, J. Liu, K.M. Kemner, Science 276 (1997) 923.
2768. T. Asefa, M. Kruk, M.J. MacLachlan, N. Coombs, H. Grondey, M. Jaroniec, G.A. Ozin, J. Am. Chem. Soc. 123 (2001) 8520.
2769. A. Sayari, Chem. Mater. 8 (1996) 1840.
2770. A. Taguchi, F. Schuth, Microporous Mesoporous Mater. 77 (2004) 1.
2771. W.H. Zhang, X.-B. Lu, J.-H. Xiu, Z.-L. Hua, L.-X. Zhang, M. Robertson, J.-L. Shi, D.-S. Yan, J.D. Holmer, Adv. Funct. Mater. 14 (2004) 544.
2772. A. Bhaumik, T. Tatsumi, J. Catal. 189 (2000) 31.
2773. N. Igarashi, S. Kidani, R. Ahemaito, K. Hashimoto, T. Tatsumi, Microporous Mesoporous Mater. 81 (2005) 97.
2774. M. Kruk, M. Jaroniec, Y. Sakamoto, O. Terasaki, R. Ryoo, C.H. Ko, J. Phys. Chem. B 104 (2000) 292.
2775. T. Sun, M.S. Wong, J.Y. Ying, Chem. Commun. (2000)2057.
2776. M. Ogawa, N. Yamamoto, J. Porous Mater. 6 (1999) 19.
2777. M. Kruk, T. Asefa, M. Jaroniec, G.A. Ozin, J. Am. Chem. Soc. 124 (2002) 6383.
2778. S.A. Bagshaw, A.R. Hayman, Adv. Mater. 13 (2001) 1011.
2779. T. Sun, J.Y. Ying, Nature 389 (1997) 704.
2780. R. Ryoo, I.-S. Park, S. Jun, C.W. Lee, M. Kruk, M. Jaroniec, J. Am. Chem. Soc. 123 (2001) 1650.
2781. C.G. Sonwane, S.K. Bhatia, N. Calos, Ind. Eng. Chem. Res. 37 (1998) 2271.
2782. M. Ogawa, Chem. Commun. (1996)1149.
2783. H. Naono, M. Hakuman, T. Shiono, J. Colloid Interface Sci. 186 (1997) 360.
2784. S. Shylesh, S.P. Mirajkar, A.P. Singh, J. Mol. Catal. A. Chem. 239 (2005) 57.
2785. H. Yang, G. Zhang, X. Hong, Y. Zhu, Microporous Mesoporous. Mater. 68 (2004) 119.
2786. E.J. Sullivan, D.B. Hunter, R.S. Bowman, Environ. Sci. Technol. 32 (1998) 1948.
2787. K.J. Chao, C.N. Wu, H. Chang, L.J. Lee, S. Hu, J. Phys. Chem. 101 (1997) 6341.
2788. O. Dag, G.A. Ozin, Adv. Mater. 13 (2001) 1182.
2789. Z. Zhu, Z. Chang, L. Kevan, J. Phys. Chem. B 103 (1999) 2680.
2790. Y. Wang, Y. Ohishi, T. Shishido, Q. Zhang, W. Yang, Q. Guo, H. Wan, K. Takehira, J. Catal. 220 (2003) 347.
2791. B.M. Weckhuysen, I.E. Wachs, R.A. Schoonheydt, Chem. Rev. 96 (1996) 3327.
2792. Y.-S. Li, P.B. Wright, R. Puritt, T. Tran, Spectrochim. Acta A 60 (2004) 2759.

2793. Y.K. Gun'ko, T.S. Perova, S. Balakrishnan, D.A. Potapova, R.A. Moore, E.V. Astrova, Phys. Stat. Sol. A (2003) 492.
2794. S.L. Burkett, S.D. Sims, S. Mann, Chem. Commun. (1996)1367.
2795. A. Sakthivel, P. Selvam, J. Catal. 211 (2002) 134.
2796. N. Nowotny, Y.L.N. Pedersen, U. Hanefeld, T. Maschmayer, Chem.Eur. J. 8 (2002) 3724.
2797. A. Bhaumik, M.P. Kapoor, S. Inagaki, Chem. Commun. (2003)470.
2798. B.E. Logan, D. LaPoint, Water Res. 36 (2002) 3647–3653.
2799. B. Gu, W. Dong, G.M. Brown, D.R. Cole, Environ. Sci. Technol. 37 (2003) 2291–2295.
2800. J. Yoon, Y. Yoon, G. Amy, J. Cho, D. Foss, T.H. Kim, Water Res. 37 (2003) 2001–2012.
2801. R. Parette, F.S. Cannon, Water Res. 39 (2005) 4020–4028.
2802. B. Gu, G.M. Brown, P.V. Bonnesen, L. Liang, B.A. Moyer, R. Ober, S.D. Alexandratos, Environ. Sci. Technol. 34 (2000) 1075.
2803. R. Parette, Department of Civil and Environmental Engineering, The Pennsylvania State University, University Park, 2003.
2804. S.D. Alexandratos, L.A. Hussain, Ind. Eng. Chem. Res. 34 (1995) 251.
2805. B. Lee, L.L. Bao, H.J. Im, S. Dai, E.W. Hagaman, J.S. Lin, Langmuir 19 (2003) 4246.
2806. A.U. Baes, T. Okuda, W. Nishijima, E. Shoto, M. Okada, Water Sci. Technol. 35 (1997) 89.
2807. L. Groisman, C. Rav-Acha, Z. Gerstl, U. Mingelgrin, J. Environ. Qual. 33 (2004) 1930.
2808. H. Yoshitake, T. Yokoi, T. Tatsumi, Chem. Mater. 14 (2002) 4603.
2809. M. Jang, E.W. Shin, J.K. Park, S.I. Choi, Environ. Sci. Technol. 37 (2003) 5062.
2810. Z. Luan, E.M. Maes, M.A.W. Van der Heide, D. Zhao, R.S. Czernuszewicz, L. Keven, Chem. Mater. 11 (1999) 3680.
2811. Z. Luan, J.A. Fournier, J.B. Wooten, D.E. Miser, Microporous Mesoporous Mater. 83 (2005) 150–158.
2812. M. Kruk, M. Jaroniec, Langmuir 13 (1997) 6267–6273.
2813. Y.S. Ho, G. McKay, Water Res. 34 (2000) 735–742.
2814. H. Zhu, D.J. Jones, J. Zajac, R. Dutartre, M. Rhomari, J. Roziere, Chem. Mater. 14 (2002) 4886.
2815. E.W. Shin, J.S. Han, M. Jang, S.-H. Min, J.K. Park, R.M. Rowell, Environ. Sci. Technol. 38 (2004) 912.
2816. E.M. Wong, M.A. Markowitz, S.B. Qadri, S.L. Golledge, D.G. Castner, B.P. Gaber, Langmuir 18 (2002) 972.
2817. P. Wu, T. Tatsumi, Chem. Mater. 14 (2002) 1657.
2818. M. Jang, J.K. Park, E.W. Shin, Microporous Mesoporous Mater. 75 (2004) 159.
2819. B. Gu, G.M. Brown, S.D. Alexandratos, R. Ober, V. Patel, Oak Ridge National Laboratory, 1999.
2820. R.L. Wadlinger, G.T. Keller, E.J. Rosinski, US Patent 3,308,069 (1967).
2821. M.M.J. Treacy, J.M. Newswam, Nature 332 (1988) 249.
2822. J.B. Higgins, R.B. Lapierre, J.L. Schlenker, A.C. Rohrman, J.D. Wood, G.T. Keller, W.J. Rohrbaugh, Zeolites 8 (1988) 446.
2823. M. Sun, J. Sun, Q. Li, Chem. Lett. (1998)519.
2824. A. Corma, A. Martinez, P.A. Arroyo, J.L.F. Monteiro, E.F. Sousa-Aguiar, Appl. Catal. 142 (1996) 139.
2825. J.M. Bennett, J.W. Richardson, J.J. Pluth, J.V. Smith, Zeolites 7 (1987) 160.
2826. L. Yang, Y. Aizhen, Q. Xu, Appl. Catal. 67 (1991) 169.
2827. R. Khouzami, G. Coudurier, B.F. Mentzen, J.C. Vedrine, Stud. Surf. Sci. Catal. 37 (1988) 355.
2828. J. Das, C.V.V. Satyanarayana, D.K. Chakrabarty, S.N. Piramanayagam, S.N. Shringi, J. Chem. Soc., Faraday Trans. 88 (1992) 3255.
2829. A.M. Prakash, C.V.V. Satyanarayana, R.P. Bagwe, S. Ashtekar, D.K. Chakrabarty, Microporous Mater. 6 (1996) 89.
2830. X. Zhang, J. Wang, J. Zhong, Petrochem. Technol. 33 (8) (2004) 781.
2831. Y. Lu, M. He, X. Shu, B. Zong, Appl. Catal. A 255 (2003) 345.
2832. B. Wang, Q. Gao, J. Gao, D. Ji, X. Wang, J. Suo, Appl. Catal. A 274 (2004) 167.
2833. X. Zhu, S. Liu, Y. Song, L. Xu, Appl. Catal. A 288 (2005) 134.
2834. X. Zhu, S. Liu, Y. Song, S. Xie, L. Xu, Appl. Catal. A 290 (2005) 191.

2835. G. Zhao, J. Teng, Y. Zhang, Z. Xie, Y. Yue, Q. Chen, Y. Tang, Appl. Catal. A 299 (2005) 167.
2836. R. Xu, W. Pang, The Chemistry of Molecular Sieve and Multiporous Materials (Fen Zi Shai yu Duo Kong Cai Liao Hua Xue), Science Press (Ke Xue Chua Ban She), Beijing, 2004.
2837. E.M. Flanigen, H. Khatami, A.H. Seymenski, in: E.M. Flanigen, L.B. Sand (Eds.), Advanced Chemistry Series 101, American Chemical Society, Washington, DC, 1971.
2838. E. Geidel, H. Bohlig, C. Peuker, Stud. Surf. Sci. Catal. 65 (1991) 511.
2839. G. Bellussi, A. Carati, M.G. Clerici, G. Maddinelli, R. Millini, J. Catal. 133 (1992) 220.
2840. Z. Liu, R.J. Davis, J. Phys. Chem. 98 (1994) 1253.
2841. Z. Wu, Y. Zhao, D. Liu, Microporous Mesoporous Mater. 68 (2004) 127.
2842. B. Rakshe, V. Ramashe, V. Ramaswam, A.V. Ramaswam, J. Catal. 163 (1996) 501.
2843. Z. Dang, A.G. Anderson, Y. Amenomiya, B.A. Morrow, J. Phys. Chem. 99 (1995) 14437.
2844. X. Wang, F. Lefebvre, J. Patarin, J.M. Basset, Microporous Mesoporous Mater. 42 (2001) 269.
2845. M. Briend, A. Lamy, M.J. Peltre, P.P. Man, D. Barthomeuf, Zeolites 13 (1993) 201.
2846. Z. Zhang, B. Zong, Chinese J. Catal. 24 (11) (2003) 856.
2847. T.R. Hughes, H.M. White, J. Phys. Chem. 71 (1967) 2192.
2848. P.E. Eberly, J. Phys. Chem. 72 (1968) 1042.
2849. J.C. Vedrine, A. Aurox, V. Bolis, J. Catal. 59 (1979) 248.
2850. N.Y. Topsøe, K. Pdersen, E. Derouane, J. Catal. 70 (1981) 41.
2851. H. Liu, X. Bao, W. Wei, G. Shi, Microporous. Mesoporous Mater. 66 (2003) 117.
2852. W. Guo, L. Huang, P. Deng, Z. Xue, Q. Li, Microporous Mesoporous Mater. 44–45 (2001) 427.
2853. K Tanabe, M. Misono, Y. Ono, H. Hattori, New Solid Acids and Bases their Catalytic Properties, Kodausha, Tokyo, 1989.
2854. Y.V. Kissin, Catal. Rev. 43 (1& 2) (2001) 85.
2855. S.-H. Park, J.B. Parise, H. Gies, H. Liu, C.P. Grey, B.H. Toby, J. Am. Chem. Soc. 122 (44) (2000) 11023.
2856. S.-H. Park, P. Daniels, H. Gies, Microporous Mesoporous Mater. 37 (2000) 129.
2857. S.-H. Park, H. Gies, B.H. Toby, J.B. Parise, Chem. Mater. 14 (2002) 3187.
2858. S.-H. Park, H. Liu, M. Kleinsorge, C.P. Grey, B.H. Toby, J.B. Parise, Chem. Mater. 16 (2004) 2605.
2859. S.-H. Park, H. Boysen, J.B. Parise, Acta Cryst. B 62 (2005) 42.
2860. F. Liebau, Microporous Mesoporous Mater. 58 (2003) 15.
2861. U. Simon, M.E. Franke, Microporous Mesoporous Mater. 41 (2000) 1.
2862. M.E. Franke, U. Simon, Solid State Ionics 118 (1999) 311.
2863. U. Simon, U. Flesch, J. Porous Mater. 6 (1999) 33.
2864. P.M. de Wol., Acta Cryst. A 30 (1974) 777–785.
2865. H. Jobic, J. Kräger, M. Bée, Phys. Rev. Lett. 82 (1999) 4260.
2866. H. Jobic, Phys. Chem. Chem. Phys. 1 (1999) 525.
2867. P. Norby, Mater. Sci. Forum 228–231 (1996) 147.
2868. S. Merlino, Eur. J. Mineral. 2 (1990) 809.
2869. T.S. Ercit, J. Van Velthuizen, Can. Mineral. 32 (1994) 855.
2870. M.J. Annen, M.E. Davis, J.B. Higgins, J.L. Schlenker, Mater. Res. Soc. Symp. Proc. 233 (1991) 245.
2871. C. Röhrig, H. Gies, B. Marler, Zeolites 14 (1994) 498.
2872. C. Röhrig, H. Gies, Angew. Chem. Int. Ed. Engl. 34 (1995) 63.
2873. W.J. Kabsch, J. Appl. Cryst. 26 (1993) 795.
2874. M.A. Estermann, W. Steurer, Phase Transit. 67 (1996) 165.
2875. G.M. Sheldrick, SHELXTL, Siemens Analytical X-ray Instruments Inc., Madison, Wisconsin, USA, 1990.
2876. J.R. Macdonald, Impedance Spectroscopy, Wiley, New York, 1987.
2877. M.J. Rice, W.L. Roth, J. Solid State Chem. 4 (1972) 294.
2878. Y.T. Tsai, D.H. Whitmore, Sold State Ionics 7 (1982) 129.
2879. G. Kelemen, W. Lortz, G. Schön, J. Mater. Sci. 24 (1989) 333.
2880. G. Kelemen, G. Schön, J. Mater. Sci. 27 (1992) 6036.

2881. M. Nakayama, Y. Ishida, H. Ikuta, M. Wakihara, Solid State Ionics 117 (1999) 265.
2882. R. Nagel, Th.W. Gross, H. Günther, H.D. Lutz, J. Solid State Chem. 165 (2002) 303.
2883. R.T. Yang, Adsorbents: Fundamentals and Applications, Wiley, New York, 2003.
2884. R.W. Rousseau, Handbook of Separation Process Technology, Wiley, New York, 1987.
2885. G. Wilkinson, F.G.A. Stone, E.W. Abel, Comprehensive Organometallic Chemistry, Pergamon, Oxford, 1982.
2886. C.-Z. Tang, Y.-F. Xu, Z. Pang, X.-F. Hou, J. Organomet. Chem. 682 (2003) 123.
2887. D. Byun, J. Zink, Inorg. Chem. 42 (2003) 4308.
2888. N.V. Choudary, P. Kumar, T.S.G. Bhat, S.H. Cho, S.S. Han, J.N. Kim, Ind. Eng. Chem. Res. 41 (2002) 2728.
2889. J.H. Kim, J. Won, Y.S. Kang, J. Membr. Sci. 237 (2004) 199.
2890. W.J. Mortier, Compilation of Extra-Framework Sites in Zeolites, Butterworth Scientic Limited, Guildford, 1982.
2891. A.A. Verberckmoes, B.M. Weckhuysen, J. Pelgrims, R.A. Schoonheydt, J. Phys. Chem. 99 (1995) 15222.
2892. K. Pierloot, A. Delabie, C. Ribbing, J. Phys. Chem. B 102 (1998) 10789.
2893. D. Bae, K. Se., Microporous Mesoporous Mater. 40 (2000) 219.
2894. D. Bae, K. Se., Microporous Mesoporous Mater. 33 (1999) 265.
2895. A.M. Beale, G. Sankar, Chem. Mater. 18 (2006) 263.
2896. C. Weidenthaler, W. Schmidt, Chem. Mater. 12 (2000) 3811.
2897. R. Singh, P.K. Dutta, Microporous Mesoporous Mater. 21 (1998) 103.
2898. P. Gallezot, B. Imelik, J. Chim. Phys. 71 (1974) 155.
2899. V. Rakic, V. Dondur, R. Hercigonja, J. Serb. Chem. Soc. 68 (2003) 409.
2900. M. Briend-Fauré, J. Jeanjean, M. Kermarec, D. Delafosse, J. Chem. Soc., Faraday Trans. 1 (74) (1978) 1538.
2901. M. Briend-Fauré, M.F. Guilleux, J. Jeanjean, D. Delafosse, G. Gjega-Mariadossu, M. Bureau-Tardy, Acta Phys. Chem. 24 (1978) 99.
2902. A.M. Rigby, M.V. Frash, J. Mol. Catal. A: Chem. 126 (1997) 61.
2903. P.A. Jacobs, Carboniogenic Activity of Zeolites, Elsevier, 1977.
2904. R. Bakhtiar, J.J. Drader, R.K. Arneson, D.B. Jacobson, Rapid Commun. Mass Spectrom. 10 (1996) 1405.
2905. Q. Chen, C. Sioma, S.Z. Kan, B.O.S. Freiser, Int. J. Mass Spectrom. 179–180 (1998) 231.
2906. M. Lautens, W. Klute, W. Tam, Chem. Rev. 96 (1996) 49.
2907. D. Schröder, D. Sulzle, J. Hrus¢ak, D.K. Böhme, H. Schwarz, Int. J. Mass Spectrom. Ion Process. 110 (1991) 145.
2908. N.E. Schore, Chem. Rev. 88 (1988) 1081.
2909. K. Schroeter, C.A. Schalley, R. Wesendrup, D. Schröder, H. Schwarz, Organometallics 16 (1997) 986.
2910. R.M. Barrer, Hydrothermal Chemistry of Zeolites, Academic Press, New York, 1982.
2911. G.T. Kokotailo, C.A. Fyfe, Y. Feng, H. Grondey, H. Gies, B. Marler, D.E. Cox (Eds.), Catalysis by Microporous Materials, Studies Surface Science and Catalysis, vol. 94, Elsevier, Amsterdam, 1995.
2912. Ch. Baerlocher, W.M. Meier, D.H. Olson, Atlas of Zeolite Framework Types, fifth revised ed., Elsevier, Amsterdam, 2001.
2913. B.K. Petersen, J. Phys. Chem. B 103 (1999) 3145.
2914. D. Bae, K. Se., Microporous Mesoporous Mater. 40 (2000) 233.
2915. K. Pierloot, A. Delabie, C. Ribbing, A.A. Verberckmoes, R.A. Schoonheydt, J. Phys. Chem. B 102 (1998) 10789.
2916. J.A. McCleverty, T.J. Thomas, Comprehensive Coordination Chemistry: From Biology to Nanotechnology, Elsevier, Oxford, 2004.
2917. D.L. Vanoppen, D.E. De Vos, M.J. Genet, P.G. Rouxhet, P.A. Jacobs, Angew. Chem., Int. Ed. Engl. 34 (1995) 560.
2918. Q. Tang, Q. Zhang, P. Wang, Y. Wang, H. Wan, Chem. Mater. 16 (2004) 1967.
2919. C.J. Blower, T.D. Smith, Zeolites 13 (1993) 394.
2920. C. Ronay, K. Se., Zeolites 13 (1993) 97.

2921. D.T. Richens, The Chemistry of Aqua Ions: Synthesis, Structure and Reactivity: A Tour Through the Periodic Table of the Elements, Wiley, Chichester, 1997.
2922. J.J. Pluth, J.V. Smith, Mater. Res. Bull. 7 (1972) 1311.
2923. M.A. Heilbronn, J.C. Vickermann, J. Catal. 33 (8) (1974) 434.
2924. H. Praliaud, G. Coudrier, J. Chem. Soc., Faraday Trans. I 75 (1979) 2601.
2925. A.A. Verberckmoes, B.M. Weckhuysen, R.A. Schoonheydt, Microporous Mesoporous Mater. 22 (1998) 165.
2926. H. Hoser, S. Krzyzanowski, F. Trifiro, J. Chem. Soc., Faraday Trans. I 71 (1975) 665.
2927. T.A. Egerton, A. Hagen, F.S. Stone, J.C. Vickermann, J. Chem. Soc., Faraday Trans. I 68 (1972) 723.
2928. P.J. Hutta, J.H. Lunsford, J. Chem. Phys. 66 (1977) 4716.
2929. T. Yanagisawa, T. Shimizu, K. Kuroda, C. Kato, Bull. Chem. Soc. Jpn. 63 (1990) 988.
2930. S. Inagaki, Y. Fukushima, K. Kuroda, J. Chem. Soc., Chem. Commun. (1993)680.
2931. S. Inagaki, A. Koiwai, N. Suzuki, Y. Fukushima, K. Kuroda, Bull. Chem. Soc. Jpn. 69 (1996) 1449.
2932. D. Trong On, D. Desplantier-Giscard, C. Danumah, S. Kaliaguine, Appl. Catal. 253 (2003) 545.
2933. Y. Murakami, S. Yamakita, T. Okubo, S. Maruyama, Chem. Phys. Lett. 375 (2003) 393.
2934. S.H. Joo, S.J. Choi, I. Oh, J. Kwak, Z. Liu, O. Terasaki, R. Ryoo, Nature 412 (2001) 169.
2935. F. Hoffmann, M. Cornelius, J. Morell, M. Fröba, Angew. Chem. Int. Ed. 45 (2006) 3216.
2936. A. Firouzi, F. Atef, A.G. Oertli, G.D. Stucky, B.F. Chmelka, J. Am. Chem. Soc. 119 (1997) 3596.
2937. K. Sakamoto, in: J. Xia, I.A. Nnanna (Eds.), Surfactant Science Series, Protein-Based Surfactants, vol. 101, Marcel Dekker, New York, 2001.
2938. T. Yokoi, H. Yoshitake, T. Tatsumi, Chem. Mater. 15 (2003) 4536.
2939. S. Che, A.E. Garcia-Bennett, T. Yokoi, K. Sakamoto, H. Kunieda, O. Terasaki, T. Tatsumi, Nature Mater. 2 (2003) 801.
2940. E. Garcia-Bennett, O. Terasaki, S. Che, T. Tatsumi, Chem. Mater. 16 (2004) 813.
2941. A.E. Garcia-Bennett, S. Che, K. Miyasaka, Y. Sakamoto, T. Ohsuna, Z. Liu, O. Terasaki, Stud. Surf. Sci. Catal. 156 (2005) 11.
2942. A.E. Garcia-Bennett, N. Kupferschmidt, Y. Sakamoto, S. Che, O. Terasaki, Angew. Chem. Int. Ed. 44 (2005) 5317.
2943. C. Gao, Y. Sakamoto, K. Sakamoto, O. Terasaki, S. Che, Angew. Chem. Int. Ed. 45 (2006) 4295.
2944. T. Yokoi, H. Yoshitake, T. Yamada, Y. Kubota, T. Tatsumi, J. Mater. Chem. 16 (2006) 1125.
2945. S. Che, Z. Liu, T. Ohsuna, K. Sakamoto, O. Terasaki, T. Tatsumi, Nature 429 (2004) 281.
2946. T. Ohsuna, Z. Liu, S. Che, O. Terasaki, Small 1 (2005) 233.
2947. H. Jin, Z. Liu, T. Ohsuna, O. Terasaki, Y. Inoue, K. Sakamoto, T. Nakanishi, K. Ariga, S. Che, Adv. Mater. 18 (2006) 593.
2948. X. Wu, H. Jin, Z. Liu, T. Ohsuna, O. Terasaki, K. Sakamoto, S. Che, Chem. Mater. 18 (2006) 241.
2949. B. Wang, C. Chi, W. Shan, Y. Zhang, N. Ren, W. Yang, Y. Tang, Angew. Chem. Int. Ed. 45 (2006) 2088.
2950. T. Hattori, N. Kitamura, N. Yamato, H. Yokota, Eur. Pat. Appl. (2003)13.
2951. Y. Shinko, H. Yoshihara, T. Hattori, Jpn. Kokai Tokkyo Koho (2004) 13.
2952. H. Marsh, F. Rodrýguez-Reinoso, Activated Carbon, Elsevier, Oxford, 2006.
2953. F.S. Baker, C.E. Miller, A.J. Repik, E.D. Tolles, Kirk-Othmer Encyclopedia of Chemical Technology, Wiley, New York, 1992.
2954. M. Molina-Sabio, F. Rodrýguez-Reinoso, Colloids Surf. A: Physicochem. Eng. Asp. 241 (2004) 15.
2955. F. Caturla, M. Molina-Sabio, F. Rodrýguez-Reinoso, Carbon 29 (1991) 999.
2956. M. Molina-Sabio, F. Rodrýguez-Reinoso, F. Caturla, M.J. Sellés, Carbon 33 (1995) 1105.
2957. M. López, M. Labady, J. Laine, Carbon 34 (1996) 825.
2958. M. Jagtoyen, F. Derbyshire, Carbon 31 (1993) 1185.
2959. M. Jagtoyen, F. Derbyshire, Carbon 36 (1998) 1085.
2960. F. Rodrýguez-Reinoso, M. Molina-Sabio, C. Almansa, Spanish Patent 2,165,784 (2003).

2961. M. Molina-Sabio, C. Almansa, F. Rodrýguez-Reinoso, Carbon 41 (2003) 2113.
2962. C. Almansa, M. Molina-Sabio, F. Rodrýguez-Reinoso, Microporous Mesoporous Mater. 76 (2004) 185.
2963. F. Rodrýguez-Reinoso, J.M. Martin-Martinez, C. Prado-Burguete, B. McEnaney, J. Phys. Chem. 91 (1987) 515.
2964. ASTM Standard Test Method for Apparent Density of Activated Carbon 1989, Designation: D 2854–2889.
2965. R. Denoyel, J. Fernandez-Colinas, Y. Grillet, J. Rouquerol, Langmuir 9 (1993) 515.
2966. F. Kraehenbuehl, H.F. Stoeckli, Carbon 24 (1986) 483.
2967. H. Benaddi, D. Legras, J.N. Rouzaud, F. Beguin, Carbon 36 (1998) 306.
2968. A.M. Puziy, O.I. Poddubnaya, A. Martýnez-Alonso, F. Suárez-Garcýa, J.M.D. Tascón, Carbon 40 (2002) 1493.
2969. R. Kimmich, Chem. Phys. 284 (2002) 253.
2970. B. Lefevre, A. Saugey, J.L. Barrat, L. Bocquet, E. Charlaix, P.F. Gobin, G. Vigier, Colloids Surf. A 241 (2004) 265.
2971. V.D. Borman, A.A. Belogorlov, A.M. Grekhov, G.V. Lisichkin, V.N. Tronin, V.I. Troyan, Tech. Phys. Lett. 30 (2004) 973.
2972. X. Kong, Y. Qiao, Appl. Phys. Lett. 86 (2005) 151919.
2973. F.B. Surani, X. Kong, D.B. Panchal, Y. Qiao, Appl. Phys. Lett. 87 (2005) 163111.
2974. X. Kong, F.B. Surani, Y. Qiao, J. Mater. Res. 20 (2005) 1042.
2975. F.B. Surani, X. Kong, Y. Qiao, Appl. Phys. Lett. 87 (2005) 251906.
2976. D.T. Wasan, A.D. Nikolov, Nature 423 (2002) 156.
2977. D.J. Mann, M.D. Halls, Phys. Rev. Lett. 90 (2003) 195503.
2978. M. Majumder, N. Chopra, R. Andrews, B.J. Hinds, Nature 438 (2005) 44.
2979. F.A.L. Dullien, Porous Media – Fluid Transport and Pore Structure, Academic Press, New York, 1992.
2980. X. Kong, Y. Qiao, J. Appl. Phys. 100 (2006) 014308.
2981. L. Pizzio, P. Vázquez, C. Cáceres, M. Blanco, E. Alesso, M.I. Erlich, R. Torviso, L. Finkielsztein, B. Lantaño, G. Moltrasio, J. Aguirre, Catal. Lett. 93 (2004) 67.
2982. L.R. Pizzio, C.V. Cáceres, M.N. Blanco, Appl. Catal. 167 (1998) 283.
2983. D. Lapham, J.B. Mo.at, Langmuir 7 (1991) 2273.
2984. H. Hayashi, J.B. Moffat, J. Catal. 83 (1983) 192.
2985. L.R. Pizzio, M.N. Blanco, Appl. Catal. A: Gen. 255 (2003) 265.
2986. T. Okuhara, N. Mizuno, M. Misono, Adv. Catal. 41 (1996) 221.
2987. W. Huang, D. He, J. Liu, Q. Zhu, Appl. Catal. 199 (2000) 93.
2988. G.P. Romanelli, J.C. Autino, M.N. Blanco, L.R. Pizzio, Appl. Catal. A: General 295 (2005) 209.
2989. R.Sh. Mikhail, S. Brunauer, E.E. Bodor, J. Colloid Interface Sci. 26 (1968) 45.
2990. A. Lecloux, J.P. Pirard, J. Colloid Interface Sci. 70 (1979) 265.
2991. P.D. Metelski, T.W. Swaddle, Inorg. Chem. 38 (1999) 301.
2992. J.B. Moffat, Polyhedron 5 (1986) 261.
2993. G.M. Brown, M.R. Noe-Spirlet, W.R. Busing, H.A. Levy, Acta Cryst. B 33 (1977) 1038.
2994. A. Vakulenko, Y. Dobrovolsky, L. Leonova, A. Karelin, A. Kolesnikova, N. Bukun, Solid State Ionics 136 (7) (2000) 285.
2995. J.B. McMonagle, J.B. Moffat, J. Colloid Interface Sci. 101 (1984) 479.
2996. C. Rocchiccioli-Deltche., R. Thouvenot, R. Franck, Spectrochim. Acta 32 A (1976) 587.
2997. N. Essayem, G. Coudurier, M. Fournier, J. Védrine, Catal. Lett. 34 (1995) 223.
2998. M.A. Parent, J.B. Moffat, Catal. Lett. 48 (1997) 135.
2999. V.M. Mastikhin, S.M. Kulikov, A.V. Nosov, I.V. Khozevnikov, I.L. Mudrakovsky, M.N. Timofeeva, J. Mol. Catal. 60 (1990) 65.
3000. P.M. Rao, A. Wolfson, S. Kababya, S. Vega, M.V. Landau, J. Catal. 232 (2005) 210.
3001. N. Essayem, Y.Y. Tong, H. Jobic, J.C. Vedrine, Appl. Catal. A: Gen. 194 (5) (2000) 109.
3002. M.A. Gao, J.B. Moffat, Catal. Lett. 42 (1996) 135.
3003. R. Cid, G. Pecci, Appl. Catal. 14 (1985) 15.
3004. I.V. Khozhevnikov, Chem. Rev. 98 (1998) 171.

3005. D.B. Taylor, J.B. McMonagle, J.B. Moffat, J. Colloid Interface Sci. 108 (1985) 278.
3006. C. Rocchiccioli-Deltche., M. Fournier, R. Franck, R. Thouvenot, Inorg. Chem. 22 (1983) 207.
3007. D. He, W. Huang, J. Liu, Q. Zhu, J. Mol. Catal. A: Chem. 145 (1999) 335.
3008. Y. Izumi, M. Ogawa, W. Nohora, K. Urabe, Chem. Lett. (1992) 1987.
3009. A. Corma, A. Martýnez, C. Martýnez, J. Catal. 164 (1996) 422.
3010. J.C. Edwards, C.Y. Thiel, B. Benac, J.F. Knifton, Catal. Lett. 51 (1998) 77.
3011. D. He, W. Huang, J. Liu, Q. Zhu, Catal. Today 51 (1999) 127.
3012. F. Cavani, F. Trifro, A. Vaccari, Catal. Today 11 (1991) 2.
3013. D. Tichit, B. Coq, Cattech 7 (2003) 206.
3014. D. Tichit, M. Hassane Lhouty, A. Guida, B. Huong Chiche, F. Figueras, A. Auroux, D. Bartalini, E. Garrone, J. Catal. 151 (1995) 50.
3015. J.I. Di Cosimo, V.K. Diez, M. Xu, E. Iglesia, C.R. Apesteguia, J. Catal. 178 (1998) 499.
3016. J.S. Valente, F. Figueras, M. Gravelle, P. Khumbar, J. Lopez, J.-P. Besse, J. Catal. 189 (2000) 370.
3017. B.F. Sels, D.E. De Vos, P.A. Jacobs, Catal. Rev. 43 (2001) 443.
3018. A. de Roy, C. Forano, J.-P. Besse, in: V. Rives (Ed.), Layered Double Hydroxides: Present and Future, Nova Sci., New York, 2001.
3019. D. Tichit, A. Rolland, F. Prinetto, G. Fetter, M.J. Martinez Ortiz, M.A. Valenzuela, P. Bosch, J. Mater. Chem. 12 (2002) 3832.
3020. M.J. Climent, A. Corma, S. Iborra, K. Epping, A. Velty, J. Catal. 225 (2004) 316.
3021. P. Benito, F.M. Labajos, V. Rives, J. Solid State Chem. 179 (2006) 3725.
3022. D. Tichit, O. Lorret, B. Coq, F. Prinetto, G. Ghiotti, Microporous Mesoporous Mater. 80 (2005) 213.
3023. F. Prinetto, G. Ghiotti, P. Graffin, D. Tichit, Microporous Mesoporous Mater. 39 (2000) 229.
3024. M.A. Aramendia, V. Borau, C. Jiménez, J.M. Marinas, J.R. Ruiz, F.J. Urbano, J. Solid State Chem. 168 (2002) 156.
3025. M. Jitianu, M. Balasoiu, M. Zaharescu, A. Jitianu, A. Ivanov, J. Sol-Gel Sci. Technol. 19 (2000) 453.
3026. T. Lopez, P. Bosch, E. Ramos, R. Gomez, O. Novaro, D. Acosta, F. Figueras, Langmuir 12 (1996) 189.
3027. P. Hollins, Surf. Sci. Rep. 16 (1992) 51.
3028. Yu A. Lokhov, A.A. Davidov, Kinet. Katal. 21 (1980) 1523.
3029. V.B. Kazansky, V.Yu. Borovkov, E.G. Derouane, Catal. Lett. 19 (1993) 327.
3030. V.B. Kazansky, V.Yu. Borokov, A.I. Serykh, Catal. Lett. 49 (1997) 35.
3031. Z. Gandao, B. Coq, L.C. de Ménorval, D. Tichit, Appl. Catal. A: Gen. 147 (1996) 395.
3032. F. Prinetto, M. Manzoli, G. Ghiotti, M.J. Martinez Ortiz, D. Tichit, B. Coq, J. Catal. 222 (2004) 238.
3033. B.L. Mojet, J.T. Miller, D.E. Ramaker, D.C. Koningsberger, J. Catal. 186 (1999) 373.
3034. D.E. Ramaker, J. de Graaf, J.A.R. van Veen, D.C. Koningsberger, J. Catal. 203 (2001) 7.
3035. H. Knözinger, H. Krietenbrink, J. Chem. Soc., Faraday Trans. I 71 (1975) 2421.
3036. J. Yarwood, in: Spectroscopy and Structure of Molecular Complexes, Plenum, London and New York, 1973.
3037. K.F. Purcell, R.S. Drago, J. Am. Chem. Soc. 88 (1966) 919.
3038. J.C. Lavalley, Catal. Today 27 (1996) 377.
3039. C. Binet, A. Jadi, J.C. Lavalley, J. Chim. Phys. 89 (1992) 31.
3040. D. Tichit, B. Coq, S. Cernaux, R. Durand, Catal. Today 75 (2002) 97.
3041. M.A. Aramendia, V. Boreau, C. Jimenez, J.M. Marinas, J.R. Ruiz, F.J. Urbano, J. Colloid Interface Sci. 238 (2001) 385.
3042. M.A. Aramendia, V. Boreau, C. Jimenez, J.M. Marinas, J.R. Ruiz, F.J. Urbano, Appl. Catal. A: Gen. 249 (2003) 1.
3043. S. Nishiyama, M. Yamamoto, J. Chem. Eng. Jpn. 37 (2004) 310.
3044. G. Boxhoorn, J. Moolhuysen, J.P.G. Coolegem, R.A. van Santen, J. Chem. Soc., Chem. Commun. (1985)1305.
3045. K.R. Dunbar, R.A. Heintz, Prog. Inorg. Chem. 45 (1997) 283.

3046. J. Balmaseda, E. Reguera, J. Rodrýguez-Hernández, L. Reguera, M. Autie, Microporous. Mesoporous Mater. 96 (2006) 222.
3047. S.S. Kaye, J.R. Long, J. Am. Chem. Soc. 127 (2005) 6506.
3048. K.W. Chapman, P.D. Southon, C.L. Weeks, C.J. Kepert, Chem. Commun. (2005) 3322.
3049. M.R. Hartman, V.K. Peterson, Y. Liu, S.S. Kaye, J.R. Long, Mater. Chem. 18 (2006) 3221.
3050. R. Martinez-Garcia, M. Knobel, E. Reguera, J. Phys.: Condens. Matter 18 (2006) 11243.
3051. S. Ferlay, T. Mallah, R. Ouahes, P. Veillet, M. Verdaguer, Nature 378 (1995) 701.
3052. S.M. Holmes, G.S. Girolami, J. Am. Chem. Soc. 121 (1999) 5593.
3053. O. Sato, T. Iyoda, A. Fujishima, K. Hashimoto, Science 272 (1996) 704.
3054. S. Ohkoshi, K. Arai, Y. Sato, K. Hashimoto, Nat. Mater. 3 (2004) 857.
3055. W.E. Buschmann, J. Ensling, P. Gutlich, J.S. Miller, Chem. Eur. J. 5 (1999) 3019.
3056. D. Louer, R. Vargas, J. Appl. Crystallogr. 15 (1982) 542.
3057. D.R. Lide (Ed.), CRC Handbook of Chemistry and Physics, eighty fourth ed., Boca Raton, 2003–2004.
3058. J. Balmaseda, E. Reguera, J. Fernandez, A. Gordillo, H. Yee-Madeira, J. Phys. Chem. Solids 64 (2003) 685.
3059. M.M. Dubinin, in: D.A. Cadenheat (Ed.), Progress in Surface Science and Membrane Science, Academic Press, New York, 1975.
3060. F. Stoeckli, Russ. Chem. Bull. Int. Ed. 50 (2001) 2265.
3061. M.H. Simonot-Grange, J. Chim. Phys. 84 (1987) 1161.
3062. P. Cartraud, A. Cointot, A. Renaud, J. Chem. Soc., Faraday Trans. 1 (77) (1981) 1561.
3063. B.P. Bering, V.V. Serpinski, Izv. Akad. Nauk SSSR, Ser. Khim. 11 (1974) 2427.
3064. R. Roque-Malherbe, Microporous Mesoporous Mater. 41 (2000) 227.
3065. A. Ludi, A. Gudel, Struct. Bond. 14 (1973) 1.
3066. A. Ludi, H.U. Gudel, Helv. Chim. Acta 51 (1968) 2006.
3067. D.F. Mullica, W.O. Milligan, Acta Crystallogr. B 34 (1978) 3558.
3068. G.W. Beall, W.O. Milligan, Inorg. Chem. 16 (1977) 2715.
3069. E. Reguera, J. Rodrýguez-Hernández, A. Champi, J.G. Duque, E. Granados, C. Rettori, Z. Phys. Chem. 220 (2006) 1609.
3070. A. Gomez, J. Rodrýguez, E. Reguera, J. Chem. Crystallogr. 34 (2004) 893.
3071. Y. Zhang, Inorg. Chem. 21 (1992) 3889.
3072. P. Gravereau, E. Garnier, A. Hardy, Acta Crystallogr. B 35 (1979) 2843.
3073. R. Martinez-Garcýa, M. Knobel, E. Reguera, J. Phys. Chem. B 110 (2006) 7296.
3074. K. Nakamoto, in: Infrared and Raman Spectra of Inorganic and Coordination Compounds, Wiley, New York, 1986.
3075. A.G. Sharpe, in: P. Maitlis, F.A.G. Stone, R. West (Eds.), The Chemistry of Cyano Complexes of the Transition Metals, Academic Press, New York, 1976.
3076. A. Gomez, E. Reguera, Int. J. Inorg. Mater. 3 (2001) 1045.
3077. E. Reguera, J. Balmaseda, G. Quintana, J. Fernández, Polydedron 17 (1998) 2353.
3078. I.D. Brown, Acta Crystallogr. A 32 (1976) 24.
3079. E. Reguera, J. Fernandez, Hyperfine Interact. 88 (1994) 49.
3080. F. Helfferich, Ion Exchange, Dover Publications, New York, 1962.
3081. R.H. Perry, D. Green, Perry's Chemical Engineer's Handbook, McGraw-Hill, International ed., 1984.
3082. D. Kratochvil, B. Volesky, G. Demopoulos, Water Res. 31 (1997) 2327.
3083. C.P. Gomes, M.F. Almeida, J.M. Loureiro, Sep. Purif. Technol. 24 (2001) 35.
3084. A.K. Singhal, AIChE Symp. Ser. 74 (1978) 36.
3085. D.M. Ruthven, Chem. Eng. Sci. 28 (1973) 701.
3086. T.F. Speth, Masters Thesis, Michigan Technology University, 1986.
3087. K.R. Franklin, R.P.J. Townsend, J. Chem. Soc., Faraday Trans. I 81 (1985) 1071.
3088. R.P. Townsend, Pure Appl. Chem. 58 (1986) 1259.
3089. B.K. Kaui, N.H. Sweed, Fundamentals of Adsorption, AIChE, New York, 1984.
3090. D.M. Ruthven, M. Goddard, Zeolites 6 (1986) 275.
3091. J. Lehto, R. Harjula, React. Funct. Polym. 27 (1995) 121.
3092. V.J. Inglezakis, H.P. Grigoropoulou, Microporous Mesoporous Mater. 61 (2003) 273.

3093. A.H. Englert, J. Rubio, Int. J. Miner. Process. 75 (2005) 21.
3094. E.L. Cooney, N.A. Booker, D.C. Shallcross, G.W. Stevens, Sep. Sci. Technol. 34 (1999) 2741.
3095. M. Matosic, I. Mijatovic, E. Hodzic, Chem. Biochem. Eng. Q 14 (2000) 141.
3096. A. Demir, A. Gunay, E. Debik, Water SA 28 (2002) 329.
3097. A.S. Sheta, A.M. Fatalah, M.S. Al-Sewailem, E.M. Khaled, A.S.H. Sallam, Microporous Mesoporous Mater. 61 (2003) 127.
3098. G.X.S. Zhao, J.L. Lee, P.A. Chia, Langmuir 19 (2003) 1977.
3099. N. Bektas, S. Kara, Sep. Purif. Tech. 39 (2004) 189.
3100. A. Chojnacki, K. Chojnacka, J. Hoffmann, H. Gorecki, Miner. Eng. 17 (2004) 933.
3101. M.J. Rivero, O. Primo, M. Inmaculada Ortiz, J. Chem. Technol. Biotechnol. 79 (2004) 822.
3102. G.O. Wood, Carbon 39 (2001) 343.
3103. S. Kleineidam, C. Schuth, P. Grathwohl, Environ. Sci. Technol. 36 (2002) 4689.
3104. C.T. Hsieh, H. Teng, J. Chem. Technol. Biotechnol. 75 (2000) 1066.
3105. A. Bembnowska, R. Pelech, E. Milchert, J. Colloid Interface Sci. 265 (2003) 276.
3106. T.S. Singh, K.K. Pant, Separ. Purif. Tech. 36 (2004) 139.
3107. E. Erdem, N. Karapinar, R. Donat, J. Colloid Interface Sci. 280 (2004) 309.
3108. S. Rengaraj, Y. Kim, C.K. Joo, K. Choi, J. Yi, Korean J. Chem. Eng. 21 (2004) 187.
3109. J. Peric, M. Tro, N.V. Medvidovic, Water Res. 38 (2004) 1893.
3110. A. Erdem-Senatalar, M. Tatller, A. Sirkecioglu, Colloids Surf. Physicochem. Eng. Asp. 173 (2000) 51.
3111. S. Basu1, P.F. Henshaw, N. Biswas, H.K. Kwan, Can. J. Chem. Eng. 80 (2002) 1.
3112. W. Rudzinski, D.H. Everett, Adsorption of Gases on Heterogeneous Surfaces, Academic Press, London, 1992.
3113. M. Horsfall Jr., A.I. Spiff, Acta Chim. Slov. 52 (2005) 174.
3114. O. Ceyhan, D. Baybas, Turk. J. Chem. 25 (2001) 193.
3115. Y.S. Ho, J.F. Porther, G. Mc Kay, Water, Air, Soil Pollut. 141 (2002) 1.
3116. F. Carrasco-Marin, M.A. Alvarez-Merino, C. Moreno-Castilla, Fuel 75 (1996) 966.
3117. V.K. Dobruskin, Langmuir 14 (1998) 3840.
3118. S.W. Davis, S.E. Powers, J. Environ. Eng. 126 (2000) 354.
3119. M.M. Dubinin, H.F. Stoeckli, J. Colloid Interface Sci. 75 (1980) 34.
3120. G. Finger, M. Bulow, Carbon 17 (1979) 87.
3121. A. Gil, P. Grange, Colloids Surf. Physicochem. Eng. Asp. 113 (1996) 39.
3122. M. Suzuki, Adsorption Engineering, Kodansha, Tokyo, 1990.
3123. M.M. Dubinin, in: D.A. Cadenhead, J.F. Danieli, M.D. Rosenberg (Eds.), Prog. Surf. Membr. Sci., Academic, 1975, p. 1.
3124. S. Brezovska, B. Marina, B. Napanova, D. Burevski, V. Bosevska, L. Stoianovska, J. Serb. Chem. Soc. 69 (2004) 145.
3125. C. Falamaki, A. Mohammadib, M. Sohrabib, Colloids Surf. Physicochem. Eng. Aspect 246 (2004) 31.
3126. B. Buczek, L. Czepirski, Pol. J. Chem. 77 (2003) 1191.
3127. H. Shojibara, Y. Sato, S. Takishima, H. Masuoka, J. Chem. Eng. Jpn. 28 (1995) 245.
3128. M. Domingo-Garcýa, I. Fernandez-Morales, F.J. Lopez-Garzon, C. Moreno-Castilla, Langmuir 13 (1997) 1218.
3129. S. Karahan, M. Yurdakoç, Y. Seki, K. Yurdakoç J. Colloid Interface Sci. 293 (2006) 36.
3130. R. Donat, A. Akdogan, E. Erdem, H. Cetisli, J. Colloid Interface Sci. 286 (2005) 43.
3131. S.A. Khan, R. Rehman, M.A. Khan, Waste Manage. 15 (1995) 271.
3132. J.K. Moon, E.H. Lee, H.T. Kim, Korean J. Chem. Eng. 21 (2004) 1026.
3133. A. Abusafa, H. Yucel, Sep. Purif. Technol. 28 (2002) 103.
3134. K.E. Noll et al., Adsorption Technology for Air and Water Pollution Control, Lewis Publishers Inc., Chelsea, Michigan, 1992.
3135. V.J. Inglezakis, J. Colloid Interface Sci. 281 (2005) 68.
3136. V.J. Inglezakis, S.G. Poulopoulos, Adsorption, Ion Exchange and Catalysis: Design of Operations and Environmental Applications, Elsevier, Amsterdam, 2006.
3137. V.J. Inglezakis, M.D. Loizidou, H.P. Grigoropoulou, J. Colloid Interface Sci. 275 (2004) 570.

3138. M.S. Berber-Mendoza, R. Leyva-Ramos, P. Alonso-Davila, L. Fuentes-Rubio, R.M. Guerrero-Coronado, J. Colloid Interface Sci. 301 (2006) 40.
3139. D. Karadag, Y. Koc, M. Turan, B. Armaganc, J. Hazard. Mater. B 136 (2006) 604.
3140. M. Sprynskyy, B. Buszewski, A.P. Terzyk, J. Namiesnik, J. Colloid Interface Sci. 304 (2006) 21.
3141. H. Faghihian, M. Ghannadi-Marageh, H. Kazemian, Appl. Radiat. Isot. 50 (1999).
3142. I. Smiciklas, S. Dimović, I. Plecás, Appl. Clay Sci. 35 (2007) 139.
3143. A. Corma, M.S. Grande, V. Gonzalez-Alfaro, A.V. Orchilles, J. Catal. 159 (1996) 375.
3144. D. Brunel, A.C. Blanc, A. Galarneau, F. Fajula, Catal. Today 73 (2002) 139.
3145. T.R. Pauly, V. Petkov, Y. Liu, S.J.L. Billinge, T.J. Pinnavaia, J. Am. Chem. Soc. 124 (2002) 97.
3146. T. Linssen, K. Cassiers, P. Cool, E.F. Vansant, Adv. Colloid Interface Sci. 103 (2003) 121.
3147. S.C. Shen, S. Kawi, J. Phys. Chem. B 103 (1999) 8870.
3148. R. Mokaya, J. Phys. Chem. B 104 (2000) 8279.
3149. X.S. Zhao, G.-Q. Lu, X. Hu, Microporous Mesoporous Mater. 41 (2000) 37.
3150. R. Mokaya, W. Jones, Chem. Commun. (1998) 1839.
3151. R. Mokaya, Chem. Commun. (2001) 633.
3152. S.C. Shen, S. Kawi, Langmuir 18 (2002) 4720.
3153. J.P. Lourenço, A. Fernandes, C. Henriques, M.F. Ribeiro, Microporous Mesoporous Mater. 94 (2006) 56.
3154. N. Coustel, F. Di Renzo, F. Fajula, J. Chem. Soc., Chem. Commun. (1994) 967.
3155. X.S. Zhao, F. Audsley, G.-Q. Lu, J. Phys. Chem. B 102 (1998) 4143.
3156. X.S. Zhao, G.-Q. Lu, J. Phys. Chem. B 102 (1998) 1556.
3157. C.P. Jaroniec, M. Kruk, M. Jaroniec, A. Sayari, J. Phys. Chem. B 102 (1998) 5503.
3158. V. Antochshuk, M. Jaroniec, Chem. Commun. (1999) 2373.
3159. A. Matsumoto, H. Misran, K. Tsutsumi, Langmuir 20 (2004) 7139.
3160. J.M. Kisler, M.L. Gee, G.W. Stevens, A.J. O'Connor, Chem. Mater. 15 (2003) 619.
3161. M.M.L. Ribeiro Carrott, A.J.E. Candeias, P.J.M. Carrott, K.S.W. Sing, K.K. Unger, Langmuir 16 (2000) 9103.
3162. A.E. Candeias, M.M.L. Ribeiro Carrott, P.J.M. Carrott, K. Schumacher, M. Grun, K.K. Unger, in: F. Rodriguez-Reinoso, B. McEnaney, J. Rouquerol, K.K. Unger (Eds.), Characterisation of Porous Solids VI, Studies in Surface Science and Catalysis, vol. 144, Elsevier, Amsterdam, 2002.
3163. P.J. Branton, P.G. Hall, K.S.W. Sing, Adsorption 1 (1995) 77.
3164. P.L. Llewellyn, F. Schüth, Y. Grillet, F. Rouquerol, J. Rouquerol, K.K. Unger, Langmuir 11 (1995) 574.
3165. R.C. Reid, J.M. Prausnitz, B.E. Poling, The Properties of Gases and Liquids, McGraw-Hill, New York, 1986.
3166. P.I. Ravikovitch, A.V. Neimark, Colloids Surf. A 187–188 (2001) 11.
3167. P.I. Ravikovitch, A. Vishnyakov, A.V. Neimark, M.M.L. Ribeiro Carrott, P.A. Russo, P.J.M. Carrott, Langmuir 22 (2006) 513.
3168. B.V. Derjaguin, Acta Phys. Chim. 12 (1940) 181.
3169. J.C. Broekhof, J.H. de Boer, J. Catal. 10 (1968) 368.
3170. A.V. Neimark, P.I. Ravikovitch, Microporous Mesoporous Mater. 44 (2001) 697.
3171. J. Hagymassy Jr., S. Brunauer, R.Sh. Mikhail, J. Colloid Interface Sci. 29 (1969) 485.
3172. A. Raoof, J-P. Guilbaud, H. Van Damme, P. Porion, P. Levitz, J. Colloid Interface Sci. 206 (1998) 1.
3173. H. Naono, M. Hakuman, J. Colliod Interface Sci. 158 (1993) 19.
3174. V.B. Fenelonov, A.Yu. Derevyankin, S.D. Kirik, L.A. Solovyov, A.N. Shmakov, J.-L. Bonardet, A. Gedeon, V.N. Romannikov, Microporous Mesoporous Mater. 44–45 (2001) 33.
3175. N. Muroyama, T. Ohsuna, R. Ryoo, Y. Kubota, O. Terasaki, J. Phys. Chem. B 110 (2006) 10630.
3176. P.I. Ravikovitch, A.V. Neimark, Langmuir 22 (2006) 11171.
3177. A. Galarneau, D. Desplantier, R. Dutartre, F. Di Renzo, Microporous. Mesoporous Mater. 27 (1999) 297.

3178. P.L. Llewellyn, C. Sauerland, C. Martin, Y. Grillet, J.-P. Coulomb, F. Rouquerol, J. Rouquerol, in: B. McEnaney, T.J. Mays, J. Rouquerol, F. Rodriguez-Reinoso, K.S.W. Sing, K.K. Unger (Eds.), Characterisation of Porous Solids IV, Royal Society of Chemistry, London, 1997.
3179. M. Kruk, M. Jaroniec, Langmuir 15 (1999) 5683.
3180. H. Naono, M. Hakuman, T. Tanaka, N. Tamura, K. Nakai, J. Colloid Interface Sci. 225 (2000) 411.
3181. M. Rozwadowski, M. Lezanska, J. Wloch, K. Erdmann, R. Golembiewski, J. Kornatowski, Langmuir 17 (2001) 2112.
3182. F.S. Baker, K.S.W. Sing, J. Colloid Interface Sci. 55 (1976) 605.
3183. J.D. Carruthers, D.A. Payne, K.S.W. Sing, L.J. Stryker, J. Colloid Interface Sci. 36 (1971) 205.
3184. J. Rouquerol, F. Rouquerol, C. Pérés, Y. Grillet, M. Boudellal, in: S.J. Gregg, K.S.W. Sing, H.F. Stoeckli (Eds.), Characterisation of Porous Solids, Society of Chemical Industry, London, 1979, p. 107.
3185. L.J.M. Schlangen, L.K. Koopal, M.A.C. Stuart, J. Lyklema, M. Robin, H. Toulhoat, Langmuir 11 (1995) 1701.
3186. D.R. Dunphy, S. Singer, A.W. Cook, B. Smarsly, D.A. Doshi, C.J. Brinker, Langmuir 19 (2003) 10403.
3187. W. Lutz, W. Gessner, R. Bertram, I. Pitsch, R. Fricke, Microporous Mater. 12 (1997) 131.
3188. H.K. Beyer, in: H.G. Karge, J. Weitkamp (Eds.), Post-synthesis Modi.cation I, Molecular Sieves – Science and Technology, vol. 3, Springer-Verlag, Heidelberg, 2002.
3189. D. Trong On, S.M.J. Zaidi, S. Kaliaguine, Microporous Mesoporous Mater. 22 (1998) 211.
3190. L. Bonetto, M.A. Camblor, A. Corma, J. Perez-Pariente, Appl. Catal. A 82 (1992) 37.
3191. A. Corma, A. Martinez, P.A. Arroyo, J.L.F. Monteiro, E.F. Sousa-Aguiar, Appl. Catal. A 142 (1996) 139–150.
3192. J. Weitkamp, Zeolites and catalysis, Solid State Ionics 131 (2000) 175.
3193. M.V. Landau, L. Vradman, V. Valtchev, J. Lezervant, E. Liubich, M. Talianker, Ind. Eng. Chem. Res. 42 (2003) 2773.
3194. S.M. Auerbach, K.A. Carrado, P.K. Dutta, Handbook of Zeolite Science and Technology, Marcel Dekker Inc., New York, 2003.
3195. P. Botella, A. Corma, J.M. Lopez-Nieto, S. Valencia, R. Jacquot, J. Catal. 195 (2000) 161.
3196. M.A. Camblor, A. Corma, A. Misud, J. Perez-Pariente, S. Valencia, Stud. Surf. Sci. Microporous Mater. 105 (1997) 341.
3197. M.A. Camblor, A. Corma, S. Valencia, Microporous Mesoporous Mater. 25 (1998) 59.
3198. P. Prokesova, S. Mintova, J. Cejka, T. Bein, Microporous Mesoporous Mater. 64 (2003) 165.
3199. M.V. Landau, D. Tavor, O. Regev, M.L. Kaliya, M. Herskowitz, Chem. Mater. 11 (1999) 2030.
3200. L. Ding, Y. Zheng, Z. Zhang, Z. Ring, J. Chen, Microporous Mesoporous Mater. 94 (2006) 1.
3201. S.B. Kulkarni, V.P. Shiralkar, A.N. Kotasthane, R.B. Borade, P. Ratnasamy, Zeolites 2 (1982) 313.
3202. A. Iwasaki, M. Hirata, I. Kudo, T. Sano, S. Sugawara, M. Ito, M. Watanabe, Zeolites 15 (1995) 308.
3203. R.M. Barrer, F.R.S. Hydrothermal Chemistry of Zeolites, Academic Press, New York, 1982.
3204. T. Twomey, M. Mackay, H. Kuipers, R. Thompson, Zeolite 14 (1994) 162.
3205. A. Cizmek, B. Subotica, D. Kralj, V. Babic-Ivancic, Microporous Mater. 12 (1997) 267.
3206. V. Nikolakis, E. Kokkoli, M. Tsapatis, D.G. Vlachos, Chem. Mater. 12 (2000) 845.
3207. P.-P.E.A. de Moor, T.P.M. Beelen, B.U. Komanschek, O. Diat, R.A. van Santen, J. Chem. Phys. B 101 (1997) 11077.
3208. P.-P.E.A. de Moor, T.P.M. Beelen, R.A. van Santen, J. Chem. Phys. B 103 (1999) 1639.
3209. O. Regev, Y. Cohen, E. Kehat, Y. Talmon, Zeolites 14 (1994) 314.
3210. M. Tsapatis, M. Lovallo, M.E. Davis, Microporous Mater. 5 (1996) 381.
3211. S. Bosnar, B. Subotic, Microporous Mesoporous Mater. 28 (1999) 483.
3212. T. Antinic, B. Subotic, N. Stubicar, Zeolites 18 (1997) 291.

3213. H. KacireK, H. Lechert, J. Phys. Chem. 79 (1975) 1589.
3214. J. Perez-Pariente, J.A. Martens, P.A. Jacobs, Appl. Catal. 31 (1987) 35.
3215. M.A. Camblor, A. Mifsud, J. Perez-Pariente, Zeolites 11 (1991) 792.
3216. T. Lindner, H. Lechert, Zeolites 14 (1994) 582–587.
3217. A. Iwasaki, I. Kudo, T. Sano, in: H. Chon, S.-K. UH (Eds.), Progress in Zeolite and Microporous Materials, Studies in Surface Science and Catalysis, vol. 105, Elsevier, Amsterdam, 1997.
3218. R.B. Borade, A. Clearfield, Microporous Mater. 5 (1996) 289–297.
3219. B. Su, V. Norberg, Zeolites 19 (1997) 65–74.
3220. R.B. Borade, A. Clearfield, Microporous Mater. 2 (1994) 167.
3221. F. Vaudry, F. Di Renzo, P. Espiau, F. Fajula, Zeolites 19 (1997) 253.
3222. X. Gu,J. Dong, T.M. Nenoff, Ind. Eng. Chem. Res. 44 (2005) 937.
3223. F. Bonhomme, M.E. Welk, T.M. Neno.,Microporous Mesoporous Mater. 66 (2003) 181.
3224. J. Sterte, S. Mintova, G. Zhang, B.J. Schoeman, Zeolites 18 (1997) 387.
3225. K. Okada, Y. Kameshima, C.D. Madhussodana, R.N. Das, Sci. Technol. Adv. Mater. 5 (2004) 479.
3226. M. Pan, Y.S. Lin, Microporous Mesoporous Mater. 43 (2001) 319.
3227. L. Tosheva, V.P. Valtchev, C. R. Chim. 8 (2005) 475.
3228. L.C. Boudreau, J.A. Kuck, M. Tsapatsis, J. Membr. Sci. 152 (1999) 41.
3229. X. Xu, W. Yang, J. Liu, L. Lin, Microporous Mesoporous Mater. 43 (2001) 299.
3230. B.-H. Jeong, K.-I. Sotowa, K. Kusakabe, J. Membr. Sci. 224 (2003) 151.
3231. Z. Wang, A. Mitra, H. Wang, L. Huang, Y. Yan, Adv. Mater. 13 (2001) 1463.
3232. S. Mintova, T. Bein, Adv. Mater. 13 (2001) 1880.
3233. W. Yang, X. Wang, Y. Tang, Y. Wang, C. Ke, S. Fu, J. Macromol. Sci. – Pure Appl. Chem. 39 (2002) 509.
3234. Y. Wang, Y. Tang, X. Wang, W. Shan, C. Ke, Z. Gao, J. Hu, W. Yang, J. Mater. Sci. Lett. 20 (2001) 2091.
3235. C. Ke, W.L. Yang, Z. Ni, Y.J. Wang, Y. Tang, Y. Gu, Z. Gao, Chem. Commun. (2001) 783.
3236. T. Seike, M. Matsuda, M. Miyake, Solid State Ion. 151 (2002) 123.
3237. B.J. Schoeman, J. Sterte, J.-E. Otterstedt, Zeolites 14 (1994) 110.
3238. K. Shqau, M.L. Mottern, D. Yu, H. Verweij, J. Am. Ceram. Soc. 89 (2006) 1790.
3239. M.M.J. Treacy, J.B. Higgins (Eds.), Collection of Simulated XRD Powder Patterns for Zeolites, Elsevier, Amsterdam, London, New York, Oxford, Paris, Shannon, Tokyo, 2001.
3240. S. Caratzoulas, D.G. Vlachos, M. Tsapatsis, J. Phys. Chem. B 109 (2005) 10429.
3241. G. Xu, J. Zhang, G. Song, Powder Technol. 134 (2003) 218.
3242. P. Lightfoot, D.A. Woodcock, M.J. Maple, L.A. Villaescusa, P.A. Wright, J. Mater. Chem. 11 (2001) 212.
3243. C.A. Kennedy, M.A. White, Solid State Commun. 134 (2005) 271.
3244. C.A. Kennedy, PhD thesis, Department of Chemistry, Dalhousie University, Halifax, Canada, 2005.
3245. R.K. Vyas, S. Kumar, Ind. Eng. Chem. Res. 34 (1995) 4058.
3246. H. Sahnoune, PH. Grenier, Chem. Eng. J. 40 (1989) 45.
3247. H. Mori, Y. Hamamoto, S. Yoshida, Nippon Reito Kucho Gakkai Ronbunshu 17 (2000) 171.
3248. A. Griesinger, K. Spindler, E. Hahne, Int. J. Heat Mass Transfer 42 (1999) 4363.
3249. K. Spindler, A. Griesinger, E. Hahne, High Temp. – High Pressures 29 (1997) 45.
3250. V.V. Murashov, M.A. White, Mater. Chem. Phys. 75 (2002) 178.
3251. V.V. Murashov, J. Phys.: Condens. Matter 11 (1999) 1261.
3252. A.J.H. McGaughey, M. Kaviany, Int. J. Heat Mass Transfer 47 (2004) 1799.
3253. B.-Z. Zhan, M.A. White, K.N. Robertson, T.S. Cameron, M. Gharghouri, Chem. Commun. (2001) 1176.
3254. B.-Z. Zhan, M.A. White, M. Lumsden, J. Mueller-Neuhaus, K.N. Robertson, T.S. Cameron, M. Gharghouri, Chem. Mater. 14 (2002) 3636.
3255. L. Qiu, P.A. Laws, B.-Z. Zhan, M.A. White, Can. J. Chem. 84 (2006) 134.
3256. P. Laws, MSc thesis, Department of Chemistry, Dalhousie University, Halifax, Canada, 2002.
3257. R. Berman, Thermal Conduction in Solids, Oxford University Press, Oxford, 1976.

3258. M.B. Jakubinek, C. Samarasekera, M.A. White, J. Mater. Res. 21 (2006) 287.
3259. P.G. Klemens, High Temp. – High Pressures 23 (1991) 241.
3260. D.G. Cahill, R.O. Pohl, Annu. Rev. Phys. Chem. 39 (1988) 93.
3261. S. Förster, M. Antonietti, Adv. Mater. 10 (1998) 195.
3262. C.J. Brinker, Y. Lu, A. Sellinger, H. Fan, Adv. Mater. 11 (1999) 579.
3263. G.J. de A.A. Soler-Illia, P. Innocenzi, Chem. Eur. J. 12 (2006) 4478.
3264. G.J. de A.A. Soler-Illia, E.L. Crepaldi, D. Grosso, D. Durand, C. Sanchez, Chem. Commun. (2002) 2298.
3265. N.A. Melosh, P. Davidson, B.F. Chmelka, J. Am. Chem. Soc. 122 (2000) 823.
3266. N.A. Melosh, P. Lipic, F.A. Bates, F. Wudl, G.D. Stucky, B.F. Chmelka, Macromolecules 32 (1999) 4332.
3267. J.H. Smatt, S. Schunk, M. Linden, Chem. Mater. 15 (2003) 2354.
3268. K. Vallé, P. Belleville, F. Pereira, C. Sanchez, Nat. Mater. 5 (2006) 107.
3269. K. Suzuki, K. Ikari, H. Imai, J. Mater. Chem. 13 (2003) 1812.
3270. H. Amenitsch, M. Rappolt, M. Kriechbaum, H. Mio, P. Laggner, S.J. Bernstor.,J. Synchrotron Radiat. 5 (1998) 506.
3271. M. Cestelli Guidi, M. Piccinini, A. Marcelli, A. Nucara, P. Calvani, E. Burattini, J. Opt. Soc. Am. A 22 (2005) 2810.
3272. P. Falcaro, S. Costacurta, G. Mattei, H. Amenitsch, A. Marcelli, M.C. Guidi, M. Piccinini, A. Nucara, L. Malfatti, T. Kidchob, P. Innocenzi, J. Am. Chem. Soc. 127 (2005) 3838.
3273. P. Innocenzi, L. Malfatti, T. Kidchob, P. Falcaro, S. Costacurta, M. Guglielmi, G. Mattei, H. Amenitsch, J. Synchrotron Radiat. 12 (2005) 734.
3274. P. Falcaro, D. Grosso, H. Amenitsch, P. Innocenzi, J. Phys. Chem. B 108 (2004) 10942.
3275. D. Grosso, A.R. Balkenende, P.A. Albouy, A. Ayral, H. Amenitsch, F. Babonneau, Chem. Mater. 13 (2001) 1848.
3276. M.P. Tate, V.N. Urade, J.D. Kowalski, T. Wei, B.D. Hamilton, B.W. Eggiman, H.W. Hillhouse, J. Phys. Chem. B 110 (2006) 9882.
3277. S. Besson, T. Gacoin, C. Jacquiod, C. Ricolleau, D. Babonneau, J.-P. Boilot, J. Mater. Chem. 10 (2000) 1331.
3278. P. Innocenzi, J. Non-Cryst. Solids 316 (2003) 309.
3279. K.M. Davis, M. Tomozawa, J. Non-Cryst. Solids 201 (1996) 177.
3280. G.E. Walrafen, S.R. Samanta, J. Chem. Phys. 69 (1978) 493.
3281. J.G. Bayly, V.B. Kartha, W.H. Stevens, Infrared Phys. 3 (1963) 211.
3282. Y. Shen, P. Wu, J. Phys. Chem. B 107 (2003) 4224.
3283. P. Innocenzi, P. Falcaro, J. Mio Bertolo, A. Bearzotti, H. Amenitsch, J. Non-Cryst. Solids 351 (2005) 1980.
3284. P. Falcaro, J. Mio Bertolo, P. Innocenzi, H. Amenitsch, A. Bearzotti, J. Sol-Gel Sci. Technol. 32 (2004) 107.
3285. A. Bearzotti, J. Mio Bertolo, P. Innocenzi, P. Falcaro, E. Traversa, J. Eur. Ceram. Soc. 24 (2004) 1969.
3286. J. Mio Bertolo, A. Bearzotti, P. Falcaro, E. Traversa, P. Innocenzi, Sens. Lett. 1 (2003) 64.
3287. P. Innocenzi, A. Martucci, M. Guglielmi, A. Bearzotti, E. Traversa, J.C. Pivin, J. Eur. Ceram. Soc. 21 (2001) 1985.
3288. I. Noda, Appl. Spectrosc. 47 (1993) 1329.
3289. I. Noda, A.E. Dowrey, C. Marcott, G.M. Story, Y. Ozaki, Appl. Spectrosc. A 54 (2000) 296.
3290. I. Noda, Y. Ozaki, Two-Dimensional Correlation Spectroscopy Applications in Vibrational and Optical Spectroscopy, Wiley, New York, 2004.
3291. P. Innocenzi, M.O. Abdirashid, M. Guglielmi, J. Sol-Gel Sci. Technol. 3 (1994) 47.
3292. R.E. Galindo, A. Van Veen, H. Schut, S.W.H. Eijt, C.V. Falub, A.R. Balkenende, F.K. De Theije, Mater. Eng. B 102 (2003) 403.
3293. D.M. Antonelli, Microporous Mesoporous Mater. 30 (1999) 315.
3294. P. Yang, D. Zhao, D.I. Margolese, B.F. Chmelka, G.D. Stucky, Chem. Mater. 11 (1999) 2813.
3295. P. Yang, D. Zhao, D.I. Margolese, B.F. Chmelka, G.D. Stucky, Nature 396 (1998) 152.
3296. K. Cassiers, T. Linssen, V. Meynen, P. Van Der Voort, P. Cool, E.F. Vansant, Chem. Commun. (2003) 1178.

3297. D.M. Antonelli, J.Y. Ying, Angew. Chem. Int. Ed. Engl. 34 (1995) 2014.
3298. P.C.A. Alberius, K.L. Frindell, R.C. Hayward, E.J. Kramer, G.D. Stucky, B.F. Chmelka, Chem. Mater. 14 (2002) 3284.
3299. E.L. Crepaldi, G.J. de Soler-Illia, D. Grosso, F. Cagnol, F. Ribot, C. Sanchez, J. Am. Chem. Soc. 125 (2003) 9770.
3300. E.L. Crepaldi, G.J.D.A.A. Soler-Illia, D. Grosso, P.-A. Albouy, H. Amenitsch, C. Sanchez, Stud. Surf. Sci. Catal. 141 (2002) 235.
3301. D. Grosso, G.J.A.A. Soler-Illia, F. Babonneau, C. Sanchez, P.-A. Albouy, A. Brunet-Bruneau, A.R. Balkenende, Adv. Mater. 13 (2001) 1085.
3302. D. Grosso, G.J.A.A. Soler-Illia, E.L. Crepaldi, F. Cagnol, C. Sinturel, A. Bourgeois, A. Brunet-Bruneau, H. Amenitsch, P.A. Albouy, C. Sanchez, Chem. Mater. 15 (2003) 4562.
3303. S.Y. Choi, M. Mamak, N. Coombs, N. Chopra, G.A. Ozin, Adv. Func. Mater. 14 (2004) 335.
3304. B. Smarsly, D. Grosso, T. Brezesinski, N. Pinna, C. Boissie, M. Antonietti, C. Sanchez, Chem. Mater. 16 (2004) 2948.
3305. J.A. Knowles, M.J. Hudson, J. Chem. Soc. Chem. Commun. (1995) 2083.
3306. E. Zhao, O. Hernandez, G. Pacheco, S. Hardcastle, J.J. Fripiat, J. Mater. Chem. 8 (1998) 1635.
3307. G. Pacheco, E. Zhao, A. Garcia, A. Sklyarov, J.J. Fripiat, Chem. Commun. (1997) 491.
3308. D. Khushalani, G.A. Ozin, A. Kuperman, J. Mater. Chem. 9 (1999) 1491.
3309. U. Ciesla, M. Froba, G. Stucky, F. Schuth, Chem. Mater. 11 (1999) 227.
3310. M. Linden, J. Blanchard, S. Schacht, S.A. Schunk, F. Schuth, Chem. Mater. 11 (1999) 3002.
3311. U. Ciesla, S. Schacht, G.D. Stucky, K.K. Unger, F. Schueth, Angew. Chem. Int. Ed. Engl. 35 (1996) 541.
3312. F. Kleitz, S.J. Thomson, Z. Liu, O. Terasaki, F. Schuth, Chem. Mater. 14 (2002) 4134.
3313. J. Fung, I. Wang, Appl. Catal. A 166 (1998) 327.
3314. B.M. Reddy, A. Khan, Catal. Rev. Sci. Eng. 47 (2005) 257.
3315. J.H. Schattka, D.G. Shchukin, J. Jia, M. Antonietti, R.A. Caruso, Chem. Mater. 14 (2002) 5103.
3316. A. Kitiyanan, S. Ngamsinlapasathian, S. Pavasupree, S. Yoshikawa, J. Solid State Chem. 178 (2005) 1044.
3317. J.N. Kondo, T. Yamashita, K. Nakajima, D.L. Lu, M. Hara, K. Domen, J. Mater. Chem. 15 (2005) 2035.
3318. S.H. Elder, Y. Gao, X. Li, J. Liu, D.E. McCready, C.F. Windisch Jr., Chem. Mater. 10 (1998) 3140.
3319. E.F. Marques, H.D. Burrows, M. da Graca Miguel, J. Chem. Soc., Faraday Trans. 94 (1998) 1729.
3320. S. Sato, S. Oimatsu, R. Takahashi, T. Sodesawa, F. Nozaki, Chem. Commun. (1997) 2219.
3321. R. Takahashi, S. Takenaka, S. Sato, T. Sodesawa, K. Ogura, K. Nakanishi, J. Chem. Soc., Faraday Trans. 94 (1998) 3161.
3322. S. Takenaka, R. Takahashi, S. Sato, T. Sodesawa, J. Sol-Gel Sci. Technol. 19 (2000) 711.
3323. E.R. Vance, D.J. Cassidy, C.J. Ball, G.J. Thorogood, J. Nucl. Mater. 190 (1992) 295.
3324. M. Aldissi, S.J. Henderson, J.W. White, T. Zemb, Mater. Sci. Forum 27–28 (1988) 437.
3325. P. Debye, H.R. Anderson Jr., H. Brumberger, J. Appl. Phys. 28 (1957) 679.
3326. J. Teixeira, J. Appl. Cryst. 21 (1988) 781.
3327. A. Guinier, G. Fournet, Small Angle Scattering of X-rays, Wiley, New York, 1955.
3328. T.M. Sabine, W.K. Bertram, Acta Crystallogr. Sect. A 55 (1999) 500.
3329. T.J. Pinnavaia, W. Zhang, Stud. Surf. Sci. Catal. 117 (1998) 23.
3330. R. Gopalan, Y.S. Lin, Ind. Eng. Chem. Res. 34 (1995) 1189.
3331. M. Hirano, C. Nakahara, K. Ota, M. Inagaki, J. Am. Ceram. Soc. 85 (2002) 1333.
3332. G.B. Deacon, F. Huber, R.J. Phillips, Inorg. Chim. Acta 104 (1985) 41.
3333. K.S. Finnie, J.R. Bartlett, J.L. Woolfrey, Langmuir 14 (1998) 2744.
3334. S. Pawsey, K. Yach, J. Halla, L. Reven, Langmuir 16 (2000) 3294.
3335. S.J. Hug, D. Bahnemann, J. Electron Spectrosc. Relat. Phenom. 150 (2006) 208.
3336. V. Torma, H. Peterlik, U. Bauer, W. Rupp, N. Husing, S. Bernstor.,M. Steinhart, G. Goerigk, U. Schubert, Chem. Mater. 17 (2005) 3146.

3337. J. Blanchard, F. Ribot, C. Sanchez, P.-V. Bellot, A. Trokiner, J. Non- Cryst. Solids 265 (2000) 83.
3338. R. Takahashi, K. Nakanishi, N. Soga, J. Sol-Gel Sci. Technol. 33 (2005) 159.
3339. E.R. Vance, B.A. Hunter, R.L. Davis, K.D. Hawkins, J. Solid State Chem. 150 (2000) 209.
3340. H.J. Rossell, J. Solid State Chem. 99 (1992) 52.
3341. K.L. Fujdala, T.D. Tilley, Chem. Mater. 14 (2002) 1376.
3342. R.L. Brutchey, J.E. Goldberger, T.S. Koffas, T.D. Tilley, Chem. Mater. 15 (2003) 1040.
3343. M.Z.C. Hu, J.T. Zielke, C.H. Byers, J.S. Lin, M.T. Harris, J. Mater. Sci. 35 (2000) 1957.
3344. E.L. Crepaldi, G.J.A.A. Soler-Illia, D. Grosso, M. Sanchez, New J. Chem. 27 (2003) 9.
3345. T.L. Hanley, V. Luca, I. Pickering, R.F. Howe, J. Phys. Chem. B 106 (2002) 1153.
3346. W. Makowski, T. Lojewski, Appl. Catal. B. Environ. 13 (1997) 205.
3347. J. Barrault, A. Derouault, G. Courtois, J.M. Maissant, J.C. Dupin, C. Guimon, H. Martinez, E. Dumitriu, Appl. Catal. A: Gen. 262 (2004) 43.
3348. O. Lebedeva, D. Tichit, B. Coq, Appl. Catal. A: Gen. 13 (1999) 61.
3349. M. Turco, G. Bagnasco, U. Costantino, F. Marmottini, T. Montanari, G. Ramis, G. Busca, J. Catal. 228 (2004) 43.
3350. D. Kishore, S. Kannan, J. Mol. Catal. A 244 (2006) 83.
3351. G. Fornasaria, R. Glfcklerb, M. Livia, A. Vaccari, Appl. Clay Sci. 29 (2005) 258.
3352. G. Centi, G. Fornasari, C. Gobbi, M. Livi, F. Tri.ro, A. Vaccari, Catal. Today 73 (2002) 287.
3353. G. Fornasari, F. Triffiro, A. Vaccari, F. Prinetto, G. Ghiotti, G. Centi, Catal. Today 75 (2002) 421.
3354. F. Basile, G. Fornasari, M. Livi, F. Tinti, F. Triffiro, A. Vaccari, Top. Catal. 30/31 (2004) 223.
3355. K. Katoh, T. Kihara, T. Asanuma, M. Gotoh, N. Shibaki, Toyota Technol. Rev. 44 (1995) 27.
3356. S. Matsumoto, Y. Ikeda, H. Suzuki, M. Ogai, N. Miyoshi, Appl. Catal. B: Environ. 25 (2000) 115.
3357. N. Das, D. Tichit, R. Durand, P. Graffin, B. Coq, Catal. Lett. 71 (2001) 181.
3358. F. Basile, G. Fornasari, M. Gazzano, A. Vaccari, Appl. Clay Sci. 16 (2000) 185.
3359. M. Intissar, F. Malherbe, V. Prévot, F. Leroux, J. Colloid Interface Sci. 299 (2006) 747.
3360. M. Intissar, J.-C. Jumas, J.-P. Besse, F. Leroux, Chem. Mater. 15 (2003) 4625.
3361. M. Masato, H. Shin, Chem. Commun. 15 (2003) 1962.
3362. P. Beaudot, M.E. de Roy, J.P. Besse, J. Solid State Chem. 161 (2001) 332.
3363. P. Beaudot, M.E. de Roy, J.P. Besse, Chem. Mater. 16 (2004) 935.
3364. V. Prevot, C. Forano, J.P. Besse, J. Solid State Chem. 2 (2000) 153.
3365. S. Miyata, Clays Clay Miner. 31 (1983) 305.
3366. G.R. Eugene, Inorg. Synth. 6 (1960) 1.
3367. S. Bonnet, C. Forano, A. DeRoy, J.P. Besse, P. Maillard, M. Momenteau, Chem. Mater. 8 (1996) 952.
3368. M. Meyn, K. Beneke, G. Lagaly, Inorg. Chem. 32 (1993) 1209.
3369. J. Fujita, A.E. Martell, K. Nakamoto, J. Chem. Phys. 36 (1962) 324.
3370. H.G.M. Edwards, P.H. Hardman, J. Mol. Struct. 273 (1992) 73.
3371. H.G.M. Edwards, D.W. Farwell, S.J. Rose, D.N. Smith, J. Mol. Struct. 249 (1991) 233.
3372. V. Prevot, C. Forano, J.P. Besse, Appl. Clay Sci. 183 (2001) 15.
3373. S. Carlino, M.J. Hudson, Solid State Ionics 110 (1998) 153.
3374. S. Carlino, M.J. Hudson, S. Waqif Husain, J.A. Knowles, Solid State Ionics 84 (1996) 117.
3375. E. Traversa, P. Nunziante, G. Ghozzini, Thermochim. Acta 199 (1992) 25.
3376. J. Sanchez Valente, F. Figueras, M. Gravelle, P. Kumbhar, J. Lopez, J.P. Besse, J. Catal. 189 (2000) 370.
3377. F. Rouquerol, J. Rouquerol, K.S.W. Sing, Adsorption by Powders and Porous Solids Principles Methodology and Applications, Academic Press, London, 1999.
3378. M. Jaroniec, M. Kruk, A. Sayari, in: L. Bonneviot, F. Beland, C. Danumh, S. Giasson, S. Kaliaguine (Eds.), Studies in Surface Science Catalysis, vol. 117, Elsevier, Amsterdam, 1994.
3379. J. Muscat, V. Swamy, N.M. Harrison, Phys. Rev. B 65 (2002) 224112.
3380. H. Zhang, J.F. Banfield, J. Mater. Chem. 8 (1998) 2073.
3381. D.A. Antonelli, Microporous Mesoporous Mater. 30 (2–3) (1999) 315.
3382. E. Beyers, P. Cool, E.F. Vansant, Microporous Mesoporous Mater. 99 (2007) 112.
3383. M.M. Yusuf, H. Imai, H. Hirashima, J. Sol-gel. Sci. Technol. 28 (2003) 97.

3384. D. Khushalani, G.A. Ozin, A. Kuperman, J. Mater. Chem. 7 (1999) 1491.
3385. J. Aguado-Serrano, M.L. Rojas-Cervantes, Microporous Mesoporous Mater. 88 (2006) 205.
3386. K.M.S. Khalil, M.I. Zaki, Powder Technol. 120 (2001) 256.
3387. S. Kambe, K. Murakoshi, I. Kitamura, Y. Wada, S. Yanagida, H. Kominami, Y. Kera, Sol. Energy Mater. Sol. Cells 61 (2000) 427.
3388. J.C. Yu, J. Yu, L. Zhang, W. Ho, J. Photochem. Photobiol. A 148 (2002) 263.
3389. M.M. Mohamed, W.A. Bayoumy, M. Khairy, M.A. Mousa, Microporous Mesoporous Mater. 97 (2006) 66.
3390. M. Schneider, A. Baiker, Catal. Today 35 (1997) 339.
3391. J.H. Clark, Green Chem. 1 (1999) 1.
3392. A. Karlsson, M. Stocker, R. Schmidt, Microporous Mesoporous Mater. 27 (1999) 181.
3393. P. Van Der Voort, P.I. Ravikovitch, K.P. De Jong, M. Benjelloun, E. Van Bavel, A.H. Janssen, A.V. Weimark, B.M. Weckhuysen, E.F. Vansant, J. Phys. Chem. B 106 (2002) 5873.
3394. S.R. Yoganarasimhan, C.N.R. Rao, Trans. Faraday Soc. 58 (1962) 1579.
3395. G. Genli, Appl. Catal. A 147 (1996) 267.
3396. H.P. Klug, L.E. Alexander, X-ray Diffraction Procedures, Wiley, New York, 1970.
3397. Y. Yang, Appl. Catal. A 273 (2004) 201.
3398. P. Gallezot, C. Leclercq, Catalyst Characterization: Physical Techniques for Solid Materials, Quoted by B. Imelik, J.C. Vedrine, Plenum Press, New York, 1994.
3399. L.I. Bekkermann, I.P. Dobrovolskii, A.A. Ivakin, Russ. J. Inorg. Chem. 21 (1976) 223.
3400. M.M. Mohamed, F.I. Zidan, M.A. Fodeal, J. Mater. Sci. 42 (2007) 712.
3401. M.M. Mohamed, J. Colloid Interface Sci. 272 (1) (2004) 28.
3402. S.D. Mahanti, W. Hammond, T.J. Pinnavaia, E. Prouzet, Microporous Mesoporous Mater. 27 (1999) 19.
3403. I.R. Beattie, T.R. Gilson, J. Chem. Soc. A (1969) 2322.
3404. M. Primet, P. Pichat, M.-V. Mathieu, J. Phys. Chem. 75 (9) (1971) 1216.
3405. M.A. Hasan, M.I. Zaki, L. Pasupulety, J. Phys. Chem. B 106 (2002) 12747.
3406. M. Primet, P. Pichat, M.V. Mathieu, J. Phys. Chem. B 267 (1968) 799.
3407. L.H. Little, Infrared Spectra of Adsorbed Species, Academic Press, New York, 1966 .
3408. M.M. Mohamed, M.M. Al-Esaimi, J. Mol. Catal. A 255 (1–2) (2006) 53.
3409. J. Kornatowski, B. Wichterlova, J. Jirovsky, E. Loppler, W. Pilz, J. Chem. Soc., Faraday Trans. 92 (6) (1996) 1067.
3410. Z.-W. Qu, G.-J. Raoes, J. Phys. Chem. B 110 (2006) 8998.
3411. B. Jonson, B. Rebenstorf, R. Larsson, S.L.T. Andersson, J. Chem. Soc., Faraday Trans. 1 84 (10) (1988) 3547.
3412. G. Carja, G. Delahay, Appl. Catal. B 47 (2004) 59.
3413. A.J. Lecloux, in: J.R. Anderson, M. Boudast (Eds.), Catalysis Science and Engineering, vol. 2, Springer, Berlin, 1981.
3414. M. Primet, P. Pichat, M.-V. Mathieu, J. Phys. Chem. 75 (9) (1971) 1221.
3415. M.I. Zaki, M.A. Hasan, F.A. Al-sagheer, L. Pasupulety, Colloids Surf. A 190 (2001) 261.
3416. E.P. Parry, J. Catal. 2 (1963) 371.
3417. G.A.M. Hussein, N. Sheppasd, M.I. Zaki, R.B. Fahim, J. Chem. Soc., Faraday Trans. 1 85 (1989) 1723.
3418. C. Martin, I. Martin, C. Del Moral, V. Rives, J. Catal. 146 (1994) 415.
3419. M.I. Zaki, G.A.M. Hussein, S.A.A. Mansour, H.A. El-Ammawy, J. Mol. Catal. 51 (1989) 209.
3420. C. Morterra, J. Chem. Soc., Faraday Trans. 1 84 (5) (1988) 1617.
3421. F. Hoffmann, M. Cornelius, J. Morell, M. Fröba, J. Nanosci. Nanotechnol. 6 (2006) 265.
3422. W.J. Hunks, G.A. Ozin, J. Mater. Chem. 15 (2005) 3716.
3423. G. Kickelbick, Angew. Chem. Int. Ed. 43 (2004) 3102.
3424. K. Landskron, B.D. Hatton, D.D. Perovic, G.A. Ozin, Science 302 (2003) 266.
3425. K. Landskron, G.A. Ozin, Science 306 (2004) 1529.
3426. O. Olkhovyk, M. Jaroniec, J. Am. Soc. Chem. 127 (2005) 60.
3427. O. Olkhovyk, S. Pikus, M. Jaroniec, J. Mater. Chem. 15 (2005) 1517.
3428. J. Morell, M. Güngerich, G. Wolter, J. Jiao, M. Hunger, P.J. Klar, M. Fröba, J. Mater. Chem. 16 (2006) 2809.

3429. R.M. Grudzien, B.E. Grabicka, S. Pikus, M. Jaroniec, Chem. Mater. 18 (2006) 1722.
3430. J. Liu, J. Yang, Q. Yang, G. Wang, Y. Li, Adv. Func. Mater. 15 (2005) 1297.
3431. M. Benitez, D. Das, R. Ferreira, U. Pischel, H. Garcýa, Chem. Mater. 18 (2006) 5597.
3432. B.M. Lok, C.A. Messina, R.L. Patton, R.T. Gajek, T.R. Cannan, E.M. Flanigen, J. Am. Chem. Soc. 106 (1984) 6092.
3433. G. Cao, M. Afeworki, G.J. Kennedy, K.G. Strohmaier, D.L. Dorset, Acta Cryst. B63 (2007) 56.
3434. R.W. Grosse-Kunstleve, L.B. McCusker, Ch. Baerlocher, J. Appl. Cryst. 30 (1997) 985.
3435. R.E. Morris, A. Burton, L.M. Bull, S.I. Zones, Chem. Mater. 16 (2004) 2844.
3436. A. Madek, J.S. Harwood, L. Frydman, J. Am. Chem. Soc. 117 (1995) 12779.
3437. C. Fernandez, J.P. Amoureux, J.M. Chezeau, L. Delmotte, H. Kessler, Microporous Mater. 6 (1996) 331.
3438. C. Fernandez, C. Morais, J. Rocha, M. Pruski, Solid State Nucl. Magn. Reson. 21 (2002) 61.
3439. J.-P. Amoureux, M. Pruski, in: D.M. Grant, R.K. Harris (Eds.), Encyclopedia of Nuclear Magnetic Resonance, vol. 9, Wiley, New York, 2002.
3440. M. Afeworki, D.L. Dorset, G.J. Kennedy, K.G. Strohmaier, Chem. Mater. 6 (2005) 1697.
3441. M. Afeworki, G.J. Kennedy, D.L. Dorset, K.G. Strohmaier, Chem. Mater. 6 (2005) 1705.
3442. G.J. Kennedy, M. Afeworki, D.C. Calabro, C.E. Chase, R.J. Smiley, Appl. Spectrosc. 58 (2004) 698.
3443. D. Barthomeuf, Zeolites 14 (1994) 394.
3444. H.G. Karge, in: H.V. Bekkum, E.M. Flanigen, J.C. Jansen (Eds.), Introduction to Zeolite Science and Practice, Studies in Surface Science and Catalysis, vol. 58, Elsevier, Amsterdam, 1991.
3445. E.G. Derouane, J.P. Gilson, J.B. Nagy, Zeolites 2 (1982) 42.
3446. J.P. van der Berg, J.P. Wolthuizen, A.D.H. Claque, G.R. Hays, R. Huis, J.H.C. van Hooff, J. Catal. 80 (1983) 130.
3447. S. Maixner, C.Y. Chen, P.J. Grobet, P.A. Jacobs, J. Weitkamp, New developments in zeolite science and technology, in: Y. Murakami, A. Iijima, J.W. Ward (Eds.), Proceedings of the Seventh International Zeolite Conference, Tokyo, Japan, August 17–22, Kodansha/Elsevier, Tokoyo/Amsterdam, 1986.
3448. A.V. Kucherov, A.A. Slinkin, Kinet. Katal. 23 (1982) 1172.
3449. A.V. Kucherov, A.A. Slinkin, Kinet. Katal. 24 (1983) 947.
3450. A.V. Kucherov, A.A. Slinkin, D.A. Kondratyev, T.N. Bodarenko, A.M. Rubinstein, Kh.M. Minachev, J. Mol. Catal. 37 (1986) 107.
3451. J.C. Vedrine, P. Dejaifve, E.D. Garbowski, Catalysis by zeolites, in: B. Imelik et al. (Eds.), Proceedings of the International Symposium, Ecully (Lyon), France, September 9–11, Elsevier, Amsterdam, 1980.
3452. J.C. Vedrine, P. Dejaifve, E.D. Garbowski, Stud. Surf. Sci. Catal. 5 (1980) 29.
3453. H. Förster, S. Franke, J. Seebode, J. Chem. Soc., Faraday Trans. I 79 (1983) 373–382.
3454. H.S. Cerquiera, P. Ayrault, J. Datka, M. Guisnet, Microporous Mesoporous Mater. 38 (2000) 197.
3455. H.S. Cerquiera, A. Rabeharitsara, P. Ayrault, J. Datka, P. Magnoux, M. Guisnet, Stud. Sci. Catal. 139 (2001) 29.
3456. H.S. Cerquiera, P. Ayrault, J. Datka, P. Magnoux, M. Guisnet, J. Catal. 196 (2000) 149.
3457. B. Gil, K. Mierzýnska, M. Szczerbińska, J. Datka, Microporous Mesoporous Mater. 99 (2007) 328.
3458. J. Bellamy, Infrared Spectra of Complex Molecules, Chapman and Hall, London/New York, 1980.
3459. G. Shirane, S. Hoshina, J. Phys. Soc. Jpn. 6 (1951) 265.
3460. D.A. Woodcock, P. Lightfoot, C. Ritter, Chem. Commun. 1 (1998) 107.
3461. J.S.O. Evans, T.A. Mary, A.W. Sleight, J. Solid State Chem. 137 (1998) 148.
3462. D.A. Woodcock, P. Lightfoot, P.A. Wright, L.A. Villaescusa, M.J. Dýaz-Cabañas, M.A. Camblor, J. Mater. Chem. 9 (1999) 349.
3463. D.A Woodcock, P. Lightfoot, L.A. Villaescusa, M.J. Dýaz-Cabañas, M.A. Camblor, D. Engberg, Chem. Mater. 11 (1999) 2508.
3464. L.A. Villaescusa, P. Lightfoot, S.J. Teat, R.E. Morris, J. Am. Chem. Soc. 123 (2001) 5453.

3465. M.P. Attfield, A.W. Sleight, J. Chem. Soc. Chem. Commun. 5 (1998) 601.
3466. S.H. Park, R.W. Grosse Kunstleve, H. Graetsch, H. Gies, Stud. Surf. Sci. Catal. 105 (1997) 1989.
3467. D.S. Bhange, V. Ramaswamy, Mater. Res. Bull. 41 (2006) 1392.
3468. I. Bull, P. Lightfoot, L.A. Villaescusa, L.M. Bull, R.K.B. Gover, J.S.O. Evans, R.E. Morris, J. Am. Chem. Soc. 125 (2003) 4342.
3469. B.A. Marinkovic, P.M. Jardim, A. Saavedra, L.Y. Lau, C. Baehtz, R.R. de Avillez, F. Rizzo, Microporous Mesoporous Mater. 71 (2004) 117.
3470. P. Tschaufeser, S.C. Parker, J. Phys. Chem. 99 (1995) 10609.
3471. J.W. Couvest, R.H. Jones, S.C. Parker, P. Tschaufeser, C.R.A. Catlow, J. Phys. Condens. Matter 5 (1993) L329.
3472. J.D. Gale, J. Phys. Chem. B 102 (1998) 5423.
3473. P.M. Jardim, B.A. Marinkovic, A. Saavedra, L.Y. Lau, C. Baehtz, F. Rizzo, Microporous Mesoporous Mater. 76 (2004) 23.
3474. R. Szoztak, T.L. Thomas, J. Catal. 100 (1986) 555.
3475. H.M. Rietveld, Acta. Crystallogr. 22 (1967) 151.
3476. H.M. Rietveld, J. Appl. Crystallogr. 2 (1969) 65.
3477. A.C. Larson, R.B. von Dreele, GSAS Generalized strucure analysis system, Laur 86–748, Los Alamos National Laboratory, Los Alamos, New Mexico, 1994.
3478. B. Toby, J. Appl. Cryst. 34 (2001) 210.
3479. H. van Koningsveld, H. van Bekkum, J.C. Jansen, Acta Cryst. B 43 (1987) 127.
3480. G. Artioli, C. Lamberti, G.L. Marra, Acta Cryst. B 56 (2000) 2.
3481. A.W. Sleight, Inorg. Chem. 37 (1998) 2854.
3482. A. Bieniok, K.D. Hammonds, Microporous Mesoporous Mater. 25 (1998) 193.
3483. S. Sen, R.R. Wusirika, R.E. Youngman, Microporous Mesoporous Mater. 87 (2006) 217.
3484. F. Caruso, E. Donath, H. Möhwald, J. Phys. Chem. B 102 (1998) 2011.
3485. A.L. Doadrio, E.M.B. Sousa, J.C. Doadrio, J. Pérez-Pariente, I. Izquierdo-Barba, M. Vallet-Regý, J. Controlled Release 97 (2004) 125.
3486. J. Xu, Z. Luan, H. He, W. Zhou, L. Kevan, Chem. Mater. 10 (1998) 3690.
3487. L. Li, J.L. Shi, L.X. Zhang, L.M. Xiong, J.N. Yan, Adv. Mater. 16 (2004) 1079.
3488. L. Ramqvist, K. Hamrin, G. Johansson, A. Fahlmann, C. Nordling, J. Phys. Chem. Solids 30 (1969) 1835.
3489. M.A. Arillo, M.L. López, C. Pico, M.L. Veiga, A. Jiménez-López, E. Rodrýguez-Castellón, J. Alloys Compd. 317–318 (2001) 160.
3490. M. Murata, K. Wakino, S. Ikeda, J. Electron. Spectrosc. Relat. Phenom. 6 (1975) 459.
3491. V. Di Castro, G. Polzonetti, J. Electron. Spectrosc. Relat. Phenom. 48 (1989) 117.
3492. M.L. Rojas, J.L.G. Fierro, L.G. Tejuca, A.T. Bell, J. Catal. 124 (1990) 41.
3493. V.R. Galakhov, M. Demeter, S. Bartkowski, M. Neumann, N.A. Ovechkina, E.Z. Kurmaev, N.I. Lobachevskii, J. Mitchell, D.L. Ederer, Phys. Rev. B 65 (2002) 113102.
3494. D.A. Shirley, Phys. Scr. 11 (1975) 117.
3495. G.G. Xia, Y.G. Yin, W.S. Willis, J.Y. Wang, S.L. Suib, J. Catal. 185 (1999) 91.
3496. Y.F. Shen, R.P. Serger, R.N. Deguzman, S.L. Suib, L. Mccurdy, D.I. Potter, C.L. O'Young, Science 260 (1993) 511.
3497. O. Giraldo, S.L. Brock, M. Marquez, S.L. Suib, H. Hillhouse, M. Tsapatsis, Nature 405 (2000) 38.
3498. Q. Feng, H. Kanoh, K. Ooi, J. Mater. Chem. 9 (1999) 319.
3499. Z.M. Wang, S. Tezuka, H. Kanoh, Chem. Mater. 13 (2001) 530.
3500. L.Y. Li, D.L. King, Chem. Mater. 17 (2005) 4335.
3501. Y.C. Son, V.D. Makwana, A.R. Howell, S.L. Suib, Angew. Chem., Int. Ed. 40 (2001) 4280.
3502. M. Abecassis-Wolfovich, R. Jothiramalingam, M.V. Landau, M. Herskowitz, B. Viswanathan, T.K. Varadarajan, Appl. Catal. B 59 (2005) 91.
3503. J.X. Dai, S.F.Y. Li, K.S. Siow, Z.Q. Gao, Electrochim. Acta 45 (2000) 2211.
3504. L.D. Noailles, C.S. Johnson, J.T. Vaughey, M.M. Thackeray, J. Power Sources 81–82 (1999) 259.

3505. X. Chen, Y.F. Shen, S.L. Suib, C.L. O'Young, Chem. Mater. 14 (2002) 940.
3506. P. Strobel, J.C. Charenton, Rev. Chim. Miner. 23 (1986) 125.
3507. D.C. Golden, C.C. Chen, J.B. Dixon, Clays Clay Miner. 35 (1987) 271.
3508. R. Giovanoli, M. Faller, Chimia 43 (1989) 54.
3509. J. Liu, V. Makwana, J. Cai, S.L. Suib, M. Aindow, J. Phys. Chem. B 107 (2003) 9185.
3510. N. Duan, S.L. Suib, C.L. O'Young, J. Chem. Soc., Chem. Commun. 13 (1995) 1367.
3511. J.C. Villegas, L.J. Garces, S. Gomez, J.P. Durand, S.L. Suib, Chem. Mater. 17 (2005) 1910.
3512. J. Liu, Y.C. Son, J. Cai, X. Shen, S.L. Suib, M. Aindow, Chem. Mater. 16 (2004) 276–285.
3513. M. Polverejan, J.C. Villegas, S.L. Suib, J. Am. Chem. Soc. 126 (2004) 7774.
3514. L. Balducci, D. Bianchi, R. Bortolo, R. D'Aloisio, M. Ricci, R. Tassinari, R. Ungarelli, Angew. Chem. Int. Ed. 42 (2003) 4937.
3515. F. Kleitz, F. Marlow, G.D. Stucky, F. Schüth, Chem. Mater. 13 (2001) 3587.
3516. C.W. Guo, Y. Cao,S.H. Xie, W.L. Dai, K.N. Fan, Chem. Commun. (2003)700.
3517. Q.S. Huo, D.Y. Zhao, J.L. Feng, K. Weston, S.K. Buratto, G.D. Stucky, S. Schacht, F. Schüth, Adv. Mater. 9 (1997) 974.
3518. F. Marlow, B. Splietho.,B. Tesche, D.Y. Zhao, Adv. Mater. 11 (1999) 632.
3519. J. Loerke, F. Marlow, Adv. Mater. 14 (2002) 1745.
3520. H.P. Lin, C.Y. Mou, Science 273 (1996) 765.
3521. H.P. Lin, S. Cheng, C.Y. Mou, Chem. Mater. 10 (1998) 581.
3522. F. Kleitz, U. Wilczok, F. Schüth, F. Marlow, Phys. Chem. Chem. Phys. 3 (2001) 3486.
3523. F. Marlow, F. Kleitz, Microporous Mesoporous Mater. 44 (2001) 671.
3524. Y. Ono, K. Nakashima, M. Sano, Y. Kanekiyo, K. Inoue, J. Hojo, S. Shinkai, Chem. Commun. (1998)1477.
3525. J.H. Jung, Y. Ono, S. Shinkai, Langmuir 16 (2000) 1643.
3526. F. Miyaji, S.A. Davis, J.P.H. Charmant, S. Mann, Chem. Mater. 11 (1999) 3021.
3527. W. Zhu, Y. Han, L. An, Microporous Mesoporous Mater. 84 (2005) 69.
3528. Q. Lu, D. Chen, X. Jiao, Chem. Mater. 17 (2005) 4168.
3529. S. Braun, S. Shtelzer, S. Rappoport, D. Avnir, M. Ottolenghi, J. Non- Cryst. Solids 147 (1992) 739.
3530. K.M. Roth, Y. Zhou, W. Yang, D.E. Morse, J. Am. Chem. Soc. 127 (2005) 325.
3531. N. Kröger, R. Deutzmann, M. Sumper, Science 286 (1999) 1129.
3532. J. Cha, K. Shimizu, Y. Zhou, S.C. Christiansen, F. Bradley, G.D. Stucky, D.E. Morse, Proc. Natl. Acad. Sci. USA 96 (1999) 361.
3533. Y. Zhou, K. Shimizu, J. Cha, G.D. Stucky, D.E. Morse, Angew. Chem., Int. Ed. 38 (1999) 779.
3534. K.M. Delak, N. Sahai, Chem. Mater. 17 (2005) 3221.
3535. S.A. Bagshaw, Chem. Commun. (1999) 1785.
3536. S.A. Bagshaw, J. Mater. Chem. 11 (2001) 831.
3537. Q. Xiao, J. Wang, D. Gao, P. Sun, Z. Yuan, T. Chen, B. Li, D. Ding, Chem. J. Chin. Univ. 27 (2006) 1395.
3538. R.K. Rana, V.S. Murthy, J. Yu, M.S. Wong, Adv. Mater. 17 (2005) 1145.
3539. R. Ryoo, J.M. Kim, C.H. Ko, C.H. Shin, J. Phys. Chem. 100 (1996) 17718.
3540. S.H. Tolbert, A. Firouzi, G.D. Stucky, B.F. Chmelka, Science 278 (1997) 264.
3541. L. Pasqua, F. Testa, R. Aiello, F.D. Renzo, F. Fajula, Microporous Mesoporous Mater. 44–45 (2001) 111.
3542. M.R. Porter, Handbook of Surfactants, second ed., Blackie, London, 1994.
3543. T. Imae, T. Kohsaka, J. Phys. Chem. 96 (1992) 10030.
3544. M.A. Cassidy, G.G. Warr, J. Phys. Chem. 100 (1996) 3337.
3545. T.R. Pauly, T.J. Pinnavaia, Chem. Mater. 13 (2001) 987.
3546. W. Lin, H. Frei, J. Phys. Chem. B 109 (2005) 4929.
3547. W. Lin, H. Frei, J. Am. Chem. Soc. 127 (2005) 1610.
3548. R. Nakamura, H. Frei, J. Am. Chem. Soc. 128 (2006) 10668.
3549. P. Hoare, in: A.J. Bard, R. Parson, J. Jordan (Eds.), Standard Potentials in Aqueous Solution, Marcel Dekker, New York, 1985.
3550. B.J. Hathaway, D.G. Holah, J. Chem. Soc. (1964)2400.

3551. F. Blatter, F. Moreau, H. Frei, J. Phys. Chem. 98 (1994) 13403.
3552. J. Panpranot, S. Kaewkun, P. Praserthdam, J.L. Goodwin, Catal. Lett. 91 (2003) 95.
3553. Q. Tang, Q. Zhang, H. Wu, Y. Wang, J. Catal. 230 (2005) 384.
3554. A.B.P. Lever, Inorganic Electronic Spectroscopy, second ed., Elsevier, Amsterdam, 1984.
3555. S. Lim, D. Ciuparu, Y. Chen, L. Pfefferle, G.L. Haller, J. Phys. Chem. B 108 (2004) 20095.
3556. G. Busca, R. Guidetti, V. Lorenzelli, J. Chem. Soc., Faraday Trans. 86 (1990) 989.
3557. K. Schmidt, W. Hauswirth, A. Muller, J. Chem. Soc., Dalton Trans. (1975) 2199.
3558. D. Bruhwiler, H. Frei, J. Phys. Chem. B 107 (2003) 8547.
3559. J.M. Fletcher, B.F. Greenfield, C.J. Hardy, D. Scargill, J.L. Woodhead, J. Chem. Soc. (1961) 2000.
3560. M.R. Basila, T.R. Kantner, J. Phys. Chem. 71 (1967) 467.
3561. K. Brodersen, H.J. Becher, Chem. Ber. 89 (1956) 1487.
3562. K. Niwa, H. Takahashi, K. Higasi, T. Kajiura, Bull. Chem. Soc. Jpn. 44 (1971) 3010.
3563. A. Eldewik, R.F. Howe, Microporous Mesoporous Mater. 48 (2001) 65.
3564. W. Lin, H. Frei, J. Am. Chem. Soc. 124 (2002) 9292.
3565. M.S. Morey, S. O'Brien, S. Schwarz, G.D. Stucky, Chem. Mater. 12 (2000) 898.
3566. W.S. Ahn, D.H. Lee, T.J. Kim, J.H. Kim, G. Seo, R. Ryoo, Appl. Catal. A: Gen. 181 (1999) 39.
3567. M.E. Raimondi, E. Gianotti, L. Marchese, G. Martra, T. Maschmeyer, J.M. Seddon, S. Coluccia, J. Phys. Chem. B 104 (2000) 7102.
3568. M.S. Morey, J.D. Bryan, S. Schwarz, G.D. Stucky, Chem. Mater. 12 (2000) 3435.
3569. Y. Hu, G. Martra, J. Zhang, S. Higashimoto, S. Coluccia, M. Anpo, J. Phys. Chem. B 110 (2006) 1680.
3570. K. Nakashiro, Y. Ono, Bull. Chem. Soc. Jpn. 66 (1993) 9.
3571. B. Kraushaar, W.G.M. Hoogervorst, R.R. Andrea, C.A. Emeis, W.H.J. Stork, J. Chem. Soc., Faraday. Trans. 87 (1991) 891.
3572. G. Sankar, J.M. Thomas, C.R.A. Catlow, C.M. Barker, D. Gleeson, N. Kaltsoyannis, J. Phys. Chem. B 105 (2001) 9028.
3573. M.W. Andersen, O. Terasaki, T. Ohsuna, T.A. Philippou, S.P. MacKay, A. Ferreira, J. Rocha, S. Lidin, Nature 367 (1994) 347.
3574. G. Blasse, G.J. Dirksen, Chem. Phys. Lett. 77 (1981) 9.
3575. G. Blasse, Struct. Bond. 76 (1991) 153.
3576. N.S. Hush, Prog. Inorg. Chem. 8 (1967) 391.
3577. G. Blasse, P.H.M. De Korte, A. Mackor, J. Inorg. Nucl. Chem. 43 (1981) 1499.

Subject Index

FSC
www.fsc.org
MIX
Papier | Fördert gute Waldnutzung
FSC® C083411

Kolibri 360 GmbH
Ferdinand-Jühlke-Straße 7
99095 Erfurt, Deutschland
produktsicherheit@kolibri360.de